Understanding Fiber Optics

Written by: Jeff Hecht
Contributing Editor
Fiberoptic Product News,
Lasers & Applications

BNA 6-1-88

HOWARD W. SAMS & COMPANY

A Division of Macmillan, Inc.
4300 West 62nd Street
Indianapolis, Indiana 46268 USA

International Standard Book Number: 0-672-27066-8
Library of Congress Catalog Card Number: 86-6311

Acquisitions Editor: Greg Michael
Editor: Sara Black
Illustrators: T.R. Emrick, Ralph E. Lund, John Bonecutter
Cover Art: Diebold Glascock Advertising Inc.
Cover Photography: Cassell Productions, Inc.
Components Courtesy of: Naval Avionics Center
Compositor: Shepard Poorman Communications Corp.

Printed in the United States of America

Ethernet is a registered trademark of Xerox Corporation.
Kevlar is a registered trademark of DuPont Company.
Selfoc is a registered trademark of Nippon Sheet Glass Company.

Acknowledgments

My thanks for comments on earlier drafts of various chapters go to: Robert Gallawa, David Charlton, Jim Hayes, Jim Walyus, Marjorie Katz, Art Nelson, and Ron Doyle.

Table of Contents

Preface

You have undoubtedly heard of fiber optics. AT&T and GTE have saturated the airwaves and print media with advertisements heralding this bright new technology. Futurists talk about the marvels of lightwave communications and photonic technology. *Omni* magazine writes about "Fiberopolis." Long-distance telephone calls go through optical fibers. Tomorrow, fibers will span the oceans.

The enthusiasm is not mere hype; the technology is real and important. From the oceans to the prairies, phone companies are laying fiber in the ground, pulling it through manholes, and hanging it from poles. The Army is buying fiber for portable battlefield communication systems. Medical fiber-optic systems let physicians peer inside the body without surgery. Very few technologies ever come close to the fantastic growth rates that market pundits love to predict. Fiber optics have.

You have picked up this book because you're curious about fiber optics and don't know much about it. You are not alone. A decade ago, fiber optics was tucked away in the back pages of optics textbooks, and optics courses were options for senior-level physics majors. Even today, only three universities in the country have full-fledged optics programs, and no university has a program that specializes in fiber optics. Most of today's optics specialists were trained in other fields, typically electronics or physics.

This book introduces fiber optics. It is arranged somewhat like a textbook. Each chapter starts by saying what it covers, ends by saying what it has covered, and provides a short multiple-choice quiz on the chapter contents. Like other books in the series, this book builds understanding step-by-step. The first chapter provides a broad overview of fiber optics, then the two following chapters look at the fundamental technology and at its uses more closely. The introduction is intended to prepare you for chapters describing the hardware of fiber communication systems. Measurement is an important concern covered in its own chapter. Then the following chapters describe the different types of fiber communication systems. The final chapter looks at non-communication uses of fiber-optic technology and the principles behind them. Try to master each chapter before going on to the next.

Some comparisons assume a general familiarity with electronics, but that is not essential; you need not know electronics to understand fiber optics. At times the text takes detours to explain fundamental optical concepts essential to fiber optics. A glossary and index also are included so the book will be a useful reference after you have studied it.

Fiber optics is a fascinating and growing field, which is sure to have a growing impact on you. I hope you come to share my enthusiasm.

J.H.

Introduction to Fiber Optics

A PERSONAL VIEW

Light is an old friend. I've been fascinated with light and optics ever since I can remember. I started playing with lenses and prisms a quarter of a century ago, and though my box of optical toys has spent some time in storage over the years, I still get it out now and then. Over the years, I've added some new playthings, and many of them involve fiber-optic technology.

At Home

The first optical fibers I saw were in decorative lamps. A group of fibers was tied together at one end and splayed out in a fan at the other. A bulb at the tied end illuminated them, and the light emerging from the loose ends made them glitter. The effect was pretty enough that when I was in college I bought one for my sister, but useless enough that I wandered away to explore other things.

In the Community

When next I saw fiber optics, in the mid-1970s, the technology had come a long way. Fibers had been improved enough that telephone companies were looking at them for communications. Those were the days when phone companies were—with good reason—described as "traditionally conservative" in their use of technology. Cautiously, they probed and tested fiber optics, almost as if they were a bomb squad examining a package they thought might explode. It was not until 1977 that, within a month, first GTE then AT&T dared to venture down manholes and stick pieces of fiber-optic cable in telephone circuits carrying live traffic.

Looking back, that technology looks primitive. It was daring then, and it worked. Not only that, it worked flawlessly. The small armies of engineers monitoring those test beds came to countless technical meetings afterward repeating the same monotonous but thrilling conclusion: "It works. Nothing has gone wrong."

In the Business World

I was at the first fiber-optic trade show in the late 1970s and have watched the excitement spread. Each year the meetings have grown larger. For a few years, breakthroughs were almost routine. The first generation of systems was barely in the ground before a second generation was ready. Today, the telephone industry is on a third generation of fiber-optic technology. What was impossible a few years ago is routine today.

The Future

Looking back, it's been an incredible ride. I've watched a technology spring from the laboratory into the real world. Once I heard about fiber optics from research scientists; now I hear about fiber from telephone service people. But the fun isn't over yet. The fiber-optics revolution will continue until fiber comes all the way to homes. It won't come tomorrow, but when it does come, it will bring a wealth of new information services. First, the services will go to businesses; eventually they will go into homes. The visionaries who foresaw a wired city were wrong—we will have a fibered society instead. We can all watch it happen.

But that's enough of this visionary stuff. Let's get down to the nuts and bolts—and fiber.

ABOUT THIS CHAPTER

The idea of communicating by light was around long before fiber optics, as were fibers of glass. It took many years for the ideas behind fiber optics to evolve from conventional optics. Even then, people were thinking more of making special optical devices than of optical communications. In this chapter we will see how fiber-optic technology evolved and how it can solve a wide variety of problems in communications.

HOW AND WHY FIBER OPTICS EVOLVED

Light normally travels in straight lines, but sometimes it is useful to make it go around corners. Some ideas have been around for many years.

Left alone, light will travel in straight lines. Even though lenses can bend light and mirrors can deflect it, light still travels in a straight line between the optical devices. This is fine for most purposes. Cameras, binoculars, telescopes, and microscopes wouldn't form images properly if light didn't travel in a straight line.

However, there also are times when people want to look around corners or probe inside places that are not in a straight line from their eyes. Or they may just need to pipe light from place to place, for communicating, viewing, illuminating, or other purposes. That's when they need fiber optics.

Piping Light

The problem arose long before the solution was recognized. In 1880, a Concord, Massachusetts, engineer named William Wheeler patented a scheme for piping light through buildings. Evidently not believing that Thomas Edison's incandescent bulb would prove practical, Wheeler planned to use light from a bright electric arc to illuminate distant rooms. He devised a set of pipes with reflective linings and diffusing optics to carry light through a building, then diffuse it into other rooms, a concept shown in one of his patent drawings in *Figure 1-1*.

Although he was in his 20s when he received his patent, Wheeler had already helped found a Japanese engineering school. He went on to become a widely known hydraulic engineer. Nevertheless, light piping was not one of his successes. Incandescent bulbs proved so practical that they're still in use today. Even if they hadn't, Wheeler's light pipes

**Figure 1-1.
Wheeler's Vision of
Piping Light (U.S.
Patent 247,229)**

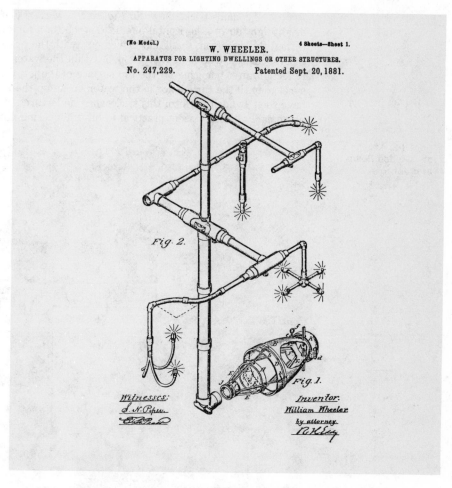

(No Model.)   4 Sheets—Sheet 1.

W. WHEELER.
APPARATUS FOR LIGHTING DWELLINGS OR OTHER STRUCTURES.
No. 247,229.   Patented Sept. 20, 1881.

Fig. 2.

Fig. 1.

Witnesses:

Inventor:
William Wheeler
by attorney

probably wouldn't have reflected enough light to do the job. However, his idea of light piping reappeared again and again until it finally coalesced into the optical fiber.

Total Internal Reflection

Ironically, the fundamental concept underlying the optical fiber was known well before Wheeler's time. A phenomenon called total internal reflection, which will be described in more detail in Chapter 2, can confine light within glass or other transparent materials denser than air. If the light in the glass strikes the edge at a glancing angle, it cannot pass out of the material and is instead reflected back inside it. Glassblowers probably saw this effect long ago in bent glass rods, but it wasn't widely recognized until the mid-1800s, when British physicist John Tyndall used it in his popular lectures on science.

Tyndall's trick, shown in *Figure 1-2*, worked like this. He shone a bright light down a horizontal pipe leading out of a tank of water. When he turned the water on, the liquid flowed out, with the pull of gravity forming a parabolic arc. The light was trapped within the water by total internal reflection, first bouncing off the top surface of the jet, then off the lower surface, until the turbulence in the water broke up the beam. Wheeler may have seen Tyndall perform this trick when he lectured in Boston, but presumably it didn't seem practical to his engineering mind.

Figure 1-2.
Light Guided Down a
Water Jet

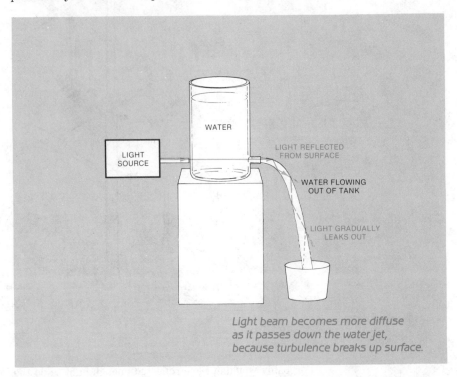

WATER

LIGHT
SOURCE

LIGHT REFLECTED
FROM SURFACE

WATER FLOWING
OUT OF TANK

LIGHT GRADUALLY
LEAKS OUT

Light beam becomes more diffuse as it passes down the water jet, because turbulence breaks up surface.

Optical Communication

Meanwhile, in Washington, a young American scientist who already had an international reputation was working on what he considered his greatest invention—the Photophone. Earlier, Alexander Graham Bell had used electricity to carry voices in the telephone. However, Bell was intrigued by the idea of sending signals without wires. He thought of optical communication, an idea that probably goes back to signal fires on prehistoric hilltops. The first "telegraph," devised by French engineer Claude Chappe in the 1790s, was an optical telegraph. Operators in towers relayed signals from one hilltop to the next by moving semaphore arms. Samuel Morse's electric telegraph put the optical telegraph out of business, but it left behind countless Telegraph Hills.

In 1880, Bell demonstrated that light could carry voices through the air without wires. Bell's Photophone reproduced voices by detecting variations in the amount of sunlight or artificial light reaching a receiver. It was the first form of "wireless" voice communications. However, it never proved practical because too many things could get in the way of the beam.

Later, others used light to carry voices through the open air, something that is now done with a few laser systems but is not used widely. In the 1930s, another engineer, Norman R. French—ironically an employee of the American Telephone & Telegraph Corp., the company built around Bell's telephone—patented the idea of communicating via light sent through pipes. Even more time passed before the optical fiber was invented.

THE CLAD FIBER

The key development in making optical fibers usable was a cladding to keep the light from leaking out.

Although Tyndall could demonstrate light guiding in his stream of water, he couldn't do so very well. The rough boundary of the water broke up the light beam. Other effects also limited light guiding in bent glass rods. The problem is that light can leak out wherever the rod touches something other than air. Because the rod cannot hang unsupported in the air, it has to touch something.

The solution to that problem seems obvious only with 20–20 hindsight. In the 1950s, Brian O'Brien, Sr., in the United States and Harry Hopkins and Narinder Kapany in England started looking for ways to guide light. The key concept was making a two-layer fiber, shown in *Figure 1-3*. Light would be carried in the inner layer or core of the fiber. The outer layer, the cladding, would confine the light within the core, because its refractive index (like that of air) was below that of the core, but it also would prevent light from leaking out.

Imaging

Many optical fibers can be bundled together to transmit images.

O'Brien, who had been dean of the prestigious school of optics at the University of Rochester, brought the idea of the clad fiber with him when he became director of research at the American Optical Co. in Southbridge, Massachusetts. His interest was imaging. A single fiber could not transmit an image because the light from different parts of the image would be blurred together. So O'Brien bundled many optical fibers together so a pattern of light formed on one end of the bundle would be recreated on the other end. This requires that all the fibers be at precisely the same position relative to each other at each end of the bundle, as we will examine in more detail in Chapter 20.

**Figure 1-3.
Light Cannot Leak out
of Clad Fibers**

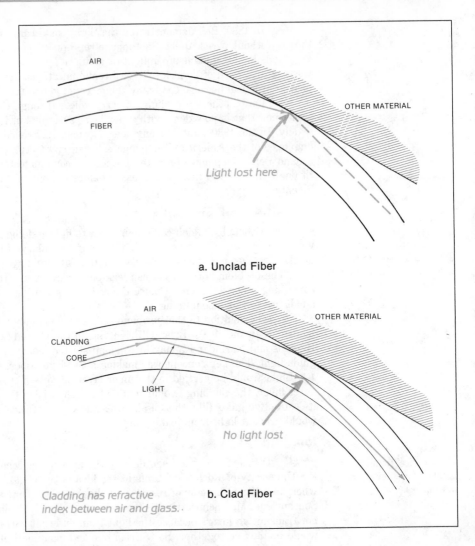

a. Unclad Fiber

b. Clad Fiber

Light lost here

No light lost

Cladding has refractive
index between air and glass.

AIR

FIBER

OTHER MATERIAL

CLADDING

CORE

LIGHT

AIR

OTHER MATERIAL

Expanded Applications

Engineers at American Optical and elsewhere soon began
extending the technology. Fibers could be grouped in flexible as well as
rigid bundles. The fibers need not be aligned with respect to each other if
the only goal was to deliver light, as in a fiber-optic lamp. Fibers could be
made from plastic as well as from glass. Applications began to emerge in
fields such as medicine and inspection.

REDUCING FIBER LOSS

Recognizing Transmission Problems

One important fundamental limitation remained: optical fibers absorbed too much light. That statement must be put in context. Even comparatively high-loss fibers are much more transparent than ordinary window glass. (Because ordinary windows are very thin, most light not transmitted by windows is reflected from the glass rather than absorbed.) Those high fiber losses were acceptable in transmitting light a few feet (ft) or a couple of meters (m), something impractical through window glass. But only 10% of the light remained after travelling 10 m (32 ft), and after 20 m only 1% was left.

At first, no one seriously thought that there was a transmission problem with fibers because they weren't trying to send light very far. Theodore H. Maiman's demonstration of the first laser in 1960 stimulated renewed interest in optical communications. However, people were slow to give optical fibers serious thought. Scientists knew that transparent materials absorbed too much light to transmit optical signals over long distances. Researchers at Bell Telephone Laboratories set to work on a new generation of light pipes. Instead of simply relying on reflective walls, they had elaborate gas lenses to focus light along the pipes' length.

Purifying Glass Fibers

Two engineers working at Standard Telecommunications Laboratories in England, Charles K. Kao and George Hockham, did take a careful look at the possibilities of optical-fiber telecommunications. Their theoretical analysis, published in 1966, indicated it was not the intrinsic properties of glass itself that made fiber loss high. Instead, it was impurities. By removing the impurities, they said, it should be possible to reduce loss to levels low enough that 10% of the light would remain after it had passed through 500 m (1600 ft) of fiber. Their prediction must have sounded fantastic then, but it proved too conservative.

Publication of Kao and Hockham's paper set off a world-wide race to make better fibers. The first to beat the theoretical prediction were Robert Maurer, Donald Keck, and Peter Schultz at the Corning Glass Works in 1970. Others soon followed, and losses were pushed down to even lower levels. In today's best optical fibers, 10% of the entering light remains after the light has passed through more than 50 kilometers (30 miles) of fiber. Losses are not quite that low in practical telecommunication systems, but as we will see in Chapter 4, impressive progress has been made. Because of that progress, fiber optics are becoming the backbone of our national long-distance telephone network.

BASICS OF FIBER OPTICS

Materials

Not all optical fibers are the same. Many distinct types, designed for specific applications, are available. Individual optical fibers are used in virtually all communications applications and for many other purposes as well. Each fiber is separate from all other fibers, although a number of separate fibers may be housed in a common cable. Most fibers are made of glass, plastic, or plastic-clad glass; some special fibers are made of other materials. The fibers are flexible but somewhat stiff; the degree of flexibility depends on the fiber diameter. Optical fibers are often compared to human hairs, but whoever thought of that comparison must have had some very stiff hairs or very thin plastic fibers. Communication fibers are much stiffer than even a man's coarse beard hairs. A better comparison would be to monofilament fishing line. Unlike wires, fibers spring back to their original straight form after being bent.

Alternatively, fibers can be bundled together in either of two forms. A flexible bundle is made up of many separate fibers, assembled together, with the ends of the bundle fixed and the rest unattached to each other (although typically encased in some overall housing). A rigid bundle is made by melting many fibers together into a single rod, which typically is bent to the desired shape during manufacture. Such rigid or fused bundles are less costly than flexible bundles, but their lack of physical flexibility makes them unsuitable for some applications.

Systems

You need more than just fiber to make a communication system. The basic elements of a system are shown in *Figure 1-4*. The signal originates from a modulated light source, which feeds it into a fiber, which delivers it to a receiver. The receiver decodes the optical signal and converts it to electronic form for use by equipment at the receiving end.

Fiber-optic communication systems include transmitter, receiver, and a cable structure (to house the fiber) as well as the fiber itself.

**Figure 1-4.
Fiber-Optic System
Components**

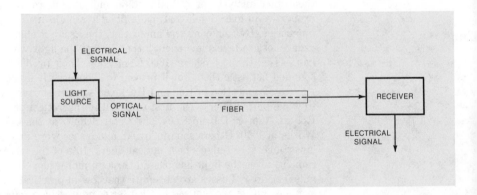

Real-world fiber-optic systems are more complex than indicated in this simple example. Most communication travels in two directions and requires a transmitter and a receiver at each end and usually separate fibers to carry signal in each direction. The fiber or fibers are housed in a cable to simplify handling and protect them from environmental stresses. Light sources must be precisely aligned with fibers so their output is collected efficiently. Likewise, if light is transferred between fibers, the two ends must be precisely aligned. Because their diameters are very small, the mechanical tolerances for proper alignment are tight. Consequently, much more attention must be paid to connectors and splices than in electrical communication over wires.

Auxiliary equipment is needed for many of these tasks. Unlike metal wires, optical fibers cannot be spliced with wire cutters and a soldering iron. Much more elaborate fiber splicers are needed, as we will see in Chapter 9. Connector installation is also a complex task and often must be performed in a factory, as we will learn in Chapter 8. Special measurement tools are needed to assess the quality of a fiber-optic link, as we will see in Chapter 13. Much of this equipment comes in various forms for different applications, because of wide differences in system requirements.

WHY FIBER OPTICS?

Light Transmission

Because fiber optics is a unique transmission medium, it has some unique advantages for certain types of communications.

The crucial operating difference between a fiber-optic communication system and other types is that signals are transmitted as light. Conventional electronic communications relies on electrons passing through wires. Radio-frequency and microwave communications (including satellite links) rely on radio waves and microwaves travelling through open space. (We will ignore free-space optical communication systems, which send a laser beam through free space or air, because there are few such systems. To prevent confusion, we will avoid the term "optical" communications.)

Different communication media are suited for different communication jobs. The choice depends on the job and the nature of the transmission medium. One important factor is how signals are to be distributed. If the same signal is to be sent from one point to many people in an area—as in broadcast television or radio—the best choice is non-directional radio transmission. Radio-frequency communication also is best if it is hard to make a physical connection, as in reaching a remote island or using a car telephone. On the other hand, a cable system is preferable for making physical interconnections among many individual fixed points, as in the telephone network. Cable also is useful in making permanent connections between two fixed points. Some of the types of transmission are shown in *Figure 1-5*.

**Figure 1-5.
Types of
Communication
Transmission**

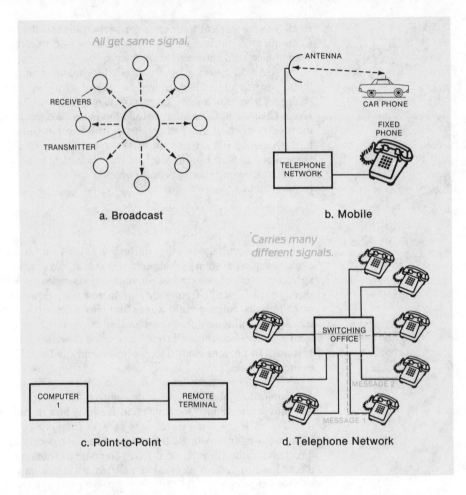

a. Broadcast

b. Mobile

c. Point-to-Point

d. Telephone Network

Capacity and Distance

High transmission capacity and long transmission distance are two major advantages of fiber-optic cables.

When choosing the type of cable, key factors to consider are how far the signal must go and the amount of information the signal is carrying. Transmission distance depends on such effects as transmitter power, receiver sensitivity, and losses in the intervening medium. The amount of information carried is measured in the passage of signals through the system, much as we measure water volume flowing through a pipe. Some transmission media, such as simple pairs of wires, work fine for low-speed signals but not for high-speed signals. Others, such as coaxial cables, can carry high-speed signals, but only over short distances. The attraction of optical fibers is that they can carry information at high speeds over long distances. (We'll attach some numbers to those terms in later chapters.)

Secondary Factors

Secondary factors also influence the choice of transmission medium. A typical example is electromagnetic interference (EMI), which like static on an AM radio can sometimes block signal transmission on wires. Optical fibers cannot pick up EMI because they carry signals as light.

Different factors can dictate the choice of fiber for other applications. For example, in military systems and financial institutions, security of the transmitted signals is a vital concern. When cables must be installed in existing buildings, size and rigidity of the cable determine ease of installation, which becomes a major economic issue. The related issues of weight and bulk are major practical concerns in communication systems designed to be portable, ranging from electronic news-gathering equipment to a battlefield communication system. Avoiding sparks is a must in systems being installed in refineries and chemical plants where the atmosphere may contain explosive gases. As we will see in Chapter 3, fiber optics can solve many of these problems.

FIBER-OPTIC APPLICATIONS

The advantages of fiber optics have led to many applications in long-haul and short-distance communications.

The wide variety of fiber-optic systems that have come into use because of the advantages of optical fibers will be described in more detail in following chapters. The major types include:

- Long-haul telecommunication systems on land and at sea to carry many simultaneous telephone calls (or other signals) over long distances. These include ocean-spanning submarine cables.
- Interoffice trunks that carry many simultaneous telephone conversations between local and regional telephone company switching facilities.
- Dropoffs for telephone lines operating above the normal speed of single telephone lines.
- Connections between microwave receivers and control facilities and the distribution networks of cable-television operators.
- Cables for remote news-gathering equipment.
- Links among computers and high-resolution video terminals used for such purposes as computer-aided design.
- Local-area networks operating at speeds too high or over distances too long for the use of metal cables.
- High-speed interconnections between computers and peripheral devices, or between computers, or even within segments of single large computers.
- Moderate-speed transmission of computer data in places where fiber is most economical to install.
- Transmission in difficult environments, especially those plagued with severe EMI.
- Portable communication equipment for battlefield use.

Meanwhile, fiber-optic technology has continued to expand into many areas outside of communications, as we will cover in more detail in Chapter 20. These include fiber-optic bundles for illumination and imaging, endoscopes to view inside the body and treat diseases with light and without surgery, and optical sensors to measure rotation, pressure, sound waves, magnetic fields, and many other quantities.

These are today's fiber-optic applications. More are sure to come as fiber optics and other technologies develop. For example, fiber optics seems an ideal technology to bring a broad array of new and existing communication services to homes and businesses. Now, however, we will concentrate on the realities of present-day technology.

WHAT HAVE WE LEARNED?

1. Light rays want to go in straight lines, but optical fibers can guide them around corners.
2. Early optical communication systems sent light through the air.
3. Optical fibers must have a cladding layer to keep light from leaking out.
4. The first applications of fiber optics were outside of communications.
5. Loss of fibers was reduced dramatically to allow their use in communications.
6. Optical fibers are best for transmitting signals at high speeds or over long distances between fixed points.
7. Fiber optics have some special advantages, including immunity to electromagnetic interference, which can block transmission on wires.

WHAT'S NEXT?

In this chapter, we examined the background of fiber-optic technology. In Chapter 2, we'll learn some of the basic physics behind fiber optics before going into some specifics about fiber-optic hardware.

Quiz for Chapter 1

1. The first use of light for communication was:
   a. Claude Chappe's optical telegraph.
   b. native American smoke signals.
   c. fiber-optic communications.
   d. prehistoric signal fires.

2. Light can be guided around corners most efficiently in:
   a. reflective pipes.
   b. hollow pipes with gas lenses.
   c. clad optical fibers.
   d. bare glass fibers.

3. The first low-loss optical fibers were made:
   a. at Corning Glass Works in 1970.
   b. at Standard Tele-communication Labs in 1966.
   c. at Bell Telephone Labs in 1960.

4. Today's best optical fibers transmit light so well that 10% of the input light remains after:
   a. 0.5 km.
   b. 4 km.
   c. 20 km.
   d. 50 km.
   e. 100 km.

5. Optical fibers are made of:
   a. glass coated with plastic.
   b. ultrapure glass.
   c. plastic.
   d. all of these.

6. Essential components of any fiber-optic communication system are:
   a. light source, fiber, and receiver.
   b. light source and cable.
   c. fiber and receiver.
   d. fiber only.

7. The small size of optical fibers makes what necessary in any device connecting them?
   a. Special glue.
   b. Tight mechanical tolerances.
   c. Low optical absorption.
   d. Small overall size.

8. What are the major advantages of optical fibers for long-distance communications?
   a. Small fiber size.
   b. Do not carry electrical current.
   c. Low loss when carrying high-speed signals.
   d. Low loss only.
   e. High-speed signal capacity only.

9. Unlike wires, optical fibers are immune to:
   a. electromagnetic interference.
   b. high-frequency transmission.
   c. signal losses.

10. Applications of fiber-optic communications include:
    a. ocean-spanning submarine cables.
    b. long-distance telephone transmission on land.
    c. connecting telephone-company facilities.
    d. transmitting data to high-resolution video terminals.
    e. all the above.

Fundamentals of Fiber Optics

ABOUT THIS CHAPTER

Fiber optics is a hybrid field. It started as a spinoff of classical optics. The basic concept of a fiber is optical, and some optical fibers and fiber bundles are used as optical components. However, as fiber became a communication medium, the field borrowed concepts and terminology from electronic communications. Transmitters and receivers convert signals from electrical to optical format and back; they are part optics and part electronics. To understand fiber-optic communications, you need to know about three fields: optics, electronics, and communications.

This chapter is a starting point. In later chapters we'll go into more detail on such topics as how light is guided in fibers and how various components and systems work. However, the first step is to learn a little about optics, how fibers work, and how optical fibers can serve as the basis of a communication system.

BASICS OF OPTICS

The light carried in fiber-optic communication systems can be viewed as either a wave or a particle.

The workings of optical fibers depend on basic principles of optics and the interaction of light with matter. The first step in understanding fiber optics is to review the relevant parts of optics. The summary that follows does not cover the entire field of optics, and some parts may seem basic, but you should read it to make sure you understand the fundamentals.

From a physical standpoint, light can be seen either as electromagnetic waves or as photons, quanta of electromagnetic energy. This is the famous wave-particle duality of modern physics. Both viewpoints are valid and valuable, but the most useful viewpoint for optics often is to consider light as rays travelling in straight lines between optical elements, which can reflect or refract (bend) them.

The Electromagnetic Spectrum

What we call "light" is only a small part of the entire spectrum of electromagnetic radiation. The fundamental nature of all electromagnetic radiation is the same: it can be viewed as photons or waves and travels at the speed of light (c), which is 300,000 kilometers per second (km/s) or 180,000 miles per second (mi/s). What makes the difference between radiation in different parts of the spectrum is a quantity that can be measured in several ways: as the length of a wave, as the energy of a photon, or as the oscillation frequency of an electromagnetic field. These three approaches are shown in *Figure 2-1*.

**Figure 2-1.
Electromagnetic
Spectrum**

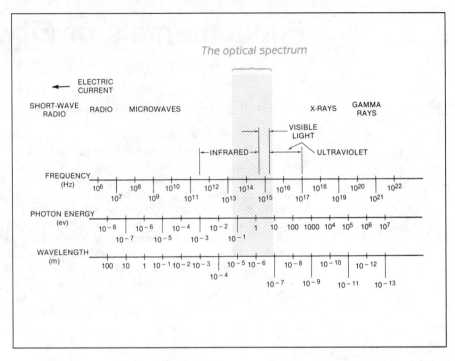

Each measurement—wavelength, energy, or frequency—has its own characteristic unit. In some parts of the spectrum, frequency is used most; in others, it is photon energy or wavelength. The optical world talks in wavelength, and wavelength is measured in metric units—meters, micrometers (μm or 10^{-6} m), and nanometers (nm or 10^{-9} m). Don't even think of wavelength in inches. (If you absolutely have to know, 1 μm is 1/25,000th of an inch.) Frequency is measured in cycles per second (cps) or hertz (Hz), with megahertz (MHz) meaning a million hertz and gigahertz (GHz) meaning a billion hertz. (If the prefixes to metric terms confuse you, their meanings are listed in *Table 2-1*. The metric system relies on prefixes to provide different units of length, weight, frequency, and other quantities. The prefix makes a unit a multiple of a standard unit. For example, a millimeter is a thousandth [10^{-3}] of a meter, and a kilometer is 1000 [10^3] meters.) Photon energy can be measured in many ways, but the most convenient here is in electron volts (eV)—the energy that an electron gains in moving through a 1-volt (V) electric field.

Table 2-1.
Metric Unit Prefixes and
Their Meanings

Prefix	Symbol	Multiple
tera	T	10^{12} (trillion)
giga	G	10^9 (billion)
mega	M	10^6 (million)
kilo	k	10^3 (thousand)
hecto	h	10^2 (hundred)
deca	da	10^1 (ten)
deci	d	10^{-1} (tenth)
centi	c	10^{-2} (hundredth)
milli	m	10^{-3} (thousandth)
micro	μ	10^{-6} (millionth)
nano	n	10^{-9} (billionth)
pico	p	10^{-12} (trillionth)
femto	f	10^{-15} (quadrillionth)

All the measurement units shown on the spectrum chart are really different rulers to measure the same thing. There are simple ways to convert between them. Wavelength is inversely proportional to frequency, according to the following formula:

$$c = \text{WAVELENGTH} * \text{FREQUENCY}$$

or

$$c = \lambda * \nu$$

where c is the speed of light, λ is wavelength, and ν is frequency. To get the right answer, all terms must be measured in the same units. Thus c must be in meters per second (m/s), λ must be in meters, and frequency must be in hertz (or cycles per second). Plugging in the number for c, we have a more useful formula:

$$\lambda * \nu = 3 \times 10^8 \text{ m/sec}$$

Not many people talk about photon energy (E) in fiber optics, but a value can be gotten from Planck's law, which states:

$$E = h\nu$$

where h is Planck's constant and ν the frequency. Because most of the interest in photon energy is in the part of the spectrum measured in wavelength, a more useful formula is:

$$E(eV) = \frac{1.2406}{\lambda(\mu m)}$$

which gives energy in electron volts when wavelength is measured in micrometers (μm).

There is one practical consequence of the wave aspect of light's personality which we will encounter later on. Light waves can interfere with each other. Normally this does not show up, because there are many different light waves present, and the effect averages out. Suppose, however, that there are only two identical light waves present, as shown in *Figure 2-2*. The total amount of light detectable is the sum of the amplitudes of the light waves squared. If the light waves are neatly lined up, what is called "in phase," they add together and give a bright spot. However, if the two light waves are aligned so the peaks of one coincide with the troughs of the other, they interfere destructively and cancel each other out. This happens when the two light waves are 180° out of phase with each other. If the two waves are out of phase by a different amount, they add together to give an intensity between the maximum and minimum possible.

Figure 2-2.
Constructive and
Destructive Interference

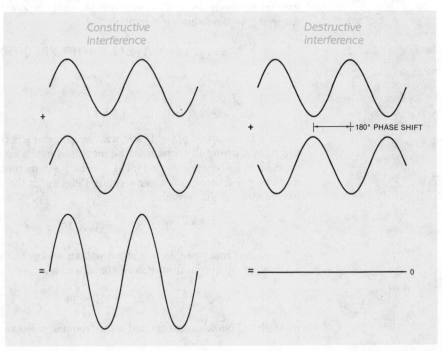

We are mainly interested in a small part of the spectrum shown in *Figure 2-1*—the optical region, where optical fibers and other optical devices work. That region includes light visible to the human eye at wavelengths of 400–700 nm, and nearby parts of the infrared and ultraviolet, which have similar properties. Roughly speaking, this means wavelengths of 200–20,000 nm (0.2–20 μm).

Fiber-optic communication systems transmit near-infrared light invisible to the human eye.

Most optical fibers used for communications transmit light in the near-infrared at wavelengths of 800–1600 nm (0.8–1.6 μm). The silica glasses used in most fibers are most transparent at those wavelengths. Plastic fibers transmit best at visible wavelengths, but they are not as transparent at those wavelengths as glass fibers are in the infrared. Special fibers now in development, made of materials other than silica, can transmit light at longer infrared wavelengths. Special grades of silica can transmit some near-ultraviolet light.

Refractive Index

The refractive index of a material is the ratio of the speed of light in a vacuum to the speed of light in the material.

The most important optical measurement for any transparent material is its refractive index (n). The refractive index is the ratio of the speed of light in vacuum to the speed of light in the medium:

$$n = \frac{c_{vac}}{c_{mat}}$$

The speed of light in a material is always slower than in a vacuum, so the refractive index is always greater than one in the optical part of the spectrum. In practice, the refractive index is measured by comparing the speed of light in the material to that in air rather than in a vacuum. That simplifies measurements but doesn't make any practical difference because the refractive index of air at atmospheric pressure and room temperature is 1.00029, so close to one that the difference is insignificant.

Refraction occurs when light passes through a surface where the refractive index changes.

Although light rays travel in straight lines through optical materials, something different happens at the surface. Light is bent as it passes through a surface where the refractive index changes—for example, as it passes from air into glass, as shown in *Figure 2-3*. The amount of bending depends on the refractive indexes of the two media and the angle at which the light strikes the surface between them. The angles of incidence and refraction are measured not from the plane of the surface but from a line normal (perpendicular) to the surface. The relationship is known as Snell's law, which is written:

$$n_i \sin I = n_r \sin R$$

where n_i and n_r are the refractive indexes of the initial medium and the medium into which the light is refracted and I and R are the angles of incidence and refraction, respectively, as shown in *Figure 2-3*.

Figure 2-3.
Light Refraction as It
Enters Glass

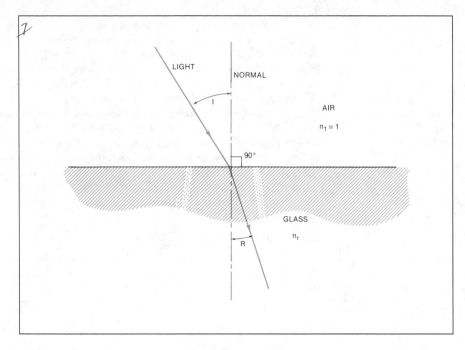

Figure 2-3 shows the standard textbook example of light going from air into glass. This bending occurs whether the surface is flat or curved. However, if both front and rear surfaces are flat, the net refractive effect is zero, as when you look through a flat window, because light emerges at the same angle that it entered (although it may be displaced). If one or both surfaces are curved, the net effect is that of looking through a lens—which in fact you are doing. That is, light rays emerge from the lens at a different angle than they entered. These overall refractive effects are shown in *Figure 2-4*.

What does this have to do with fiber optics? Stop and consider what happens when light travelling in a medium with a high refractive index (such as glass) comes to an interface with a medium having a lower refractive index (such as air). If the glass has a refractive index of 1.5 and the air an index of 1.0, the equation becomes:

$$1.5 \sin I = 1 \sin R$$

**Figure 2-4.
Light Refraction
Through a Window and
a Lens**

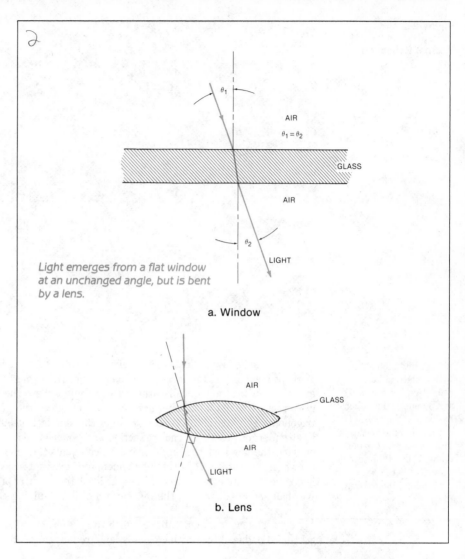

*Light emerges from a flat window
at an unchanged angle, but is bent
by a lens.*

a. Window

b. Lens

That means that instead of being bent closer to the normal, as in *Figure 2-3*, the light is bent farther from it, as in *Figure 2-5*. This isn't a problem if the angle of incidence is small. For I = 30°, sin I = 0.5, and sin R = 0.75. But a problem does occur when the angle of incidence becomes too steep. At I = 60°, sin I = 0.866, so Snell's law says that sin R = 1.299. That angle can't exist because the sine can't be a number greater than one!

**Figure 2-5.
Refraction and Total
Internal Reflection**

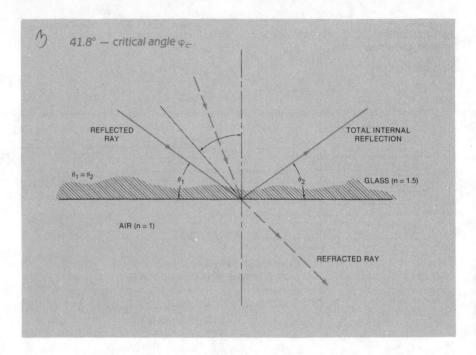

Snell's law indicates that refraction can't take place when the angle of incidence is too large, and that's true. If the angle of incidence exceeds a critical value, called the critical angle, where the sine of the angle of refraction would equal one, light cannot get out of the glass. (Recall from trigonometry that the maximum value of the sine is 1 at 90°.) Instead, the light undergoes total internal reflection and bounces back into the glass, obeying the law that the angle of incidence equals the angle of reflection, as shown in *Figure 2-5*. It is this phenomenon of total internal reflection that keeps light confined in optical fibers, at least to a first approximation. As we shall see in Chapter 4, the mechanism of light guiding can be more complex.

The critical angle above which total internal reflection takes place, ϕ_c can be deduced by turning Snell's law around, to give:

> If light hits a boundary with a material of lower refractive index at a steep enough (i.e., glancing) angle, it cannot get out and is reflected back into the high-index medium. This total internal reflection is the basic concept behind the optical fiber.

$$\phi_c = \arcsin\left(\frac{n_r}{n_i}\right)$$

For the example given, with light trying to emerge from glass into the air, the critical angle is arcsin (1/1.5) or 41.8°.

LIGHT GUIDING

Light is guided in the core of an optical fiber by total internal reflection when it hits the boundary of the lower-index cladding that surrounds it.

The two key elements of an optical fiber—from an optical standpoint—are its core and cladding. The core is the inner part of the fiber, through which light is guided. The cladding surrounds it completely. The refractive index of the core is higher than that of the cladding, so light in the core that strikes the boundary with the cladding at a glancing angle is confined in the core by total internal reflection, as shown in *Figure 2-6*.

Figure 2-6.
Light Guiding in an
Optical Fiber

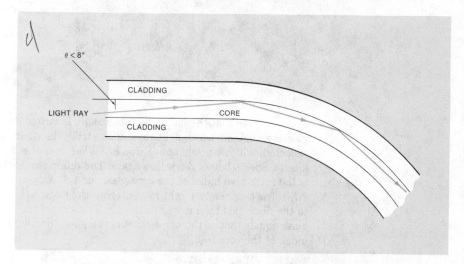

The difference in refractive index between core and cladding need not be large. In practice, it is only about 1%. This, however, still allows light guiding in fibers. For $n_r/n_i = 0.99$, the value of ϕ_c is about 82°. Thus light is confined in the core if it strikes the interface with the cladding at an angle of 8° or less to the surface. The upper limit can be considered the confinement angle in the fiber.

The angle over which a fiber accepts light depends on the refractive indexes of core and cladding glass.

Another way to look at light guiding in a fiber is to measure the fiber's acceptance angle—the angle over which light rays entering the fiber will be guided along its core, shown in *Figure 2-7*. (Because the acceptance angle is measured in air outside the fiber, it differs from the confinement angle in the glass.) The acceptance angle normally is measured as numerical aperture (NA), which for light entering a fiber from air is approximately:

$$NA = \sqrt{n_0^2 - n_1^2}$$

where n_0 is the refractive index of the core and n_1 is the index of the cladding. For a fiber with core index of 1.50 and cladding index of 1.485 (a 1% difference), NA = 0.21. An alternate but equivalent definition is the sine of the half-angle over which the fiber can accept light rays, 12° in this example, θ in *Figure 2-7*. Another alternate definition is $NA = n_0 \sin \theta_c$, where θ_c is the confinement angle in the fiber core (8° in this example).

**Figure 2-7.
Measuring the
Acceptance Angle**

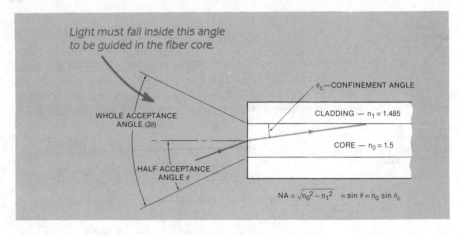

Light must fall inside this angle
to be guided in the fiber core.

θ_c—CONFINEMENT ANGLE

WHOLE ACCEPTANCE
ANGLE (2θ)

CLADDING — $n_1 = 1.485$

CORE — $n_0 = 1.5$

HALF ACCEPTANCE
ANGLE θ

$$NA = \sqrt{n_0^2 - n_1^2} = \sin\theta = n_0 \sin\theta_c$$

Note that the half acceptance angle is larger than the largest
glancing angle at which light rays must strike the cladding interface to be
reflected, which we said above was 8°. What does this mean? Go back and
look at Snell's law of refraction again. The difference is the factor n_0, which
is the refractive index of the core glass, or 1.5. As you can see in *Figure
2-7*, refraction bends a light ray entering the fiber so it is at a smaller angle
to the fiber axis than it was in the air. The sine of the angle inside the
glass equals that of the angle outside the glass, divided by the refractive
index of the core (n_0).

LIGHT COUPLING

Numerical aperture and acceptance angle measure a crucial
concern in practical fiber-optic systems: getting light into the fiber, which is
known as coupling. A century ago, British physicist Charles Vernon Boys
made fine fibers of glass comparable in size to today's optical fibers, but as
far as we know today, he didn't think of trying to transmit light along their
lengths. Even when fiber optics was first developed in the 1950s, no one
seemed to think that much light could be coupled into single fibers. Instead,
they grouped fibers into bundles to collect reasonable amounts of light.
Only when lasers made highly directional beams available did researchers
seriously begin to consider using single optical fibers.

Light Source and Fiber Size

As we saw earlier, the angles over which optical fibers can accept
light are limited, but conventional optics can readily produce such narrow
beams. Just look at a flashlight. The biggest problem is that individual
fibers are small. The commonest optical fibers used in communications are
0.25–0.5 mm in diameter, but include a plastic coating that protects the
fiber from harm. The cladding—the outer part of the fiber proper—is only
125 μm across (0.125 mm), and the inner core that carries the light is only a
fraction of that diameter (from 8 to about 85 μm). Some fibers are larger,

with cores 100–1000 μm (0.1–1 mm) in diameter. Even smaller fibers may be used in bundles; the problem of getting light into them is simplified by collecting them together.

A fiber will pick up some light from any source. Hold a single fiber so one end points at a light bulb. Now look into the other end, bending the fiber so you don't look directly at the light. You can see some light, but that is only a tiny fraction of the light from the bulb.

Efficient coupling requires a light source similar in size to the fiber core. For small-core fibers, the best match is a semiconductor diode laser, which emits light from a region a fraction of a micrometer high and a few micrometers wide. Light-emitting diodes (LEDs), with larger emitting areas, work well with larger-core fibers. The practical tradeoffs are cost and lifetime versus performance; laser diodes cost more than LEDs and have shorter lifetimes, but they can deliver higher powers into fibers and operate at higher speeds.

Alignment

Fiber-to-fiber coupling requires careful alignment and tight tolerances.

Coupling light between fibers requires careful alignment and tight tolerances. The highest efficiency comes when the ends of two fibers are permanently joined in a splice (described in Chapter 9). Temporary junctions between two fiber ends, made by connectors (described in Chapter 8) have slightly higher losses but allow much greater flexibility in reconfiguring a fiber-optic network. Special devices called couplers (described in Chapter 10) are needed to join three or more fiber ends. One of the most important functional differences between fiber-optic and wire communications is that fiber couplers are much harder to make than their metal-wire counterparts.

Coupling losses must always be considered in fiber-optic communication systems.

The losses in transferring signals between wires are so small that they normally can be neglected. This is not so for fiber optics. As we will see in Chapter 14, system designers must consider the coupling loss at every single connector, coupler, and light source. Splice losses also are significant, although they often are counted in overall fiber loss.

TRANSMISSION AND ATTENUATION

Some light is lost in transmission through a fiber. The amount of loss depends on wavelength.

Transmission of light by optical fibers is not 100% efficient. Some light is lost, in a process called attenuation of the signal. Several mechanisms are involved, including absorption by materials within the fiber, scattering of light out of the fiber core, and leakage of light out of the core caused by environmental factors. The degree of attenuation depends on the wavelength of light transmitted, as shown in *Figure 2-8.*

Figure 2-8.
Loss as a Function of
Wavelength (*Courtesy*
Corning Glass Works)

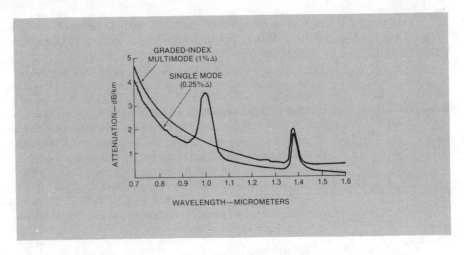

Figure 2-8.
Loss as a Function of
Wavelength (*Courtesy*
Corning Glass Works)

Attenuation measures the amount of signal lost in transmission by comparing output power with input power. Measurements are made in decibels (dB), a very useful unit, albeit a peculiar one. The decibel is a logarithmic unit measuring the ratio of output to input power. (It is actually a tenth of a unit called a "bel" after Alexander Graham Bell, but that base unit is virtually never used.) Loss in decibels is defined as:

$$\text{dB LOSS} = -10 \times \log_{10}\left(\frac{\text{POWER OUT}}{\text{POWER IN}}\right)$$

Thus, if output power is 0.001 of input power, the signal has experienced a 30-dB loss. The minus sign is there by convention to avoid negative numbers in attenuation measurements. It would not be used in systems where the signal level might increase, because then the sign of the logarithm would indicate whether the signal had decreased (minus) or increased (plus).

Attenuation of a fiber is the product of the length times the characteristic loss in decibels per kilometer.

Each optical fiber has a characteristic attenuation that is measured in decibels per unit length, normally decibels per kilometer. The total attenuation (in decibels) in the fiber equals the characteristic attenuation times the length. To understand why, consider a simple example, with a fiber having the relatively high attenuation of 10 dB/km. That is, only 10% of the light that enters the fiber emerges from a 1-km length. If that output light was sent back through the same length of fiber, only 10% of it would emerge (or 1% of the original signal) for a total loss of 20 dB.

As can be seen in *Figure 2-8*, the attenuation that light experiences in a fiber depends on its wavelength. The curves shown are fairly typical for the two major types of telecommunication fibers, which we will discuss more completely in Chapter 4. The absorption peak at 1.0 μm is caused by the peculiarities of single-mode fiber; the peaks at 1.4 μm are caused by traces of water remaining in the fiber as an impurity. Otherwise, the curve is fairly smooth in the region shown. Attenuation increases sharply at wavelengths longer than 1.6 μm, not shown in this plot, because the silica in glass begins absorbing light. Attenuation also continues increasing at shorter wavelengths.

As we will see later in this chapter, the choice of operating wavelength depends not only on fiber loss but also on the available light sources and on other fiber properties. Fiber loss in the part of the spectrum shown is low compared to other media with comparable signal-transmission capability. Attenuation is very low at 1.3 μm, and even lower at 1.55 μm. At the 0.5-dB/km attenuation typical in the 1.3-μm region, 1% of the light entering the fiber remains after 40 km, for a 20-dB loss. At the 0.2-dB/km attenuation available at 1.55 μm, 1% of the input light remains after 100 km. This allows long-distance transmission without amplifiers or repeaters, an important consideration for telecommunications.

BANDWIDTH AND DISPERSION

Optical fibers are unique in allowing high-speed signal transmission at low attenuation.

Low attenuation alone is not enough to make optical fibers invaluable for telecommunications. The thick wires used to transmit electrical power also have very low loss. Optical fibers are attractive because they are able to combine low loss with high bandwidth and to transmit high-speed signals (i.e., high information capacity).

Information Capacity

Information capacity is a very important concept in all types of communications. It is measured in different ways for analog and digital systems. In digital systems, it is measured in bits—units of information—per second. The more bits that can pass through a system in a given time, the more information it can carry. A 100-megabit per second (Mbit/s) systems can carry as much information as a hundred 1-Mbit/s systems. As long as all the information is going between the same two points, it is much cheaper to build one 100-Mbit/s system than a hundred 1-Mbit/s systems. The same principle works for analog systems, but we measure capacity as frequency in hertz.

That's fine in theory, but in practice the attenuation of wires increases with operating frequency. In coaxial cable, the loss increases sharply with frequency, as shown in *Figure 2-9*. However, in optical fibers the loss is essentially independent of signal frequency over their normal operating range. (The scale is measured in loss per kilometer of cable and does not take into account the coupling losses mentioned earlier.)

Figure 2-9.
Loss as a Function of
Frequency

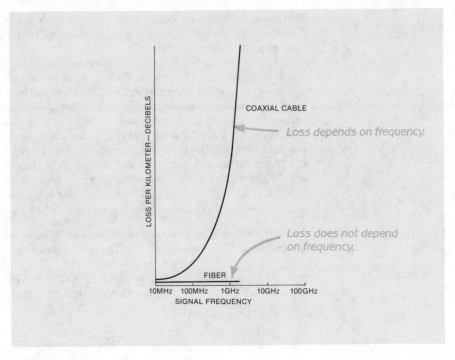

Transmission Speed Limits

Some effects limit the transmission speed of fibers. They are best seen by looking at digital transmission, although they also occur for analog transmission. Consider another view of the basic optical fiber shown in *Figure 2-10*. As long as the fiber's core diameter is much larger than the wavelength of light—i.e., well over 10 μm—rays can enter the fiber at many different angles to its axis. A ray that bounces back and forth within the core many times will travel a slightly greater distance than one that goes straight through. For instance, if one ray travelled straight through a 1-km fiber and another bounced back and forth at a 5° angle through a 100-μm core through the same fiber, the second ray would travel 3.8 m farther.

What this means for communications can be seen by considering pulses of light sent through the fiber. Some light rays would go straight down the fiber, while others would be reflected back and forth. For this example, an instantaneous pulse at the start of the fiber would spread out to 12.7 ns at the end.

This phenomenon is called pulse dispersion. Our example gives a rather oversimplified view of how light travels through the fiber in what are called different modes. As we will see in Chapter 4, this and other effects can cause various degrees of pulse dispersion. Dispersion is important because, as the pulses spread, they can overlap and interfere with each other, limiting data transmission speed. In our example, if each pulse went through 2 km of fiber, it would acquire a 25-ns tail. So the time

Pulse dispersion limits fiber transmission capacity.

Figure 2-10.
Light Rays Take
Different Paths in a
Fiber Core

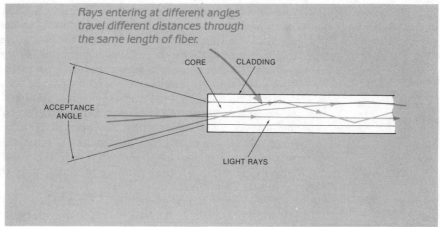

Rays entering at different angles travel different distances through the same length of fiber.

CORE   CLADDING

ACCEPTANCE
ANGLE

LIGHT RAYS

$$\frac{1}{40} = 25 \, ns/bits$$

between pulses would have to be at least that long, limiting transmission to 40 Mbit/s. Because actual input pulses would not be instantaneous, the real maximum pulse rate would be even slower.

Dispersion and Distance

Pulse dispersion increases with distance, so maximum transmission rate decreases with distance.

Because pulses stretch out a certain amount each kilometer of fiber, pulse dispersion—like attenuation measured in decibels—increases linearly with distance travelled. Typical values for fibers that carry multiple modes are measured in nanoseconds (of dispersion) per kilometer of fiber. These can be translated into an analog bandwidth limit (20 MHz-km for commercial 100-μm core fibers, for example) or a maximum data rate for digital transmission. Both bandwidth and data rate are the inverse of pulse dispersion and are quoted in units times kilometers of fiber. Thus, while pulse dispersion increases with length of the fiber, the bandwidth or data rate decreases.

$$bandwidth = \frac{1}{dispersion}$$

The bandwidths quoted so far are low by fiber-optic standards because they are for fibers that carry light in many modes. Dispersion is much smaller in fibers that carry light in only one mode, which will be described in more detail in Chapter 4. These single-mode fibers can carry hundreds of millions or billions of bits per second over many kilometers. The residual dispersion in them depends on the range of wavelengths emitted by the light source.

LIGHT SOURCES

Light sources in fiber-optic communications are semiconductor lasers or light-emitting diodes.

Fiber-optic communication systems require a light source to transmit the signal. In practical systems, these light sources are semiconductor diode lasers or LEDs. Some inexpensive short-distance systems use LEDs that emit visible light, but most systems operate at near-infrared wavelengths.

t two onsecutive pulses

Inexpensive
GaAs/GaAlAs light
sources are used for
short-distance systems,
but more costly 1300-nm
sources are used for long-
distance transmission be-
cause fiber loss is lower
at the longer wavelength.

Many fiber systems use light sources of gallium arsenide (GaAs) and gallium aluminum arsenide (GaAlAs) emitting at 800–900 nm (0.8–0.9 μm). GaAs/GaAlAs lasers and LEDs were the best sources available for the first fiber-optic systems; they remain inexpensive. However, as we can see in *Figure 2-8*, loss is lower at longer wavelengths, so GaAs and GaAlAs sources are used in systems where attenuation of a few decibels per kilometer can be tolerated. Pulse dispersion is also high.

For long-distance systems, a better wavelength is 1300 nm, where fiber loss is 0.4–0.5 dB/km, and pulse dispersion is naturally very low. Semiconductor lasers or LEDs that emit at 1300 nm are made of indium gallium arsenide phosphide (InGaAsP), which will be described in more detail in Chapter 6. Those devices are harder to make than GaAs devices and are thus much more expensive.

There is strong interest in 1550-nm light sources, because loss at that wavelength is only 0.2–0.25 dB/km. However, pulse dispersion is high at 1550 nm, as will be described in Chapter 4. InGaAsP can be used to make 1550-nm devices, but the task is harder than producing 1300-nm lasers and LEDs.

MODULATION

An important practical advantage of semiconductor lasers and LEDs is that they can be modulated directly by the same electrical current that powers them. The signal is imposed on the drive current, causing light output to vary in proportion to the signal.

Changing the current
passing through a semi-
conductor laser or LED
changes its light output,
modulating it with a
signal.

The relationships between drive current and optical output power for an LED and a semiconductor diode laser are shown in *Figure 2-11*. Changes in the drive current cause changes in the optical output. Thus a digital signal applied as a current modulation emerges as a series of light pulses. Analog modulation makes input current vary continuously over a range, so light output also rises and falls. The current–output curves of optical devices are not perfectly linear, so direct modulation can induce some distortion in analog signals. (Digital transmission is largely immune to such effects, which is one reason it is so popular.)

Direct modulation sounds so logical that it is hard to see why anyone would do anything any differently. However, other lasers cannot be modulated in the same way at high speeds; they have to be powered with a steady source and modulated with an external device that varies its transmission of light. As we will see later, external modulation may eventually be used in some ultra-high–performance laser systems, but the ability to get away from such complexity was a major reason that semiconductor lasers and LEDs were picked for fiber-optic systems.

**Figure 2-11.
Relationship Between
Drive Current and
Optical Output**

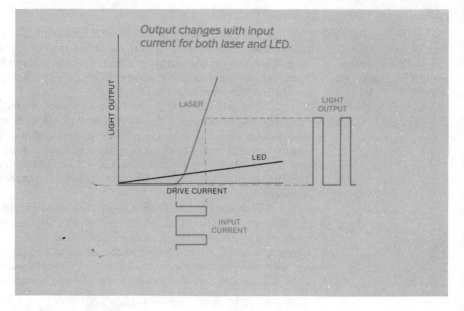

*Output changes with input
current for both laser and LED.*

MULTIPLEXING

Most fiber-optic systems
carry many signals com-
bined or multiplexed into
a single transmission.

A fiber-optic communication system may carry a single simple signal, such as a video channel. However, many systems carry several separate signals merged together or multiplexed over a single pair of optical fibers (one carrying signals in each direction). The idea is shown in *Figure 2-12*. It is particularly advantageous for fiber optics because of fiber's high transmission capacity. Hundreds or thousands of individual telephone circuits can be multiplexed onto a single pair of fibers, as we will see in Chapter 15.

**Figure 2-12.
Multiplexing Signals**

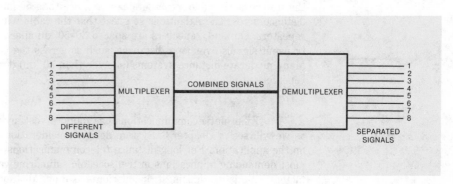

DETECTION

The final element of a fiber-optic communication system is the
receiver, which detects the optical signal and converts it to electronic form
for further transmission or processing. Typically, a receiver consists of a
semiconductor photodetector and amplifying electronics. (Its variations will
be described in more detail in Chapter 7.)

How well the detector can do its job depends on its sensitivity and
the signal level reaching it. Sensitivity, in turn, depends on operating
conditions, detector design, and how the material used in the detector
responds to different wavelengths. Silicon photodiodes, for example, work
well at 0.8–0.9 μm, but are not usable at 1.3 μm. Other detectors, of
germanium or indium gallium arsenide (InGaAs), are needed at the longer
wavelength.

The receiver electronics process the electrical signal generated by
the detector to replicate the original signal sent through the fiber. In
essence, this is a cleaning-up process. The signal itself is amplified. Pulse
edges, which had grown blurred as the signal passed through the system,
are sharpened by detecting when the signal strength passes a threshold
level. Timing of a series of pulses is checked and corrected.

REPEATERS AND REGENERATORS

By driving a transmitter with the output from a receiver,
transmission distances can be stretched far beyond what is possible in a
single length of fiber. An international consortium is using this approach to
build a transatlantic fiber-optic cable, connecting the United States with
Europe. There is nothing revolutionary about repeaters, and they don't
require fiber optics. In fact, because fiber-optic cable can carry signals
farther between repeaters than other types of cables, it is helping to
eliminate many excess repeaters.

Nonetheless, repeaters are needed in some fiber-optic systems. In
the transatlantic cable, repeaters can be more than 50 km apart, but the
distance across the Atlantic is so great that the cable will include about 125
repeaters. On land, repeaters are spaced 30–50 km apart in systems that
transmit signals over long distances, such as across the United States.
Many moderate-distance systems (shorter than 30 km) don't require any
repeaters at all.

SYSTEM CONSIDERATIONS

Fiber-optic communications has found a vast array of applications,
as we will see in Chapter 3. System design considerations depend largely
on the application. For long-distance telecommunications—including the
most demanding applications in transoceanic submarine cables—crucial
factors include maximizing transmission speed and distance and minimizing
fiber and splice loss. By contrast, connector loss becomes vital in local-area
networks that operate within buildings. In long-haul systems, it is
important to minimize the cost of cable, while the goal is to reduce the cost

of terminal equipment for local-area networks (because there are many separate terminals, rather than the two required for point-to-point telecommunications).

These system considerations make design and construction of practical fiber-optic systems a multifaceted task. Guidelines appropriate for one type of system should not be followed blindly for others, because they often lead directly to the wrong answers. Applications are diverse enough that many different components are offered.

WHAT HAVE WE LEARNED?

1. Light is guided through optical fibers by total internal reflection of light that enters them within an acceptance angle, measured directly or as the numerical aperture (NA).
2. Refractive index is a crucial property of optical materials.
3. The core of an optical fiber must have a higher refractive index than the cladding surrounding it.
4. Fiber-optic communication systems transmit near-infrared light at 800–900, 1300, or 1550 nm.
5. The small dimensions of optical fibers make tolerances tight in transferring light into fibers.
6. A major attraction of fiber optics is its ability to send high-speed signals over long distances.
7. Attenuation of optical fibers (measured in decibels) is proportional to a characteristic value for the fiber (measured in decibels per kilometer) and the length of the fiber.
8. Transmission capacity of optical fibers depends on the dispersion of light signals sent through them, which is proportional to a characteristic value (measured in nanoseconds per kilometer) times the length of the fiber.
9. Attenuation and dispersion in an optical fiber depend on wavelength.
10. Light sources in fiber-optic systems are semiconductor lasers and LEDs.

WHAT'S NEXT?

In Chapter 3, we will look at the major applications for fiber optics, particularly in communication systems.

Quiz for Chapter 2

1. If light passes from air into glass, it is:
   a. reflected.
   b. refracted.
   c. absorbed.
   d. scattered.

2. Light is confined within the core of a simple optical fiber by:
   a. refraction.
   b. total internal reflection at the outer edge of the cladding.
   c. total internal reflection at the core–cladding boundary.
   d. reflection from the fiber's plastic coating.

3. An optical fiber has a core with refractive index of 1.52 and a cladding with index of 1.45. Its numerical aperture is:
   a. 0.15.
   b. 0.20.
   c. 0.35.
   d. 0.46.
   e. 0.70.

4. The input power to a fiber-optic cable is 1 mW. The cable's loss is 20 dB. What is the output power, assuming there are no other losses?
   a. 0.10 mW.
   b. 0.05 mW.
   c. 0.01 mW.
   d. 0.001 mW.
   e. None of the above.

5. The output of a 20-km fiber-optic cable is measured at 0.005 mW. The fiber loss is 0.5 dB/km. What is the input power to the fiber?
   a. 1 mW.
   b. 0.5 mW.
   c. 0.05 mW.
   d. 0.01 mW.
   e. None of the above.

6. Optical fiber attenuation is lowest at:
   a. 800 nm.
   b. 900 nm.
   c. 1300 nm.
   d. 1400 nm.
   e. 1550 nm

7. Optical fiber attenuation can be as low as:
   a. 0.1 dB/km.
   b. 0.2 dB/km.
   c. 0.4 dB/km.
   d. 0.5 dB/km.
   e. 1.0 dB/km.

8. A pulse sent through a 20-km optical fiber is 300-ns long when it emerges. What is the fiber dispersion? Assume that the pulse was instantaneous when it entered the fiber.
   a. 35 ns/km.
   b. 30 ns/km.
   c. 25 ns/km.
   d. 20 ns/km.
   e. 15 ns/km.

9. The most common wavelength for long-distance communications is:
   a. 800 nm.
   b. 900 nm.
   c. 1300 nm.
   d. 1400 nm.
   e. 1550 nm.

10. Repeaters in present long-distance fiber-optic system can be as far apart as:
    a. 10 km.
    b. 20 km.
    c. 30 km.
    d. 50 km.
    e. 80 km.

Applications of Fiber Optics

ABOUT THIS CHAPTER

Fiber optics is best known for its communication applications, with the best-publicized ones in the telephone industry. That vision reflects reality, because the telephone industry is the largest market for fiber optics. Nonetheless, optical fibers are used in many other types of communications, including short-distance data transmission, video systems, and local-area networks. More applications are coming in areas ranging from military systems to automobiles. And some applications, such as medical endoscopy, are entirely separate from communications.

This chapter will introduce how and why fiber optics are used. We will go into more detail on applications later, but a general understanding of how fibers are used will help you better appreciate important features of fiber technology that will be described in later chapters.

TYPES OF COMMUNICATIONS

Fiber optics can be used in many—but not all—types of communications, but system designs vary widely.

Communication systems come in many sizes, shapes, and forms. For example, the telephone network differs greatly from cable-television systems, even though both provide services to individual homes from a central facility. The telephone network makes temporary two-way connections between phones in different places and provides bulk telecommunication capacity between regional centers. Cable-television systems distribute the same programming to all subscribers, and information flows almost exclusively from the network to the subscriber (who does not send signals back over the network). Cable companies generally receive most of their programs from satellites, which in turn send signals to many cable operators around the country. Cable systems also transmit much more information to homes (i.e., many television signals simultaneously) than telephone networks. As we will see in later chapters, cable and telephone systems are designed very differently, in ways that make fiber work very well for many telephone applications but not so well for cable television.

Other differences are even more obvious. Computer data communication often is over short distances within buildings or among separate buildings on a campus. Computer data-transmission speeds are much lower than long-distance telephone systems. Thus, computer data communications often cannot benefit from the main advantages of fiber optics. However, fiber optics also can solve special problems, such as electromagnetic interference with signal transmission.

In the rest of this chapter, we will look at the important types of communications and how well fiber optics meets their needs. First, however, we need to look at the crucial distinction between analog and digital communications.

ANALOG AND DIGITAL COMMUNICATIONS

Signals can be transmitted in an analog or digital format. Digital technology has many advantages, including better compatibility with fiber optics.

Communication signals can be transmitted in two fundamentally different forms, as shown in *Figure 3-1*: continuous analog signals and discrete digital signals. The level of an analog signal varies continuously. A digital signal, on the other hand, can only be at a certain number of levels. The most common number, as shown in *Figure 3-1*, is two, with signals coded in binary format, either off or on.

**Figure 3-1.
Analog and Digital
Signals**

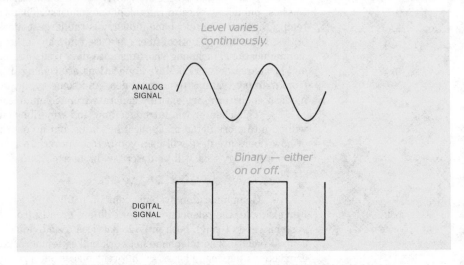

Each format has its advantages. The older analog technology is more compatible with people. Our ears, for instance, detect continuous variations in the level of sound, not just the presence or absence of sound. Our eyes likewise detect levels of brightness, not simply the presence or absence of light. For that reason, audio and video communications have traditionally been in analog form. Telephone wires deliver a continuously varying signal to your telephone handset, which converts those electronic signals into continuously varying sound waves.

Digital signals are more compatible with electronics and fiber optics. It is much simpler and cheaper to design a circuit to detect whether a signal is at a high or a low level (off or on) than to design and build one to accurately replicate a continuously varying signal. Digital signals also are much less prone to distortion, as shown in *Figure 3-2*. When an analog signal goes through a system that doesn't reproduce it exactly, the result is a garbled signal that can be unintelligible. That is exactly what happens when you get a distorted voice on the phone. However, when a digital signal is not reproduced exactly, it is still possible to tell the on from the off state.

Figure 3-2.
Distortion of Analog
and Digital Signals

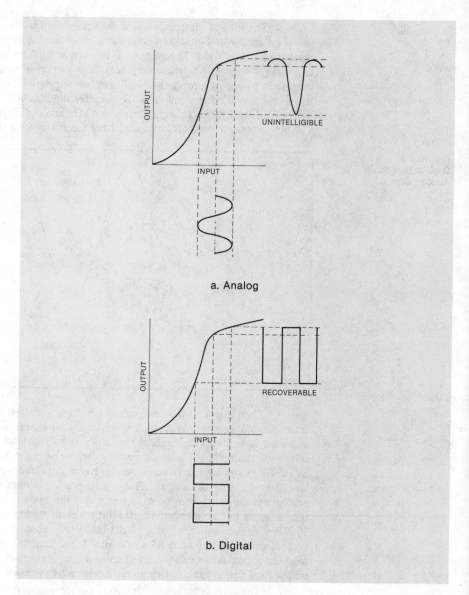

a. Analog

b. Digital

Converting Analog and Digital Signals

If people need analog signals but transmission works best with digital, what about converting between the two? That is being done increasingly in audio systems. Ultra–high-fidelity compact disc players use a laser to play back sound digitized as spots on a rapidly spinning disc. Internal electronics then convert the digitized sound back to analog form. The telephone network also can convert the analog signals a telephone generates from someone's speech into digital form.

The idea of digitization is simple, as shown in *Figure 3-3*. A circuit called an analog-to-digital converter samples an analog waveform to measure its amplitude. The samples are taken at uniform intervals (8000 times per second in a telephone circuit). The converter assigns the signal amplitude to one of a predetermined number of possible levels. For a telephone circuit, that number is 128 (exactly the number of levels that can be encoded by 7 bits). This converts the 4-kHz analog telephone signal into a digital stream of sets of 7 bits sent 8000 times a second (56,000 kbits/s).

Figure 3-3.
Digitization of an
Analog Signal

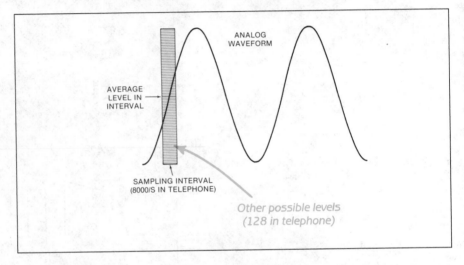

The figures in the last paragraph show one disadvantage of digital transmission. Accurate reproduction of an analog signal requires sampling at a rate faster than the highest frequency to be reproduced. In the telephone example, the sampling rate is twice the highest frequency. Many bits (7 for telephony) have to be sent per sampling interval. This requires a large transmission bandwidth. There is no precise equivalence between analog and digital transmission capacity, but the two are comparable: a transmission line able to handle 10 Mbit/s has an analog capacity of around 10 MHz. That means an analog signal takes only about a tenth of the transmission capacity as it does in digital form. That is not a problem in telephony, but as we will see in Chapter 17, it is a serious obstacle to the widespread use of fiber optics in cable television networks.

The shift from analog to digital technology is good news for fiber optics. Light sources in particular suffer from nonlinearities that induce distortion in analog optical signals at high frequencies. However, fiber-optic systems work fine in the digital mode. Moreover, they have plenty of the transmission capacity that digital systems demand. Those are major factors that have helped fiber optics grow very rapidly as the telephone industry turned to digital technology.

LONG-DISTANCE TELECOMMUNICATIONS

Telephone Network Structure

The telephone network includes subscriber loops, trunk lines, and backbone systems. Fiber optics is widely used for trunk and backbone systems.

The telephone network can be loosely divided into a hierarchy of systems, shown in simplified form in *Figure 3-4*. Your home or business telephone is part of the subscriber loop or local loop, the part between individual subscribers and telephone-company switching offices (called central offices in the telephone industry). Trunk lines run between central offices, for example, carrying telephone calls from one suburb to another or from suburbs to central cities. Long-distance telephone carriers operate long-haul backbone systems between area codes.

Figure 3-4.
Parts of the Telephone Network

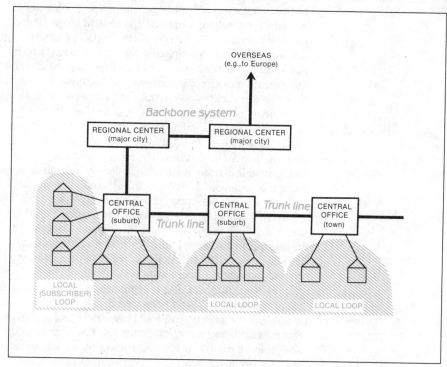

In practice, things are somewhat more complicated than shown in *Figure 3-4*. The subscriber loop may be broken down into two segments—a connection from the central office to a dropoff point and individual wires from that dropoff point to homes. Regional telephone-operating companies and long-distance carriers each have treelike networks that reach throughout the regions where they operate, making it hard to define a boundary between trunk and backbone network. Much organizational debris remains from the rapid deregulation of the telephone industry, such as central offices with tape down the middle to divide equipment that belongs to the regional telephone company from that belonging to long-distance companies.

Telephone System Hardware

The telephone system is changing from analog to digital transmission. Fiber optics are very well matched to digital requirements.

Deregulation is not the only revolution in the telephone industry. The telephone network is rapidly changing from analog to digital technology. That change is not visible to most telephone users because virtually all of it is inside telephone company facilities (e.g., the switching office and trunk and backbone communication lines). Only a handful of business users with special demands use digital transmission on their own phone lines. However, in the long term, the telephone industry is hoping to bring digital technology all the way to the subscriber.

The telephone network makes extensive use of multiplexing.

The telephone network makes extensive use of multiplexing to carry many phone circuits on single cables. Only in the local loop, between homes and central offices or remote dropoff points, is one conversation routed over a pair of wires. At the central office (or in an increasing number of cases at an intermediate dropoff point), the signals from many telephone lines are multiplexed for simultaneous transmission through a single cable. Successive levels of multiplexing raise transmission speeds and number of voice circuits carried on a single cable until the level reaches thousands in backbone systems. Originally this was done with analog technology; now more and more is done digitally.

Backbone communication systems rely on fiber optics to carry thousands of telephone circuits per fiber pair.

As we saw in Chapter 2, the advantages of fiber optics grow with transmission speed and distance. Thus, the higher the transmission speed and the longer the distance, the more likely fiber optics will be used. The nationwide backbone systems installed by long-distance telephone carriers (e.g., the American Telephone & Telegraph Co. and MCI Communications) rely largely on fiber-optic cables. Each fiber pair transmits signals at 400 Mbit/s and up, which is equivalent to thousands of 56,000-bit/s digital voice circuits. AT&T in the U.S. and the Nippon Telegraph and Telephone Corp. in Japan plan to convert their backbone systems to 1.7 Gbit/s soon. Some long-distance traffic still goes via satellite or microwave transmission on the ground, but optical fibers are taking much of the load on land.

Fiber optics are used in trunk lines carrying 45, 90, and 135 Mbit/s.

Optical fibers also carry much of the telephone trunk load. The telephone industry's first widespread use of fiber optics was for 45 Mbit/s trunks between central offices. Most urban and suburban central offices are several kilometers apart, and even the first commercial fiber systems, operating at 800–900 nm, could transmit over those distances without repeaters. When a second generation of fiber technology became available, transmitting at 1300 nm, longer and faster fiber systems came into use. Now fiber trunk cables carry signals at 45, 90, and 135 Mbit/s. Repeaters can be 30 km or more apart, depending on system requirements, which is important in transmitting signals through sparsely populated regions.

Fibers are beginning to permeate into lower-speed systems, mostly those operating at 1.5 or 6 Mbit/s in North America (standard transmission rates are different in Europe). Some carry signals from central offices to remote nodes, where the signals are split up to separate wires that carry telephone circuits to individual subscribers. Some carry signals to business and industrial users who require high-speed connections with the central office—either to drive their own switchboards or to connect directly with computers to allow more efficient data transmission than is possible if data signals have to be converted to analog form so they can go through ordinary phone lines.

By replacing one thick metal cable with four fiber cables that can fit in the same duct, telephone companies can save much money on installation costs.

Why are fiber optics used in these systems? Telephone companies cite a host of reasons. The high capacity and low attenuation of fibers make them the medium of choice for long-distance, high-speed systems. Fiber-optic cables are small enough to fit into existing underground ducts. By pulling out one old metal cable, a phone company can make room for four fiber cables. That means important savings because the cable itself is only a small part of the cost of installing a system in urban areas. (If new ducts are needed, most of the money goes for labor and construction.) What's more, the cables can be installed with extra fibers to leave room for future expansion—at minimal extra cost and without need for major construction or excavation.

Fibers have other advantages that can be important for particular applications. Fiber-optic cables can be made with no electrical conductors— so they won't carry dangerous current pulses injected by lightning strikes. Signals carried in fibers cannot pick up electromagnetic interference from power lines, generating plants, or other sources. The high capacity of fibers also leaves plenty of room for future expansion—an important consideration for telephone-industry planners looking for continued steady growth in communication demand.

SUBMARINE FIBER CABLES VERSUS SATELLITES

Undersea fiber-optic cables will compete with satellites for transatlantic communication traffic— and may win most of the telephone market.

The most dramatic example of the role fiber optics is playing in long-distance communications is development of ocean-spanning cable systems that—at least in the North Atlantic—threaten to bring much satellite communications back to earth. As we will see in Chapter 15, the coaxial cable technology used in earlier transatlantic telephone cables came to the end of its rope with a cable laid in 1983. Fiber optics offers a combination of long repeater spacing and high capacity that is just what the doctor ordered for the ailing undersea cable business. Three groups are planning to lay a total of five cables across the North Atlantic by the early 1990s. When the Office of Technology Assessment looked at these plans together with plans for several satellites serving the same route, it warned of possible "vast overcapacity in transatlantic telecommunications in the mid-1990s." That hasn't stopped anybody yet. Meanwhile, multiple cables are planned across the Pacific.

Not all submarine cables span the oceans. The Japanese fiber-optic backbone system installed by Nippon Telegraph and Telephone includes a 38.6-km undersea segment that links the country's main island with the northern island of Hokkaido. A submarine fiber cable was recently laid under the English Channel. Many shorter submarine fiber cables are being installed around the world, including one that links the Statue of Liberty and Ellis Island with a telephone office in Manhattan.

DIGITAL DATA LINKS

Fibers are used for some computer data communications but mostly only when other technologies won't do the job.

Although fiber optics are well suited for digital transmission, they are not widely used for digital data communications. The reason is that most data communication is too easy. Wires are perfectly adequate for transmitting 9600 bit/s a couple of meters from a personal computer to a printer. Fibers have found some places in digital data communications (often called data links because they link two points), but those are mostly where other technologies can't do the job.

Telephone lines often are used for digital data transmission at modest speeds, such as the 300 or 1200 bit/s of modems used with personal computers and remote terminals. For faster transmission or to maintain continual connection between devices, a better solution is a dedicated transmission line. Systems can differ widely. Distances can range from a few meters within an office to a few kilometers between buildings. Data transmission rates may be a few kilobits per second or hundreds of megabits per second.

The faster the speed or the more difficult the environment, the more likely fiber optics are to be used. If electromagnetic interference is severe enough, fiber optics may be needed to carry the equivalent of a single telephone channel. Fibers can run alongside power lines, to carry data to monitor electric power utilities or to carry data up and down an elevator shaft, which is a convenient place to string cables in a high-rise building. Their small size makes fiber cables a good choice when installation is in tight quarters or through already-crowded cable ducts. Fibers also are valuable in linking users far from the main computer or those who need unusually high data rates (e.g., users of high-resolution terminals for computer-aided design and manufacture).

LOCAL-AREA NETWORKS

The need to interconnect many terminals is one characteristic that makes local-area networks hard to build with fibers.

There is a subtle but important difference between digital data links and local-area networks. Digital data links run between two points; local-area networks interconnect a multitude of points, as shown in *Figure 3-5*. In other words, a point-to-point data link transfers information between two points at opposite ends of a transmission line. Local-area networks allow the interchange of data among many points. This distinction is crucial because fiber optics are more amenable to point-to-point transmission than networking.

Transferring optical signals from light source to fiber and between fibers is far from trivial, but it is a tractable problem. It is a straightforward task to send a signal from a light source through a fiber to a receiver. Problems begin to appear if the signal must be divided among receivers. One very serious problem is cost. As the number of terminals multiplies so does the number of transmitters and receivers. Wires can handle signals as they emerge from computers and peripheral devices, but fiber optics requires optical transmitters and receivers. Those costs add up in a multiterminal network. Complications also arise in splitting light in one fiber between two or more other fibers.

**Figure 3-5.
Transmission Systems**

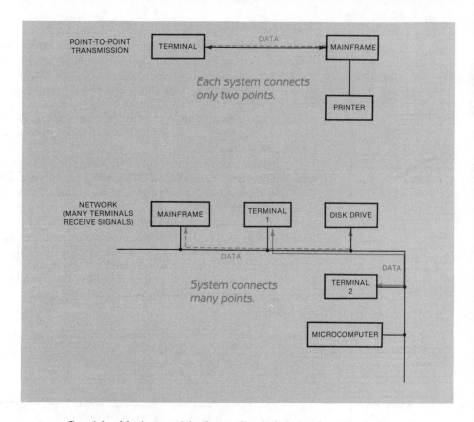

Fiber optics are used in a standard 100-Mbit/s local-area network.

Coaxial cable (or coax) is the medium of choice for today's 10-Mbit/s Ethernet-type systems. Fiber is used only in special cases, such as connections to remote nodes beyond the normal reach of coax. However, a standard fiber-optic network, the Fiber-Optic Distributed Data Interface (FDDI), is being developed for operation at 100 Mbit/s, a speed at which the advantages of fiber transmission outweigh such considerations at terminal costs. The emerging FDDI standard will be described in more detail in Chapter 18.

A less-visible data-transmission network exists inside computers. For high-speed supercomputers, interconnection becomes a crucial limitation. Optical fibers are among the technologies being studied to enhance the interconnections between integrated circuits used in large supercomputers. Fibers already connect circuit boards within some large special-purpose computers used to switch telephone signals.

VIDEO TRANSMISSION

Fiber optics is used for some video transmission, but coaxial cables are better matched to the needs of current cable-television networks.

Despite their high transmission capacity (or equivalently their broad bandwidth), optical fibers have found only limited uses in video transmission. Because they are small and lightweight, they can carry signals from portable cameras back to a mobile control room in electronic news gathering. Temporary fiber-optic cables have been laid for predictable news coverage ranging from the Olympics to space shuttle launches. However, they are little used for the most widespread application of video cables—cable television.

Design of the cable-television network is very different from that of the telephone network.

One reason they are not used widely in cable-television systems is that such systems are designed differently than the telephone network, as shown in *Figure 3-6*. There are several important differences. First, video transmission travels in only one direction. There need be no return video signal from the home. Second, video signals are not switched. Every home gets the same signal. (If extra-cost subscription-only services are sent over the cable, they go to all homes in a coded form. Only subscribers to those services get the special decoder needed to view them.) This approach has led to the design of cable systems around the tree architecture shown in *Figure 3-7*. If you compare that with *Figure 3-6*, you will see the obvious similarities.

**Figure 3-6.
Cable-Television System
Design**

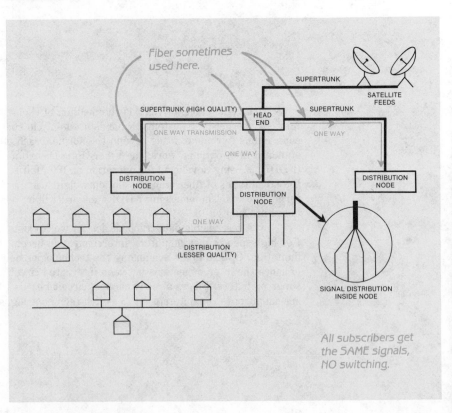

**Figure 3-7.
Tree Architecture for a
Communication System**

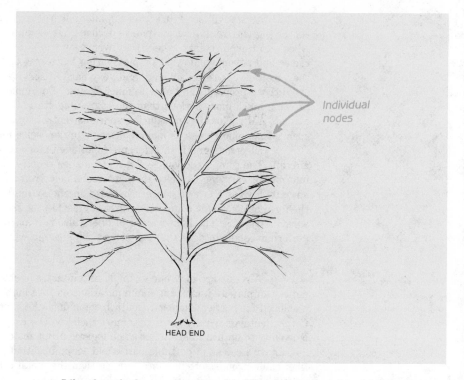

Individual nodes

HEAD END

Like the telephone network, cable-TV systems contain cables that transmit many signals simultaneously. Called supertrunks, they carry video signals from satellite feeds to the head end or center of the network and from the head end to distribution nodes. They must meet higher transmission-quality standards than the cable that distributes signals to homes. They also carry signals between two points, rather than from one point to many, as do distribution cables. As we shall see in Chapter 17, fiber optics is used in some supertrunks, but they are by no means standard. Coax rather than fiber is used for distribution cable.

One advantage of coax is its ability to drop signals off at many subscribers from a single cable. Another is that it can handle the multiplexed analog video signals transmitted to homes. All in all, fiber optics now is too expensive for cable-television networks.

AUTOMOBILES AND AIRPLANES

Key attractions of fiber optics for use in cars and planes are its light weight and immunity to electromagnetic interference.

The small size, light weight, and immunity to electromagnetic interference of optical fibers has attracted designers of automobiles and aircraft. The electronic content of cars and planes has increased steadily, and with it the need for communications. Such systems are quite different from those intended for long-distance communication because of their small overall size and the need for very low-cost components for cars.

As we will see in Chapter 19, automakers want fibers to carry control signals from a central point—perhaps in the center of the steering wheel—to remote points, where they would switch on accessories such as radios and power windows. Complex wiring harnesses do the job today, but they are bulky, easily damaged, and very hard to service. Fibers would avoid those problems and also be much lighter and immune to EMI, a problem much more serious than the spark-plug static that sometimes plagues AM radios. By introducing spurious control signals, EMI can do such threatening things as make electronic brakes seize up.

Electromagnetic interference is an even more serious problem in aircraft. The complex avionics of modern planes requires much greater information-carrying capacity than automobile electronics. In military aircraft, EMI is a crucial concern because an enemy might try to use it, in the form of electronic countermeasures, to disable the plane's sophisticated weapon systems. The light weight of fibers also is important because of the performance advantages that can come from shaving weight from a plane.

MILITARY SYSTEMS

Many military systems are beginning to use fiber-optic communications.

Aircraft are only one example of military systems where fiber optics can play an important communication role. As high technology is mobilized for battle, command, control, communications, and intelligence—C³I in Pentagon acronymese—becomes crucial on the battlefield. The Army is looking to optical fibers to ease the logistic problems of deploying bulky 26-pair wire cables in portable battlefield communication systems. And the fiber cable promises to be more reliable than the metal cable, which often suffers from broken wires. Military planners also are working on a fiber-optic system for guiding battlefield missiles to their targets, which will be described in Chapter 19.

NON-COMMUNICATION FIBER OPTICS

Optical fibers can be used for transmitting light and images and for sensing.

Most of this book is about the uses of fiber optics in communications, but fiber optics is not only communications. Optical fibers also can be used in a variety of other applications, as we will see in Chapter 20. Many of them rely on fiber-optic bundles, which were described earlier.

Illumination

A light source at one end of a bundle of optical fibers illuminates whatever is at the other end. This can make a decorative lamp, a flexible illuminator for hard-to-reach places, or a practical display, such as the one shown in *Figure 3-8*.

Figure 3-8.
Fiber-Optic Display Sign
(Courtesy Mitsubishi Rayon)

If the fibers in a bundle are arranged properly at the ends, they can transmit images. One of the most important applications for fiber-optic image transmission is endoscopy, which lets physicians view an otherwise inaccessible part of the body. For example, an endoscope can be passed down the esophagus so a physician can see into the stomach. Endoscopes also can carry laser light to treat certain conditions, such as bleeding ulcers.

Sensors

Optical fibers also can be used as sensors. Some fiber sensors use the fiber only to pick up light and bring it to a place where it can be detected. Others rely on fibers to carry light to and from a sensing element. In still others, the fiber itself is the sensing element. Examples of the latter include the sensitive acoustic detectors the Navy hopes will keep track of the whereabouts of potentially hostile submarines and rotation sensors, which can serve as gyroscopes.

WHAT HAVE WE LEARNED?

1. Fiber optics are better for transmitting digital signals than signals in the older analog format. This encourages the use of fibers in the telephone network, which is adopting the digital format.
2. The telephone system is composed of a backbone system, trunk lines, and the subscriber loop. Fiber is used mostly in the backbone system and trunk lines, where individual fiber cables carry multiplexed signals.
3. Undersea fiber-optic cables may replace many satellite channels for telephone communication across the oceans.
4. Most computer data communications are not at high enough speeds or over long enough distances to benefit greatly from fiber optics, but fibers are sometimes used to solve other problems.
5. Fiber-optic transmission has been chosen for a 100-Mbit/s local-area network standard.
6. Cable-television systems use little fiber.
7. Data networks in cars and planes may use fiber.
8. Optical fibers can be used to sense, illuminate, display, and image as well as to communicate.

WHAT'S NEXT?

In Chapter 4, we will go back and take a closer look at the characteristics of optical fibers. Later on, in Chapters 15–20, we will take a closer look at the variety of fiber-optic applications.

Quiz for Chapter 3

1. Which of the following are true for analog signals?
   a. They vary continuously in intensity.
   b. They are transmitted in parts of the telephone network.
   c. They are compatible with human senses.
   d. They can be processed electronically.
   e. All of the above.

2. Which of the following are true for digital signals?
   a. They are used in parts of the telephone network.
   b. They can be only at certain levels.
   c. They can be processed electronically.
   d. They can encode analog signals.
   e. All of the above.

3. Assume that an analog signal has to be sampled at twice its bandwidth for digitization at 128 levels. If the analog signal has bandwidth of 10,000 Hz, what is the data rate needed to transmit the digitized signal?
   a. 20 kHz.
   b. 56 kHz.
   c. 128 kHz.
   d. 140 kHz.
   e. 1.28 MHz.

4. What part of the telephone network is connected directly to your home telephone?
   a. Subscriber loop.
   b. Feeder cable.
   c. Trunk line.
   d. Backbone system.

5. Fiber optics transmits which of the following data rates in the telephone system (list all that apply)?
   a. 56 kbit/s.
   b. 1.5 Mbit/s.
   c. 6 Mbit/s.
   d. 45 Mbit/s.
   e. 400 Mbit/s and up.

6. Which of the following is not an advantage of fiber optics for telephone transmission?
   a. Will not propagate current pulses from lightning strikes.
   b. High-quality analog transmission.
   c. Small size lets four fiber cables fit in ducts that could hold only one metal cable.
   d. Immunity to electromagnetic interference.
   e. Long-distance high-speed transmission.

7. When is fiber optics used for point-to-point computer data transmission?
   a. Only when the customer is having an affair with the fiber salesperson.
   b. When personal computers need to be connected.
   c. When there is concern about electromagnetic interference.
   d. When new technology is being tested.
   e. When the data comes from a fiber-optic telephone system.

8. Fiber optics is the standard transmission medium for:
   a. the 100-Mbit/s FDDI local-area network.
   b. 10-Mbit/s Ethernet local-area networks.
   c. interconnection of computers and peripheral devices.
   d. short-distance transmission within buildings.

9. Which is not a problem for fiber optics in cable television?
   a. Attenuation of fibers is too low.
   b. Distortion of high-frequency analog signals.
   c. Splitting of signals from one cable to many subscribers.
   d. Cost of fiber systems is too high.

10. Which is not a present or potential application of fiber optics?
    a. Sensing.
    b. Power transmission to homes.
    c. Image transmission.
    d. Signal transmission in cars.
    e. Delivering laser beams inside the human body.

Types of Fibers

ABOUT THIS CHAPTER

All optical fibers are not alike. There are several major types, which are made differently, operate in different ways, have different characteristics, and serve different functions. Some differences that seem subtle can lead to large functional differences.

This chapter describes the basic types of fibers and how they work. You will learn the differences among various types of fibers and what those mean for the applications of such fibers. This is an essential groundwork to understanding the range of applications of fiber optics.

FUNCTIONAL REQUIREMENTS AND TYPES

Types of fibers have evolved in response to user needs, which themselves have changed, but many fiber types have found uses other than originally expected.

The major types of optical fibers have evolved over many years, with some types going back a decade or two. Functional requirements have evolved along with fiber technology, so interest in some types has grown and declined. For example, the first low-loss fibers were small-core single-mode types. Concern that it would be impractical to get light into their tiny cores led to the development of larger-core graded-index multimode fibers with higher transmission capacity than the simple step-index multimode fibers we examined in Chapter 2. Later, advances in coupling technology and a desire for even higher-speed transmission renewed interest in single-mode fibers, which today dominate the telecommunications market. Meanwhile, the graded-index multimode fibers, which were supplanted by single-mode fibers for telecommunications, have reappeared for short-distance communications.

Different fiber applications often have conflicting requirements.

The major reason so many fiber types are needed is that fiber applications are diverse. Different applications often have conflicting requirements. For example, the low-loss broad-bandwidth transmission essential for long-distance telecommunications is best provided by a small-core single-mode fiber. On the other hand, the needs of local-area networks for inexpensive terminal components and easy coupling are best met by a larger-core multimode fiber.

The major factors that can dictate the choice of certain fibers for specific applications include:

- Low attenuation to maximize repeater spacing or avoid repeaters altogether
- Maximization of transmission bandwidth or speed
- Ease of coupling to inexpensive large-area emitters
- Ease of making splices in the field
- Large tolerances to allow inexpensive connectors
- Cost of fiber
- Transmission wavelength
- Tolerance of high temperatures or other environmental conditions
- Flexibility of the fiber.

The rest of this chapter describes the types of fibers now available, including some in development and how they can meet these and other criteria. *Figure 4-1* gives a sampling of today's major types of single fibers, with relative dimensions in cross-section, showing variation of the key parameter of refractive index. Only core and cladding are shown for simplicity; actual fibers are coated with an outer plastic layer. We will discuss them in a rough, but somewhat arbitrary sequence of complexity, after first outlining how fibers are made. Basic fiber-optic principles are included in the discussions of the various fiber types.

FIBER MANUFACTURING

Manufacture of glass fibers starts with making a preform of highly purified glass, which is heated and drawn out into a fiber.

The manufacture of optical fibers is a precise and highly specialized process requiring special equipment. The first step in making glass fibers is to make a rod or "preform" of highly purified glass. Then that preform is heated and drawn out into a thin fiber. As the fiber is drawn, it is coated with a protective plastic layer. There are many variations in the ways preforms are made, but the other steps are similar in most processes. (Plastic fibers, however, are produced in somewhat different ways, which will not be discussed here.)

**Figure 4-1.
Major Types of Optical
Fiber**

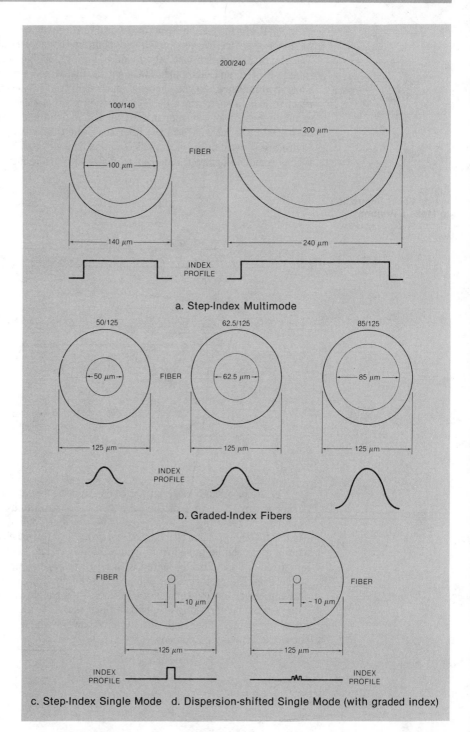

a. Step-Index Multimode

b. Graded-Index Fibers

c. Step-Index Single Mode   d. Dispersion-shifted Single Mode (with graded index)

Preforms can be made in several different ways, each of which has its advocates. One of the simpler-to-explain methods is shown in *Figure 4-2*. The starting point is a bait tube made of highly purified silica (SiO_2). The tube is heated, and ultrapurified chemicals are passed down it in a precisely controlled mixture. As the chemicals pass through the heated tube, they react to form a soot, which deposits on the inside of the tube. In this example, heat from a burner melts the soot to form a sintered glass. The deposited material will form the fiber core; the tube will form the cladding. There are several other variations, such as deposition of the soot on a rod that is removed before the deposited material is made into a preform.

Figure 4-2.
Inside Vapor Deposition
to Make a Preform
(Courtesy Corning Glass
Works)

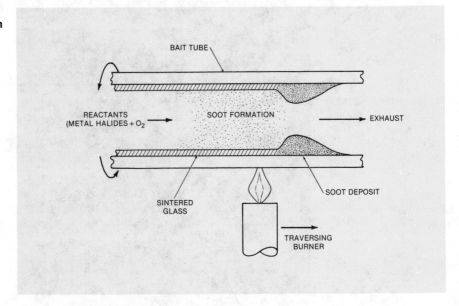

After enough material has been deposited inside the tube, the burner is turned up to heat the tube so it collapses into a solid rod. In practice, the resulting preform is cooled and removed. Later, as shown in *Figure 4-3*, the preform is mounted vertically in a furnace and heated until molten glass can be pulled from it in a fine fiber. The fiber diameter is measured and the fiber is coated with a protective plastic layer as it is pulled from the preform. The equipment used for drawing fibers is called a drawing tower, and tower it does. It is easily a couple of stories high, and it towers over everything else on the floor of a fiber factory. If most of the factory is a single floor high, it may have towers visible on the outside to house the drawing equipment.

**Figure 4-3.
Drawing Glass Fibers
from Preforms** (*Courtesy
Corning Glass Works*)

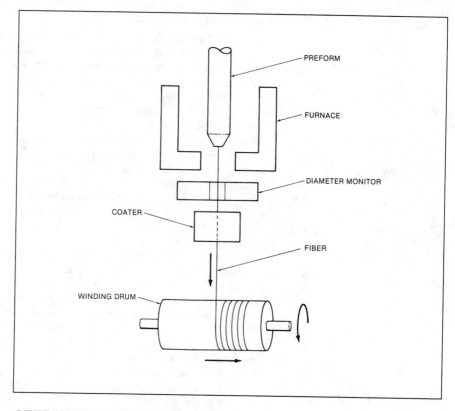

STEP-INDEX MULTIMODE FIBER

Step-index multimode fibers are conceptually the simplest fibers and were the first to find practical uses. These fibers work in the simplified way described in Chapter 2. The fiber core has a refractive index slightly higher than the cladding material, confining the light by total internal reflection to the core. The term "step-index" comes from the abrupt change in refractive index of the fiber material at the core–cladding boundary, the interface that confines light within the core. The amount of the index difference depends on the fiber design and material, but typically is small— less than 1% usually gives adequate light guiding in glass fibers.

The major attractions of step-index multimode fibers is the ease with which they can collect light. Most such fibers have cores at least 100 μm in diameter, and their numerical apertures of 0.2–0.4 are large enough to collect light efficiently. Thus they can be connected to inexpensive large-area light sources and do not require extremely precise connectors. Indeed, the largest-core fibers—all-plastic types with cores 1 mm (1000 μm) in diameter—allow such loose tolerances that signals can be transmitted even when the fibers are partly out of the connector. However, these advantages come at the cost of limited transmission bandwidth and higher losses— especially for plastic fibers—than other types of fiber.

Step-index multimode fibers confine light to the core by total internal reflection from the lower-index cladding.

Step-index multimode fibers collect light easily but have a limited bandwidth.

Modes and How They Work

The term "multimode" indicates that there are many ways in which light can travel through such fibers. The easiest way to visualize these multiple paths is by drawing light rays that enter the fiber and are reflected back and forth from the core–cladding interface, as we saw in Chapter 2. Unfortunately, reality—as usual—is considerably more complicated. Instead of talking about light rays, we should be talking about transmission modes in an optical waveguide.

What is a mode? A mode is a stable propagation state in an optical fiber. Dig into mode-propagation theory, and you will find that it is an effect caused by the wave nature of light. If light travels through an optical fiber along certain paths, the electromagnetic fields in the light waves reinforce each other to form a field distribution that is stable as it travels down the fiber. These stable operating points (standing waves) are modes. If the light tries to travel other paths, a stable wave will not propagate down the fiber—thus no mode.

The details of mode-propagation theory are far too complex to discuss here and have little relevance to the day-to-day concerns of fiber-optic users. We will take the short cut of treating modes as bundles of light rays entering the fiber at the same angle. However, some results of mode theory are worth noting. One is that the light actually penetrates slightly into the fiber cladding layer, even though it nominally undergoes total internal reflection, as shown in *Figure 4-4*. This means that attenuation in the cladding layer, while not crucial, cannot be ignored altogether. In addition, some modes may propagate partly in the cladding, where losses tend to be high because light can leak out of the fiber altogether, as well as suffer absorption in the cladding.

Figure 4-4.
Light's Path in a Step-Index Multimode Fiber

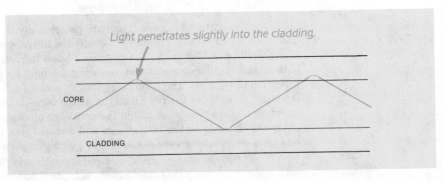

Propagation modes are standing waves that travel through the fiber. To see some of the consequences of multimode transmission, think of a mode as made up of a family of rays, all travelling down the fiber at the same angle with respect to the fiber axis. (Because the cone of rays is circularly symmetric about the fiber axis, adding them together cancels out the parts of the rays going in and out, leaving only the mode travelling down the fiber.)

Each mode has a characteristic number N. A mode N is associated with all rays travelling at an angle θ_N with respect to the fiber axis, where

$$\theta_N \approx \lambda \left(\frac{N + 1}{2Dn} \right) \text{ radians}$$

where for small values of θ, N is the mode number (the lowest-order mode is 0), λ is the wavelength, D is core diameter, and n is refractive index of the core. Note that even the rays associated with the zeroth-order mode do not travel straight along the fiber core. For a fiber with a refractive index of 1.5 and 100-μm core, the angle θ_0 (for the lowest-order mode) is approximately 0.5° for a wavelength of 0.85 μm.

The number of modes a fiber can transmit depends on its numerical aperture and core diameter, as well as the wavelength.

The number of modes that can propagate in a fiber depends on the fiber's numerical aperture (or acceptance angle) as well as on its core diameter and the wavelength of the light. For a step-index multimode fiber, the number of such modes N_m is defined by:

$$\text{MODES} = \frac{(\text{CORE DIAMETER} \times \text{NA} \times \pi/\text{WAVELENGTH})^2}{2}$$

or

$$N_m = \frac{(D \times NA \times \pi/\lambda)^2}{2}$$

where λ is the wavelength and D is the core diameter. To plug in some representative numbers, a 100-μm core step-index multimode fiber with NA = 0.29 (a typical value for such fiber) would transmit 5744 modes at 850 nm.

Modal Dispersion Effects

Modal dispersion comes from differences in the propagation angles of the ray families associated with individual modes.

Looking at the ray propagation angle θ_N gives another indication of how modal dispersion works. Remember that the mode can be viewed as a family of rays, all travelling at θ_N with respect to the fiber axis. Light travelling straight along the fiber axis would move at a speed of c/n—that is, the speed of light in a vacuum divided by the refractive index of the core material. Because the ray family for each mode has a different propagation angle, their path lengths in the fiber are different. The differences are proportional to the differences in the cosines of the propagation angles. A pulse of light enters a step-index multimode fiber in many different modes. Thus, the pulse spreads out as it travels along the fiber, causing the modal dispersion described in Chapter 2. Precise calculations are beyond the scope of this book, but typical bandwidths of step-index multimode fiber are about 20 MHz-km.

Dispersion and bandwidth depend on the fiber's internal characteristics and its length.

Modal dispersion limits pulse rate in multimode fibers because it causes successive pulses to overlap and interfere with each other. (It also distorts analog waveforms.) Total dispersion is roughly the product of the dispersion characteristic of the fiber, D_0 (measured in nanoseconds per kilometer) times the length L (measured in kilometers). To be strictly accurate, the total dispersion of multimode fibers should be calculated by the formula:

$$D = D_0 \times L^\gamma$$

where γ is a factor close to one dependent on the fiber type. For most practical purposes, however, assume $\gamma = 1$, especially because multimode fiber normally is used only over short distances.

Fiber dispersion can be converted to transmission bandwidth by the approximate formula:

$$\text{BANDWIDTH (MHz-km)} = \frac{350}{\text{DISPERSION (ns/km)}}$$

(The bandwidth in this formula—and elsewhere in the book—is defined as the frequency where output has dropped by one-half [3 dB] from its usual level.) As with dispersion, bandwidth of a fiber segment depends on length L and the fiber's characteristic bandwidth BW_0, in this case according to the formula:

$$BW = \frac{BW_0}{L^\gamma}$$

where the γ factor is the same as above and can usually be assumed to equal one except for long-distance transmission. Thus a 5-km length of 20 MHz-km fiber would have a net bandwidth of 4 MHz.

Leaky Modes

Some modes can propagate short distances in the cladding of a multimode fiber.

Measurements of numerical aperture of multimode fibers show a peculiar phenomenon: the NA appears to be highest for a short segment of fiber. Thus a 2-m length may have NA of 0.37, while a 1-km length has 0.30, which is close to the theoretical prediction. There is no room for this in the formula for NA given in Chapter 2, so what's happening? Modes that are just slightly beyond the threshold for propagating in a multimode fiber can propagate for short distances in the fiber cladding. Because this extra light appears to have been transmitted by the fiber, it artificially increases the fiber's NA.

Similarly, the highest-order modes that meet the conditions for propagation are just within the threshold. If conditions alter just a tiny bit —for example, if the fiber is bent—they might leak out. These are called leaky modes, and we will see some of their effects below.

Bending Effects

So far we've assumed that the fiber is straight, but in any real application, it will bend around corners. In practice, fiber bends are gradual relative to the diameter of the fiber, with curvature of a few centimeters or more compared to the 100-μm diameter of a typical step-index fiber core. Larger-core fibers are more rigid and have larger minimum bend radii.

To see how a bend can change a fiber's transmission properties, recall the simple ray model of transmission and look at *Figure 4-5*. When light rays travelling down the fiber strike a bend, those in higher-order modes can leak out if they hit the side of the fiber at an angle beyond the critical angle θ_c. That increases the loss in the fiber. Lower-order modes are not likely to leak out, but they can be transformed into higher-order modes, which can leak out further along the fiber at the next bend. The bends need not be large to cause losses in the fiber. Indeed, some of the most serious bending losses in multimode fibers come from microbending, when fiber in a cable develops tiny kinks.

Figure 4-5.
Light Can Leak out of Bent Fibers

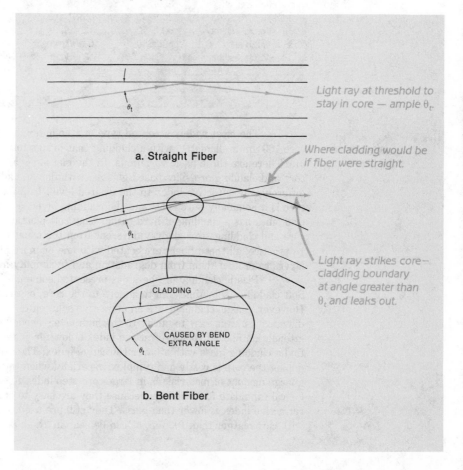

a. Straight Fiber

Light ray at threshold to stay in core — ample θ_c.

Where cladding would be if fiber were straight.

Light ray strikes core–cladding boundary at angle greater than θ_c and leaks out.

CLADDING

CAUSED BY BEND
EXTRA ANGLE

b. Bent Fiber

TYPES OF STEP-INDEX MULTIMODE FIBER

Differences in composition
and diameter of core and
cladding lead to differ-
ences in properties of
step-index multimode
fibers.

All step-index multimode fibers share many characteristics, but not all step-index multimode fibers are alike. Two key sets of parameters can vary: composition and diameter of both the core and cladding. The variation of these parameters can lead to major differences in crucial functional characteristics such as bandwidth and attenuation. Some examples are listed in *Table 4-1*, along with representative values from data sheets.

**Table 4-1.
Types of Step-Index
Multimode Fiber**

Fiber Type	Diameter, Core/Clad	Attenuation @850 nm	Bandwith, MHz-km	NA
All-glass	100/140	4 dB/km	20 MHz-km	0.21
Hard-clad silica	110/125	8 dB/km	17 MHz-km	0.30
All-glass	200/280	4 dB/km	10 MHz-km	0.21
Plastic-clad silica	200/?	10 dB/km	25 MHz-km	0.23
Plastic-clad silica	200/380	8 dB/km	10 MHz-km	0.4
Plastic-clad silica	600/750	12 dB/km	—	0.4
Hard-clad silica	600/650	6 dB/km	9 MHz-km	0.3
Plastic-clad silica	1000/1260	12 dB/km	—	0.4

The most common step-in-
dex multimode fiber has a
100-μm core and a 140-μm
cladding.

The most widely accepted type of step-index multimode fiber has a core 100 μm in diameter, with a cladding that brings total diameter to 140 μm. There are three basic variations. In the glass-on-glass fiber, both the core and cladding are silica-based glasses, with dopants added to adjust the difference in refractive index to the desired level. In plastic-clad silica, the core is a nominally pure silica and the cladding layer, a soft plastic. Hard-clad silica has a cladding of hard rather than soft plastic and is made in core and cladding sizes slightly different from the usual 100/140 dimensions. In practice, all these fibers are coated with one or more plastic buffer layers to protect them from degradation and to simplify handling.

Plastic cladding of fibers has presented some mechanical problems. Soft claddings sometimes fail to adhere to the core, especially under tension. However, plastic claddings are an attractive solution to a fiber-fabrication dilemma. Silica is easy to purify to the high levels needed for low-attenuation fibers, but its refractive index is low for glass, making it hard to find a cladding glass with a lower refractive index. The usual alternative of doping the core is costly and complex, especially when the core must contain a large amount of material, as in large-core step-index fibers. Plastics are a logical candidate for claddings because they are easy to handle and their refractive index is lower than glass. They still are used to clad many fibers with cores larger than 100 μm, as can be seen in *Table 4-1*.

The main uses of 100/140 multimode step-index fibers are in short-distance data links. Although the fibers have relatively large modal dispersion, and thus limited bandwidth, they are adequate for short-distance transmission. Their cores are large enough to accept light from inexpensive LEDs and to allow use of inexpensive connectors with large tolerances. Losses are higher than in most single-mode or multimode graded-index fibers, but this is not a problem since transmission distances are limited. As we will see later, multimode graded-index fibers are being used increasingly for short systems.

As can be seen in *Table 4-1*, multimode step-index fibers can be made with cores much larger than 100 μm. All-glass types are made with 200-μm cores and claddings 280-μm in diameter. Plastic-clad silica types are made with cores 200–1000 μm in diameter. Such fibers may be used in short-distance communications, or in other applications where efficient light collection is the main goal. That collection efficiency is offset by disadvantages including greater stiffness and higher cost than smaller-core fibers and bandwidths low enough that manufacturers avoid listing them in specification sheets.

Because of their limited transmission bandwidth and large collection area, step-index multimode glass fibers normally are used only with inexpensive GaAs LEDs emitting near 850 nm, and many manufacturers specify loss only at that wavelength. For all-glass or plastic-clad silica fibers with cores 100 or 200 μm in diameter, losses can be as low as the 3 dB/km range, comparable to that of other high-quality fibers at that wavelength. Losses generally are over 10 dB/km for larger-core plastic-clad silica fibers. An attenuation versus wavelength curve for one all-glass 100/140 step-index multimode fiber is shown in *Figure 4-6*.

**Figure 4-6.
Typical Spectral
Attenuation of 100/140
Fiber** (Courtesy SpecTran
Corp.)

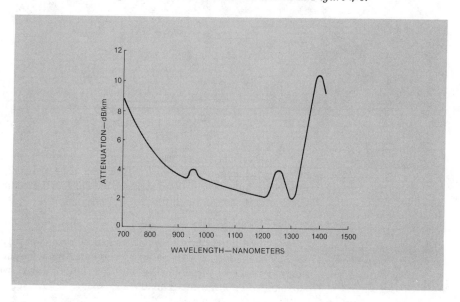

Leaving some OH in the
fiber helps alleviate radiation effects, but at a cost
of higher loss normally.

Some step-index multimode fibers are designed for special
applications, particularly to survive exposure to nuclear radiation, which
can cause temporary or permanent darkening of conventional fibers. Such
fibers have different transmission characteristics. Makers of conventional
fibers seek to remove most hydroxyl (OH) ions, which absorb strongly
between 850 and 1000 nm. However, those hydroxyl ions help alleviate
radiation effects on fibers; consequently, they are not removed from
radiation-hardened fibers. Thus, radiation-hardened fibers have higher loss
than conventional types under normal conditions but lower losses when
exposed to radiation.

Ultraviolet Fibers

Special fibers are made
for ultraviolet transmis-
sion at wavelengths as
short as 200 nm.

Extrapolation from *Figure 4-6* indicates that fiber loss should be
high in the ultraviolet. Although intrinsic losses of silica are much higher in
the ultraviolet than in the near infrared, special fibers can be made to
transmit reasonable amounts of ultraviolet light over short distances.
Figure 4-7 plots attenuation (in decibels per meter, not the usual decibels
per kilometer) of one such ultraviolet-transmitting fiber. Special ultraviolet-
transmitting fibers can transmit some light through a 1-m length at
wavelengths to 180 nm.

**Figure 4-7.
Absorption in an
Ultraviolet-Transmitting
Fiber** *(Courtesy
Fiberguide Industries)*

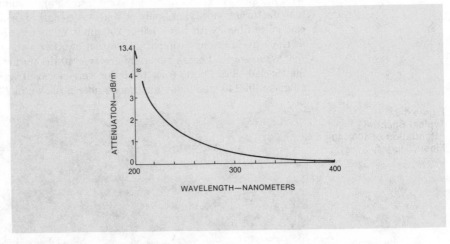

Obviously these transmission figures cannot compare with fibers
operating at longer wavelengths. However, they are adequate to carry
ultraviolet light over short distances, for illumination or measurement,
which otherwise can be difficult.

PLASTIC FIBERS

High-loss, large-core multimode step-index fibers
can be made entirely of
plastic.

Multimode step-index fibers can be made entirely of plastic as long
as they contain a core with a refractive index higher than the cladding.
Many have a core of polymethyl methacrylate (PMMA), which can be
surrounded by lower-index materials, such as fluorine-containing polymers.

This can lead to large core–cladding index differences. For example, in one commercial fiber the core has 1.495 refractive index and the cladding 1.402, corresponding to an NA of 0.47 and a full acceptance angle of 56°.

All-plastic fibers have some important attractions, including low cost, better flexibility, and ease of handling, and have been used for many years in applications such as light-outage indicators in cars and in many types of fiber-optic bundles. Flexibility and low cost are important in large-core fibers, because silica types tend to be stiff and expensive. On the negative side, all-plastic fibers have much higher attenuation and less resistance to high temperatures than glass types. The attenuation curve in *Figure 4-8* shows minimum loss measured in hundreds of decibels per kilometer, limiting transmission to short distances. Low glass transition temperatures limit most plastics to temperatures below about 85°C. New types can operate up to 125°C but have higher losses. These temperature limits have been an important problem in automobile applications.

**Figure 4-8.
Attenuation of a Plastic
Fiber** *(Courtesy
Mitsubishi Rayon)*

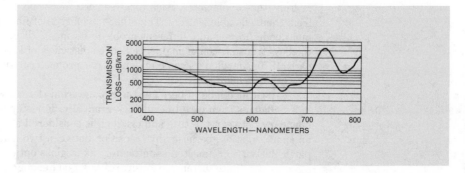

Plastic fibers have core diameters from about 85 to 3000 μm. Cladding thicknesses differ considerably; some are very thin, others are 20% or more of fiber diameter. Smaller fibers normally are used as components in flexible bundles, while the larger fibers are used separately. As with other step-index fibers, the numerical aperture and attenuation do not depend directly on fiber diameter.

All-plastic fibers have some potential communication applications in areas like automobiles, where distances are small and bandwidth requirements are modest. However, most uses of plastic fibers are likely to remain in light piping and image transmission through bundles—where the goal is to carry light no more than a few feet and bandwidth (in the communications sense) is meaningless. Typically a 2-m bundle transmits about 60% of the incoming light, although both higher and lower transmission are possible.

Uniform transmission of light throughout the visible spectrum is important in imaging.

Attenuation considerations for imaging fibers differ greatly from those for communication fibers. In communications, only a single wavelength needs to get through the fiber. For imaging, the entire visible spectrum must get through. It is important that light throughout the visible spectrum be transmitted uniformly for many imaging applications.

For instance, physicians use color in diagnosis, so endoscopes should accurately transmit the colors of tissue inside the body. Special types of fibers are made for that purpose.

Liquid Light Guides

A final variation on the step-index multimode fiber is the liquid light guide. The light guide is essentially a step-index multimode fiber. The liquid (the core) is contained in a hollow tube with a refractive index lower than the liquid, so it functions as a cladding and confines light to the liquid. Such devices have been demonstrated but have had little practical use; they won't be mentioned further here.

GRADED-INDEX MULTIMODE FIBERS

Graded-index multimode fibers were developed to offer easier coupling than small-core single-mode fibers and better bandwidth than step-index multimode fibers.

Demonstration of the first low-loss fibers did not instantly solve all problems of long-distance fiber-optic communications. The first low-loss fibers were single-mode types, with light-carrying cores only several micrometers in diameter. Such cores seemed much too small to collect light from then-available sources. The other major type of fiber—the step-index multimode fiber—had a different problem; its modal dispersion was too high to allow transmission at high speeds over distances of kilometers. In the early 1970s, developers came up with a compromise solution—the graded-index multimode fiber.

Graded-index fibers get their name from the way the refractive index changes from core to cladding—gradually. For step-index fibers, the boundary between the core and cladding is considered abrupt. That is only an approximation, because any transition between two materials—particularly two connected as intimately as in glass optical fibers—takes place over a finite distance. However, it's a good enough approximation for all practical purposes.

In graded-index fibers, the transition is deliberately made gradual. In theory, the refractive index makes a smooth drop from center of the fiber to the edge of the cladding, as shown in *Figure 4-9*. In practice, a good approximation of that smooth curve is made by depositing up to a couple hundred layers of glass with gradually changing composition in the early stages of making the preform. Heating and collapse of the preform and drawing it into fiber make the distribution reasonably smooth.

Figure 4-9.
Graded-Index Fiber
Refractive Index Profile

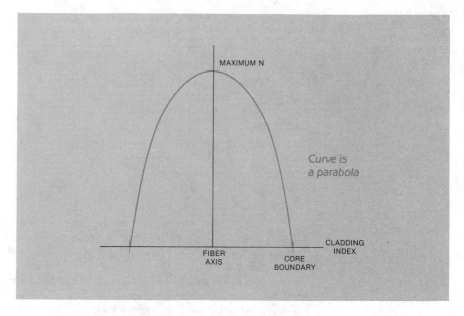

Graded-index multimode fibers have smaller cores than step-index multimode fibers—typically 50–85 μm in a fiber with 125-μm core versus 100 μm and up for step-index multimode fibers. Those cores are large enough to ease coupling tolerances, but they also can carry many modes. How does grading the refractive index help? It makes the path that light rays travel through the fiber depend on refraction rather than total internal reflection, so light rays entering the fiber at different angles travel essentially the same distances through the fiber.

Light guiding in a graded-index multimode fiber is shown in *Figure 4-10*, which can be compared with the picture of light guiding in a step-index multimode fiber in *Figure 4-4*. In the step-index fiber, the light rays zig-zag between the core–cladding boundary on each side of the fiber axis. In graded-index fiber, the gradient in the refractive index gradually bends the rays back toward the axis.

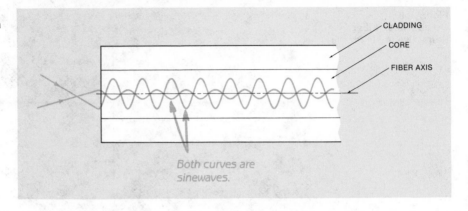

Both curves are
sinewaves.

The mathematical details are far beyond the scope of this book, but they turn out to depend on how the refractive index varies with distance from the fiber axis, the index profile. Suppose the core has maximum index of n_{core} at the center, and the cladding layer has index n_{clad}. The index changes gradually from the center of the fiber to a distance a from the center, which is the edge of the core. If you plot the change from the fiber center outward on a curve such as shown in *Figure 4-10*, it appears (for values of $r \le a$) as:

$$n(r) - n_{clad} = A \left[1 - \left(\frac{r}{a} \right)^\alpha \right]$$

in which A is a constant of proportionality and r measures distance from the fiber axis. (Distance from the axis is divided by core radius a to make the number dimensionless.) The key parameter is α, the power to which the normalized axial distance is raised. If the plot was a straight line, α would be 1; if it was a parabola, it would be 2. For an abrupt boundary, α would be infinity. The best value of α turns out to be close to 2, giving a parabolic index profile and making light rays travel a sine-wave path through the fiber. We knew that all along, which is why the index profile in *Figure 4-9* is a parabola and the paths of light through the fiber in *Figure 4-10* are sine waves.

Modal Dispersion

Index gradation balances the times that rays travelling near to and far from the fiber axis take to pass through the fiber, minimizing modal dispersion.

How does index gradation help reduce modal dispersion? It doesn't really equalize the physical distances that light rays travel through the fiber. What it does instead is to minimize the difference in time they take to pass through the fiber. This can be done because the speed at which light travels through the glass in the fiber is the speed of light in vacuum divided by the refractive index, c/n. Thus the higher the refractive index, the slower light travels. The light rays that enter the fiber at a steeper angle still have to go a longer distance through the glass. However, in the outer parts of the core, they travel through glass with a lower refractive index than the rays that go a shorter distance closer to the fiber axis. Thus, what the higher-order rays lose in speed by travelling a greater distance, they make up by going faster in the lower-index glass.

The graded-index multimode design does not eliminate modal dispersion, but it does reduce it greatly. Typical graded-index multimode fibers have bandwidths of 100–1000 MHz-km or more at their normal operating wavelengths of 850 or 1300 nm, with higher bandwidth at the longer wavelength.

Material Dispersion and Waveguide Dispersion

Although simple modal dispersion dominates the total dispersion picture for step-index multimode fibers, matters are more complex for graded-index multimode fibers, and wavelength becomes an important factor. The refractive index of glass is a function of wavelength, so the mode-equalizing effects of the fiber refractive index also depend upon wavelength.

There are two other kinds of dispersion: material dispersion and waveguide dispersion. Material dispersion occurs because a pulse of light in the fiber includes more than one wavelength. Thanks to the differences in refractive index with wavelength, different wavelengths travel through the fiber at different speeds. The range of wavelengths in a pulse also affects waveguide dispersion, which arises because of the way that light is divided between core and cladding (an effect more important for single-mode fibers).

The result shown in *Figure 4-11* is a bandwidth-versus-wavelength curve with a pronounced peak. Fiber designers can pick this peak by their choice of manufacturing parameters and dopants. If the fiber is to be used at only a single wavelength, typically 1300 nm, the peak might be placed close to that wavelength. But if operation was to be at two or more wavelengths, the fiber might be designed with peak bandwidth at an intermediate wavelength. The figure shows two different peaks, one in fiber designed for 1300 nm, the other in fiber for use at shorter wavelengths. The curve labelled "laser material limit" indicates how material dispersion limits bandwidth.

**Figure 4-11.
Range of Bandwidth in
Graded-Index
Multimode Fibers**
*(Courtesy Corning Glass
Works)*

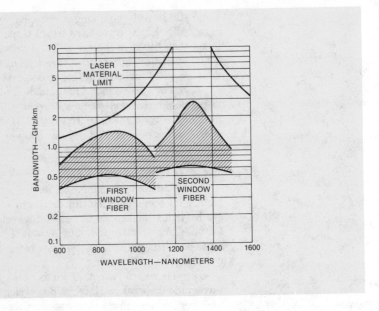

Added complications come from mode-mixing effects in long fibers, especially in fibers spliced together from shorter lengths. In essence, the light can be redistributed among modes in ways that actually can reduce pulse dispersion below the simple linear dependence on fiber length described earlier. However, other effects can cause pulse spreading to increase faster. The exact nature of these dependences are complex and still subject to some debate. Fortunately, the move away from multimode fibers in long-distance telecommunication systems has reduced their importance.

Attenuation

Attenuation of graded-index multimode fibers typically is at least as low as the best step-index multimode fibers, and often better. Most graded-index fibers are designed to operate in one (or both) of two transmission windows—at 850 or 1300 nm. Attenuation at the shorter wavelength is higher (in the 3-dB/km range), and the bandwidth is lower. However, light sources and detectors for that wavelength cost much less than those for 1300 nm. For the longer wavelength, attenuation is 1 dB/km or less, allowing transmission over longer distances, but only at a cost of more expensive terminal components. Transmission at other wavelengths is possible; *Figure 4-12* shows attenuation of a good-quality 85/125 graded-index fiber over a range of wavelengths.

Figure 4-12.
Attenuation of a
Graded-Index Fiber
(Courtesy Corning Glass
Works)

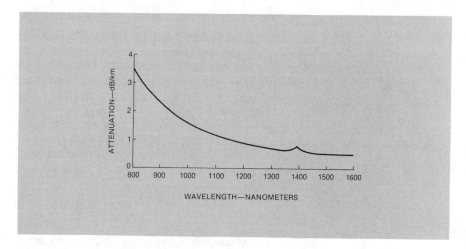

Graded-index multimode fibers have dropped from favor for long-distance communications but are finding increasing applications in shorter systems, data links, and local-area networks. Although multimode graded-index fibers cannot match the bandwidth of single-mode fibers, they offer adequate capacity for most systems no more than a few kilometers long. They also allow use of less-costly terminal components, such as LEDs rather than semiconductor lasers and connectors with looser tolerances. In applications that require many terminals, such costs are crucial concerns.

STEP-INDEX SINGLE-MODE FIBERS

If a fiber core is made
small enough, it will carry
only one mode.

Earlier in this chapter, we saw how the number of modes that a step-index fiber could carry depended on factors including the refractive-index difference between core and cladding, the operating wavelength, and the size of the fiber core. The relationship is

$$\text{MODES} = \frac{(\text{CORE DIAMETER} \times \text{NA} \times \pi/\text{WAVELENGTH})^2}{2}$$

or

$$N_m = \frac{(D \times NA \times \pi/\lambda)^2}{2}$$

where N_m is the number of modes, D is the core diameter, NA is the numerical aperture, and λ is the wavelength. So there are three ways to reduce the number of modes to get a single-mode fiber: reduce core diameter, reduce numerical aperture, and increase wavelength.

Core Size

Operating wavelength generally is fixed by considerations such as attenuation. Reducing numerical aperture too much can make it almost impossible to couple light into the fiber. That leaves reducing the core diameter as the only reasonable way to reduce the number of modes. It is not possible just to turn the equation around and solve for the D needed to limit transmission to a single mode. Proper derivation of the answer requires Bessel functions, which are the sort of things taught in third-year advanced calculus courses that flunk out engineering majors. Let's skip that step and go instead to the formula that gives conditions for single-mode operation of a step-index fiber:

$$D < \frac{2.4\lambda}{\pi \times NA}$$

For an NA of about 0.15, the core diameter must be no more than about five times the wavelength. If the NA is 0.1, the core could be up to 7.6 times the wavelength. If NA was raised to 0.2, the core would have to be smaller than 3.8 times the wavelength. These considerations lead to core diameters of 10 μm or less for single-mode fibers used in 1.3-μm communications.

The small size of a single-mode fiber core makes light coupling into the fiber difficult.

The small core size puts tight requirements on light coupling into the fiber. For handling reasons, a single-mode fiber cladding should be at least 125 μm in diameter, a dozen times (or more) the core diameter. Light sources should have output areas measured in micrometers, to match the fiber core. Connection tolerances also must be tight. If two fibers with 10-μm cores are misaligned by just 1 μm, the overlap area is reduced by 12.7%! It was such problems that made early developers wary of single-mode fibers.

Transmission Capacity and Dispersion

Step-index single-mode fibers have zero chromatic dispersion at 1300 nm.

On the other hand, fibers that can carry only a single mode banish the whole problem of modal dispersion—the main limitation on the bandwidth of multimode fibers. What remains is called chromatic dispersion (because it is dependent on the range of wavelengths being transmitted by the fiber). It is the sum of material dispersion and waveguide dispersion. These are plotted roughly in *Figure 4-13*. Here, for once, some good luck intervenes. Dispersion can be positive or negative because it measures the change in the refractive index with wavelength. That change can be an increase or decrease (i.e., positive or negative). In other words, material dispersion and waveguide dispersion can be opposite in sign. Better yet, they cancel out for step-index single-mode fibers at a wavelength close to 1300 nm, where optical fibers also have low attenuation.

**Figure 4-13.
Chromatic Dispersion in
Single-Mode Step-Index
Fibers.**

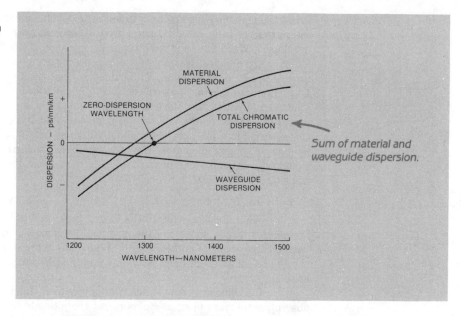

This does not mean that optical fibers have infinite transmission capacity at 1300 nm or even true zero dispersion. (Nature is not that kind to engineers.) Chromatic dispersion equals zero only at one wavelength, but all light sources emit a range of wavelengths. Other, smaller effects also keep dispersion from precisely equalling zero. But dispersion is close enough to zero that it is not a practical concern for present long-distance telephone applications using semiconductor laser sources. (Lasers are needed because the range of wavelengths emitted by LEDs is large enough that dispersion becomes significant. Also, as we shall see later, LEDs cannot couple enough power into single-mode fibers for long-distance transmission.)

Attenuation

The main limitation on 1300-nm transmission is loss as low as 0.4 dB/km in mass-produced fibers.

The main limitation on 1300-nm systems thus becomes attenuation. Losses in single-mode fibers tend to be somewhat smaller than in multimode fibers. One reason is eminently practical: the light-carrying cores that must have ultra-low loss are much smaller, simplifying production. Other factors also enter, allowing attenuation in the best commercial single-mode fibers to reach 0.4–0.5 dB/km at 1300 nm. Somewhat lower losses have been demonstrated in the laboratory but are not available in mass-produced fibers.

Long-Distance Transmission

Step-index single-mode fiber systems now can transmit 400 Mbit/s 30–50 km without repeaters.

Today's single-mode fiber systems can transmit 400 Mbit/s over distances of 30 to 50 km without repeaters at 1300 nm. Developmental systems already demonstrated can send 1.7 Gbit/s over similar distances between repeaters. That performance is good enough that single-mode fibers are now virtually the only type the telephone industry uses outside the subscriber loop. But for some purposes, such as submarine cables, developers would like to stretch repeater spacings still farther.

The simplest way to stretch repeater spacing might seem to be turning up the optical power. Unfortunately, that creates other problems. Laser lifetimes decrease, as will be shown in Chapter 6, but more fundamentally, the fiber itself has limited transmission capacity. Powers from 1300-nm lasers are only tens of milliwatts, but those powers are concentrated in a fiber core only about 10 μm across. What's more, the light travels many kilometers through the fiber. Moving to higher powers would trigger nonlinear effects that would themselves limit signal transmission through the fiber.

1550-nm Transmission

What about moving to the longer 1550-nm wavelength, where attenuation is only about half the level at 1300 nm? Laboratory researchers have pushed attenuation below 0.16 dB/km—an impressive feat and very close to the theoretical limit. Such low losses could really stretch repeater spacing.

Reaching such low losses is no minor task. Intrinsic scattering and absorption inherent in silicate glasses set the ultimate limits on lowest fiber loss. However, absorption by impurities in the glass keeps attenuation above those minimums. Some impurities are inevitable in most fiber designs, because dopants must be added to pure silica to raise the refractive index of the core glass above that of the cladding material. Recently, losses have been reduced by using pure silica as the core glass and doping the cladding glass with fluorine to reduce its refractive index below that of the core glass.

Unfortunately, dispersion, the other key parameter in single-mode fiber performance, enters the picture at 1550 nm. Recall that in *Figure 4-13* dispersion is zero only at 1300 nm. At longer wavelengths, chromatic dispersion is well above zero, large enough to become a problem at 1550 nm.

One way to reduce dispersion at 1550 nm is to use a single-frequency laser source.

One possible solution to the dispersion problem can be seen by looking closer at chromatic dispersion. The units of measurement are picoseconds of pulse spreading per kilometer of fiber per nanometer of source bandwidth. If inherent dispersion is large, the total pulse spreading could be reduced by reducing source bandwidth below the few nanometers typical of semiconductor lasers. What is needed is a new laser emitting a much narrower range of wavelengths, a so-called single-frequency laser (which also emits a range of wavelengths, but one that is much narrower than conventional lasers). Such lasers have existed in research laboratories for several years, and a few companies have claimed to have models near the market. The first commercial models came on the market in late 1986, just as this book was being finished. We will talk more about this technology in Chapter 6.

The other approach is to reduce the fiber dispersion itself. As described below, this requires a different fiber design.

DISPERSION-SHIFTED SINGLE-MODE FIBERS

We saw above that chromatic dispersion of a single-mode fiber is the sum of material dispersion and waveguide dispersion. Material dispersion depends on the glass composition, and little can be done to alter it without harming fiber transmission. Waveguide dispersion occurs because light moves faster in the low-index cladding than in the higher-index core. (The difference, like material dispersion, depends on wavelength.) The degree of waveguide dispersion depends not only on materials but also on how the light is divided between core and cladding. That is a consequence of fiber design, which can be altered. Thus, changing the waveguide dispersion offers a way to shift the zero dispersion wavelength—where waveguide dispersion and material dispersion cancel—to 1550 nm.

Changing waveguide dispersion by changing the fiber design allows shifting zero dispersion wavelength to 1550 nm.

As we have seen, the interface between core and cladding in conventional single-mode fibers is a refractive-index step, where glass composition changes abruptly. This approach originally was chosen because such fibers are simple to make, but it has proven a workhorse. Changing the waveguide dispersion to make dispersion-shifted fibers requires a more elaborate design that divides light differently between core and cladding. The graded-index segmented-core approach used in what at this writing is the only commercial dispersion-shifted fiber is shown in *Figure 4-14*. The inner core has a refractive index graded with a triangular (gradient $\alpha = 1$) profile. It is surrounded with a lower-index silica layer, which in turn is surrounded by a higher-index layer and then an outer cladding. This design makes waveguide dispersion at 1550 nm equal in magnitude but opposite in sign to material dispersion at that wavelength, so that the chromatic dispersion equals zero.

**Figure 4-14.
Dispersion-Shifted Fiber
with Segmented Core**

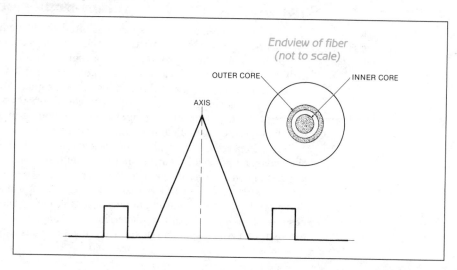

Two factors complicate the design of dispersion-shifted fiber, although neither appear insoluble. First, attenuation tends to be slightly higher in dispersion-shifted fibers than in step-index single-mode fibers because more light is carried in the lossier cladding glass. However, with care, loss of dispersion-shifted fibers can be kept within a few hundredths of a decibel per kilometer of that of step-index single-mode fiber.

A more subtle problem is concern about effects measured by the cutoff wavelength, where the fiber can start to support a second waveguide mode. As mentioned earlier, the number of modes an optical fiber can transmit depends on core diameter, numerical aperture, and wavelength. An optical fiber that is single-mode at a particular wavelength, say 1300 nm, can begin to carry two modes at some shorter cutoff wavelength. (At even shorter wavelengths, it could carry a third mode.) The practical concern is not cutoff wavelength per se, but a quantity it measures indirectly—resistance to microbending losses. The shorter the cutoff wavelength, the more a fiber suffers from microbending losses. Some dispersion-shifted fiber designs have cutoff wavelengths so short that microbending losses become serious.

Dispersion-shifted fiber is not likely to make step-index single-mode fiber obsolete. The design is inherently more complex and thus harder and more costly to produce. However, it does offer a high-performance alternative where it is desirable to minimize the number of repeaters.

The original goal in developing dispersion-shifted fibers was to avoid the need for hard-to-make single-frequency lasers emitting 1550 nm. Although the two approaches remain in competition to some extent, recent research suggests that obtaining the best performance from a long-distance fiber-optic system might require both dispersion-shifted fibers and single-frequency lasers.

DISPERSION-FLATTENED SINGLE-MODE FIBERS

Dispersion-flattened fibers in development have low dispersion at a range of wavelength.

Both step-index and dispersion-shifted single-mode fibers share one transmission-limiting characteristic. Although their attenuation is low throughout much of the 1300- to 1600-nm region (with the exception of a hydroxyl absorption peak near 1400 nm), dispersion is low at only a single wavelength. That is not a problem if high-speed signals are transmitted at only one wavelength, but it presents obstacles to efforts to expand capacity by simultaneously transmitting signals at two or more different wavelengths. The technique is called wavelength-division multiplexing.

Dispersion-shifted fibers are designed to provide zero dispersion at one wavelength. What about making material dispersion and waveguide dispersion add to zero over a broader range?

That problem turns out to be harder than it sounds because fiber designers have only a limited number of degrees of freedom in their designs. Some fibers with low dispersion at a range of wavelengths between two zero-dispersion points have been demonstrated in the laboratory, but they are not yet practical.

POLARIZING FIBERS

So far, we have oversimplified the way in which light propagates in optical fibers by totally ignoring polarization. Polarization is a consequence of the nature of electromagnetic waves. An electromagnetic wave contains two fields—one electric and one magnetic—oscillating perpendicular to each other and propagating in a direction perpendicular to both, as shown in *Figure 4-15*.

**Figure 4-15.
An Electromagnetic
Wave**

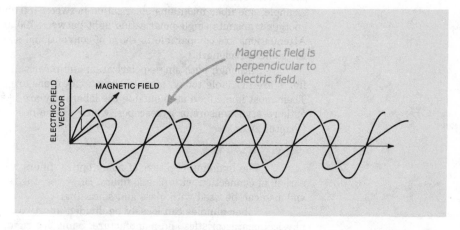

Ordinary unpolarized light is made up of many waves with their electric and magnetic fields oriented randomly (although always perpendicular to each other for each wave). If all the electric fields (and hence the magnetic fields as well) were aligned parallel to one another, the light would be linearly polarized, which is the simplest type of polarization. The two polarization directions are, logically enough, called horizontal and vertical. (Light also can be polarized circularly and elliptically, depending on the way electric and magnetic fields oscillate with respect to each other's phase, a matter beyond the scope of this chapter.)

Special fibers maintain polarization of input light or transmit only one polarization.

Normal optical fibers are insensitive to polarization because their cores are circularly symmetrical. Thus, a single-mode fiber carries light in both horizontal and vertical polarization modes. That doesn't matter for most applications, including ordinary communication systems. However, it can be a problem for some fiber-optic sensors and for advanced communication schemes under development. To deal with such concerns, two types of polarization-sensitive single-mode fibers have been developed.

The difference between the two types is subtle. One is a true single-polarization fiber that can transmit light in one linear polarization but not in the other. The other is polarization maintaining, because it maintains the polarization of light that originally entered the fiber. It does this by isolating the two orthogonal polarizations from each other even while they travel down the same single-mode fiber. (The fiber is called birefringent because it has different properties for light of different linear polarizations.) When light emerges from a polarization-maintaining fiber, it has the same polarization as when it entered.

Polarization fibers are similar in some ways to conventional single-mode fibers, but they generally operate at shorter wavelengths and thus have smaller cores (about 5 μm). Their claddings may be smaller than the 125 μm that is a de-facto standard for telecommunication fibers. They retain their polarization properties over a limited range of wavelengths. For example, one fiber maintains polarization between 780 and 850 nm, while another transmits single-polarization light between 790 and 850 nm. Attenuations are comparable to those of conventional single-mode fiber at the same wavelength.

Although both single-polarization and polarization-maintaining fibers are available today, most of their applications are in the laboratory. Their most immediate uses outside the laboratory are likely to be in fiber-optic rotation sensors or gyroscopes, which will be described briefly in Chapter 20.

BUNDLED FIBERS

The bundling together of many optical fibers was mentioned earlier in connection with plastic fibers. However, the technology is general and also can be used with glass and silica fibers.

Fiber bundles can loosely be divided into two categories based on physical characteristics and manufacture. Some are rigid, with the fibers fused together in a solid bundle. Others are flexible, with the fibers physically discrete, so the bundle can bend. Each approach has its advantages.

A second way to view fiber bundles is by application. For imaging, the fibers are arranged in the same way at each end of the bundle. The output ends of such coherent fiber bundles replicate the pattern of light at the input ends. Alternatively, for illumination, the fibers may be mixed up or randomized (sometimes with great care to make sure the pattern is truly random). Both rigid and flexible bundles can be made coherent or randomized.

One important goal in all fiber bundles is to make fiber cores as much of the surface area as possible. The cores transmit light through the bundle, but light coupled into the cladding generally is lost. Thus bundled fibers have thin claddings to maximize the packing fraction, the part of the surface occupied by fiber cores. A good fiber bundle might have an 85% packing fraction. Fiber bundle technology and its uses will be described in more detail in Chapter 20.

INFRARED FIBERS

Intense research over the past few years has gone to making optical fibers from materials that are transparent at longer infrared wavelengths than silicate glasses. The goal is fibers with much lower losses than conventional fibers—perhaps as low as 0.001 dB/km.

The basic motivation for the research is shown in *Figure 4-16*, which illustrates theoretical minimum losses for a variety of materials. Basic light-scattering processes general to all transparent materials set a lower limit on attenuation at short wavelengths, but that effect drops off sharply as wavelength increases. Absorption of light by silica makes losses increase for ordinary glass fibers at wavelengths longer than about 1600 nm in the infrared. However, other materials are very transparent at longer wavelengths, and researchers are trying to make them into ultra-low-loss fibers. For communications, that would mean much longer repeater spacings. It also might allow use of optical fibers for imaging and transmission of laser energy at wavelengths longer than now possible.

**Figure 4-16.
Theoretical Minimum
Losses of Infrared
Glasses** (*Courtesy Martin
Drexhage, Rome Air
Development Center*)

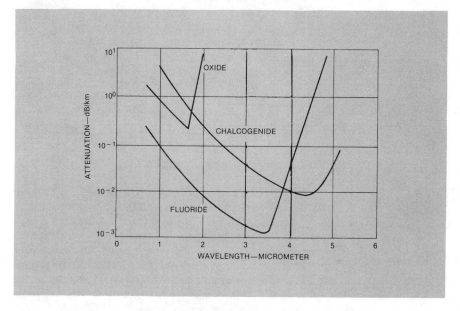

Currently the most attention is going to compounds containing zirconium fluoride and other heavy-metal fluorides, which promise good transmission at 0.2–7.5 μm. The first fibers are beginning to reach the market, mostly for laboratory use, but the losses remain far from the theoretical minima and are still higher than those for silicate glass fibers at shorter wavelengths. The fibers also are fragile and expensive.

The problems that remain in making practical infrared fibers are tough. Purification of the materials is difficult. The raw materials are more expensive than those for silicate glasses. (Contrary to occasional jokes about Saudi Arabia cornering the raw materials market for optical fibers, conventional optical fibers can't be made from raw sand like some glass products. Fiber materials require extensive purification and processing.) Infrared materials also are harder to pull into fibers because they are thinner than silicate glass when molten.

FIBER COATINGS

So far, this chapter has talked about fibers as if they were made of only two components, the light-carrying core and the cladding surrounding it. In practice, fibers are coated with one or more additional layers, typically plastic, as shown in *Figure 4-17*. This typically raises overall diameter of fibers with 125-μm cladding, to 250 or 500 μm.

**Figure 4-17.
Optical Fiber with
Plastic Coating Layers**

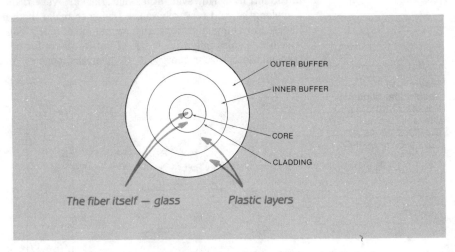

The thickness and precise composition of these coating layers depend on the type of fiber and the application for which it is designed. Their basic functions are to protect the fiber from physical and environmental damage and degradation. Researchers have found that weak points where fibers break develop from tiny microcracks in the surface of the fiber cladding. Coatings protect the fiber from damage that can produce microcracks and provide extra physical strength for the fiber. They also protect against damage from moisture or other environmental factors. They can ease handling, particularly as the fiber is stranded into the cables needed for most applications, which are described in more detail in Chapter 5.

Fibers are coated immediately after they are drawn. Thus, what you see and feel when you handle optical fibers is mostly this outer coating layer, not the cladding that is an optical part of the fiber.

WHAT HAVE WE LEARNED?

1. Optical fibers are made by depositing purified materials in a cylindrical preform, which is heated and drawn into fiber.
2. Step-index multimode fibers are simple types with cores 100–3000 μm in diameter. They collect light efficiently from large-area sources, but modal dispersion limits the bandwidth to about 20 MHz-km.
3. The number of modes a step-index fiber can transmit depends on its numerical aperture and core diameter as well as the wavelength. Single-mode types have core diameters of about 10 μm.

4. Bending can increase fiber losses by letting high-order modes leak out of the core.
5. Step-index multimode fibers come in all-glass, plastic-clad silica, and all-plastic versions of various sizes.
6. Plastic fibers have much higher attenuation than glass fibers.
7. Graded-index multimode fibers offer easier coupling than small-core single-mode fibers and better bandwidth than step-index multimode fibers. The gradation of their refractive index reduces modal dispersion to low levels, giving them a bandwidth of 100–1000 MHz-km. Core diameters typically are 50–85 μm.
8. Single-mode step-index fibers, with cores about 10 μm across, suffer no modal dispersion. The two components of their chromatic dispersion—material and waveguide dispersion—add to zero at 1300 nm, giving them extremely high bandwidth with loss around 0.4–0.5 dB/km.
9. Fiber loss at 1550 nm is only about half that at 1300 nm, but dispersion is much higher. Overall dispersion can be reduced either by using single-wavelength lasers or by making special fibers with dispersion shifted to 1550 nm.
10. Dispersion-shifted single-mode fibers have a graded-index inner core surrounded by an inner cladding and an annular step-index outer core.
11. New materials may offer much lower losses than present types at longer infrared wavelengths but are not yet practical.
12. The part of fibers we see is actually a plastic coating, not the cladding of the fiber itself.

WHAT'S NEXT

In Chapter 5, we will see what happens to fibers when they are put into cables.

Quiz for Chapter 4

1. How many modes would a step-index fiber with a core 100 μm in diameter and a numerical aperture of 0.29 transmit at 850 nm?
   a. 1.
   b. 2.
   c. About 50.
   d. Hundreds.
   e. Thousands.

2. If the measured NA of a short length of multimode fiber is higher than the theoretical prediction, it is because:
   a. the measurement was incorrect.
   b. some modes of light can travel short distances in the cladding but not in the core.
   c. theoretical values are only approximate.
   d. microbending losses were not accounted for.

3. What diameter are the cores of multimode step-index fiber?
   a. 100 μm.
   b. 200 μm.
   c. 400 μm.
   d. 1000 μm.
   e. All of the above.

4. Light is guided in multimode graded-index fibers by:
   a. total internal reflection.
   b. mode confinement in the cladding.
   c. refraction in the region where the core refractive index changes.
   d. the optics that couple light into the fiber.

5. Modal dispersion is highest in which type of fiber?
   a. Step-index multimode.
   b. Graded-index multimode.
   c. Step-index single-mode.
   d. Graded-index single-mode.

6. Which of the following will not reduce the number of modes that an optical fiber can carry?
   a. Reducing core diameter.
   b. Reducing numerical aperture.
   c. Increasing wavelength.
   d. Reducing attenuation.

7. If a fiber has numerical aperture of 0.1, what must its core diameter be less than for it to transmit only a single mode at 1.3-μm wavelength?
   a. 1.55 μm.
   b. 6.5 μm.
   c. 10 μm.
   d. 50 μm.
   e. 100 μm.

8. What makes dispersion zero at 1300 nm in step-index single-mode fibers?
   a. Waveguide and material dispersion cancel each other out.
   b. Chromatic dispersion cancels out modal dispersion.
   c. Waveguide dispersion equals the sum of material and modal dispersion.
   d. Dispersion is zero in all single-mode fibers.

9. Which of the following is needed for high-speed transmission at 1550 nm?
   a. Special fibers with zero dispersion at that wavelength.
   b. Special lasers with extremely narrow linewidth so they experience little dispersion.
   c. New technology to produce fibers with lower attenuation at 1550 nm.
   d. a and b.
   e. a and c.

10. Bandwidths of multimode graded-index fibers are:
   a. 20–100 MHz-km.
   b. 100–1000 MHz-km.
   c. 1–10 GHz-km.
   d. Over 10 GHz-km.
   e. Cover entire range.

Cabling

ABOUT THIS CHAPTER

Cabling is not glamorous, but for most communication uses of fiber optics, it is a necessity. A cable structure protects optical fibers from mechanical damage and environmental degradation, eases handling of the small fibers, and isolates them from mechanical stresses that could occur in installation or operation. The cable makes the critical difference in determining whether optical fibers can transmit signals under the ocean or just within the confines of an environmentally controlled office building.

This chapter discusses the major types of fiber-optic cable you are likely to encounter. You will see what cables do, where and why different types are installed, what cables look like on the inside, how cables are installed, and what happens to fibers in cables.

CABLING BASICS

Fiber-optic cables resemble metal-wire cables but differ because signals are transmitted as light, not electricity.

Fiber-optic cables look like conventional metal cables, and the first fiber cable designs borrowed technology from copper wire cables. Indeed, electrical and fiber cables still use many of the same materials. Black polyvinyl chloride sheaths are common on both fiber-optic cables and coaxial cables used inside buildings. Similar sheathing structures are used on metal and fiber cables to protect against the rigors of underground or aerial installation. Cut metal and fiber cables open, and the only obvious difference in internal structure is likely to be the use of fibers rather than copper wires.

Some important differences can be subtle. Because optical fibers are not conductive, they do not require electrical insulation to isolate circuits from each other. Optical cables can be made non-conductive by avoiding use of metals in their construction, which produces all-dielectric cables. Fiber-optic cables tend to be smaller because one fiber has the same capacity as many wire pairs and because fibers themselves are small.

Fibers must be isolated from tension because they break if stretched more than about 5%.

Some major differences in cable design are necessary because fibers react differently to stress than copper wires. Pull on a fiber and it will stretch slightly, then spring back to its original length. Pull the fiber hard enough, so it stretches by more than about 5%, and it will break. Pull a copper wire, applying less stress than you did to break the fiber, and it will stretch by up to about 30% and not spring back to its original length. In mechanical engineering terminology, fiber is elastic (because it contracts back to its original length), and copper is inelastic (because it stays stretched out).

These differences have major implications for cable design. Stretch a copper cable and the wire stretches with it. Stretch a fiber cable too far and the fiber breaks, even though the fiber itself is strong. Thus cable designs should protect fibers from stress along their length. As we will see later, this is done in various ways, typically by applying the strain to strength members (of metal or non-metallic materials).

REASONS FOR CABLING

Cabling is the packaging of optical fibers for easier handling and protection. Uncabled fibers work fine in the laboratory and in certain applications such as sensors and the fiber-optic system for guiding missiles, which will be described in Chapter 19. However, like wires, fibers must be cabled for most communications uses.

Ease of Handling

Cables make fibers easier to handle.

One reason for cabling fibers is to make them easier to handle. Physically, single glass optical fibers resemble monofilament fishing line, except the fibers are stiffer. Protective plastic coatings raise the outer diameter of fibers to 250–500 μm, but they are still so small they are hard to handle. They also are transparent enough to be hard to see on many surfaces. Try to pick up one loose fiber with your fingers, and you'll soon appreciate one of the virtues of cable.

Cabling also makes multiple fibers easy to handle. Most communication systems require at least two fibers, one carrying signals in each direction. Some require many fibers. Cables have been developed for up to several hundred fibers. Cabling puts the fibers in a single easy-to-see and easy-to-handle structure.

Cables also serve as mounting points for connectors and other equipment used to interconnect fibers. If you take that function too much for granted, try butting two bare fibers together with your bare hands and finding some way to hold them together permanently. Anyone for tape?

Protection

Stress Along Fibers

Cables prevent physical damage to fibers during installation and use.

Another goal of cabling is to prevent physical damage to the fiber both during installation and use. The most severe stresses along the length of cables normally come when they are pulled or laid in place. Aerial cables always experience some static stress after installation because they hang from supports. Dynamic stresses applied for short periods can be the most severe, and the most damaging to cables. The worst problems come from contractors with backhoes and other earth-moving equipment, who dig up buried cables, applying sharp forces and snapping the cables. Falling branches can break aerial cables. Cables can isolate the fibers from static stresses by applying the force to strength members. As we saw earlier, fibers are much more vulnerable to breaking when stretched than are copper wires, so strength members in fiber cables must resist stretching. However, the cable designer cannot provide absolute protection against careless contractors or heavy falling branches.

Crush-Resistance

Cables also must provide crush-resistance to prevent damage if someone steps on the cable. Requirements for crush-resistance differ greatly. Ordinary in-building cables are not made to be walked on, but some special types are designed for under-carpet installation, as shown in *Figure 5-1.* These must be flat and crush-resistant. Submarine cables must be able to withstand high static pressures underwater—and deep-sea cables must be able to withstand the pressure of several kilometers of seawater. Cables buried in the ground must withstand a different type of crushing force applied in a small area: the teeth of chewing gophers, whose front teeth—like those of other rodents—grow constantly. That is one case where the small size of fiber cables can be undesirable: small cables may be just bite-sized for a gopher.

**Figure 5-1.
Cross Sections of Two-Fiber Cables**

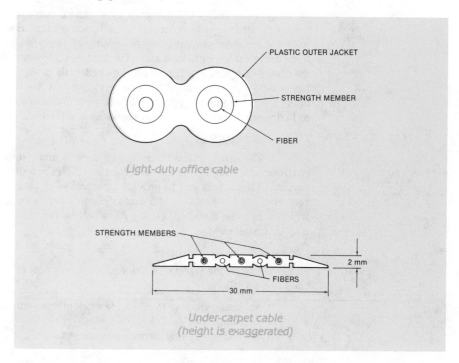

Light-duty office cable

Under-carpet cable
(height is exaggerated)

Cables provide the rigidity needed to keep fibers from being bent too tightly. They also help protect fibers from developing tiny microcracks, caused by surface nicks, which can lead to fiber breakage.

Degradation

Cabling protects fibers against degradation caused by heat, moisture, or hydrogen.

Cabling also protects fibers from more gradual degradation mechanisms. Long-term exposure to moisture and/or heat can degrade fiber strength and optical properties. Both are threats to outdoor cables. Moisture also can affect underground and underwater cables. Most cables designed for use in uncontrolled (i.e., outdoor or underground) environments include barriers to keep moisture out. Aerial cables must withstand extremes of temperature—from heating to high temperatures on a hot, sunny day in the summer to freezing in the winter. The combination of cold and moisture presents an added danger—freezing of moisture in the fiber. Because water expands when it freezes, it could apply forces on the fiber that produce microbends and increase losses.

A significant long-term concern in some fibers that transmit at 1300 or (particularly) 1550 nm is the possible influx of molecular hydrogen into the fiber. If a fiber is kept in an atmosphere with a large hydrogen content, the tiny hydrogen molecules diffuse throughout the fiber, adding significantly to losses at long wavelengths. That is done deliberately only in the laboratory. Hydrogen can accumulate in some cable structures, for example, by diffusion from or decomposition of certain plastics or by electrolytic breakdown of moisture by electrical currents in the cable (e.g., the power delivered to repeaters in undersea cables). Hydrogen effects were not discovered until many fiber-optic systems had gone into use and some early cables contained materials that could generate significant quantities of hydrogen. However, new materials have come into use, and heavy-duty cables intended for long-term use now are designed to avoid internal hydrogen build-up. In addition, new fibers are much less vulnerable to hydrogen-induced losses than earlier types.

TYPES OF CABLE

Cables are designed for particular environments.

The same optical fiber may be used in many different environments, but this is not so for cable. Cables are designed to withstand particular conditions and to provide a controlled environment for the fibers they contain. Thus, choice of a cable design depends on the environment where it is to be installed. To see what is involved, consider where cables may be installed.

Types of Environments

The major types of environments for optical cable can be loosely classified as follows:

- Inside devices (e.g., inside a telephone switching system or computer).
- Intraoffice (e.g., across a room or under a raised floor in a computer room).
- Intrabuilding (e.g., between walls or above suspended ceilings between offices in a structure).
- Plenum installations (i.e., through air ducts in a building).
- Temporary light-duty cables (e.g., remote news gathering).
- Temporary heavy-duty cables (e.g., military battlefield communications).

- Aerial cables (e.g., strung from telephone poles outdoors).
- Ducted cables (i.e., installed in plastic ducts buried underground).
- Direct-burial cables (i.e., laid directly in a trench).
- Submarine (i.e., submerged in ocean water or sometimes fresh water).
- Instrumentation cables, which may have to meet special requirements (e.g., withstand high temperatures, corrosive vapors, or nuclear radiation).
- Hybrid power-fiber cables, which carry electric power (or serve as the ground wire for an electric power system) as well as optical signals.

Those categories are not exhaustive, and some are deliberately broad and vague. Instrumentation, for example, covers cables used to log data collected while drilling to explore for oil or other minerals. Special cables are needed to withstand the high temperatures and severe physical stresses experienced within deep wells.

Cable Design Considerations

Each environment has special requirements leading to the design of many types of cables. In the following list of cables, their strengths and weaknesses are briefly summarized.

- Intra-device cables should be small, simple, and low-cost because the device protects the cable.

Standard light-duty cables are for use within buildings.

- Intraoffice cables are the standard light-duty cables used for fiber-optic data links inside normal buildings (but not in factories or other buildings where interior conditions may be hostile to the cable). Typically they contain one or two fibers (but they may contain more). They can take various forms, including the two examples shown in *Figure 5-1*: a two-fiber duplex cable that from the outside looks like the zip cord used for electric lamps and the special flat, crush-resistant types designed to run under carpeting. Other types are round or oval in cross-section. Low cost, small size, durability, ease of installation, and compatibility with connectors and existing devices are the usual goals.
- Intrabuilding cables are similar to light-duty cables for installation within offices, but they are more likely to be installed within existing walls, making ease of installation more important. They also may contain more fibers. Heavy-duty intrabuilding cables also can be used in more hostile factory environments, where they may be subjected to more stresses and more severe physical conditions.

Breakout cables are intrabuilding cables that contain subcables.

- Breakout cables are intrabuilding cables that contain subcables, so the cable can be divided into individual fibers or groups of fibers for distribution to separate end points in the office. An example is shown in *Figure 5-2*.

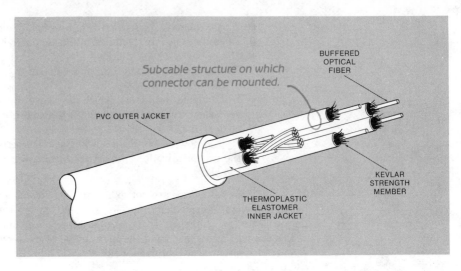

**Figure 5-2.
Breakout Cable**
(*Courtesy Math
Associates, Inc.*)

- Plenum cables are special intrabuilding cables made of materials that retard the spread of flame and produce little smoke. Because they are made of such materials (typically fluoropolymers), fire codes allow their installation in heating and cooling air ducts without special conduits to contain them. The need for special materials makes these cables relatively expensive, but installation savings offset that extra cost. Small size is important because of the high cost of the materials. Their construction is shown in *Figure 5-3*.
- Temporary light-duty cables are portable and rugged enough to withstand reasonable wear and tear. They may contain only a single fiber (e.g., a video feed from a camera) and should be durable enough to be laid and reused a few times. In general, they are comparable to interior cable.
- Temporary military cables are made rugged for military field use. Prototypes have survived such abuse as being run over by cars in a company parking lot. They are special-purpose cables made to withstand both hostile conditions and unskilled users. Because they are unusual, little more attention will be given to them.

**Figure 5-3.
Single-Fiber and
Duplex-Plenum Cables**
*(Courtesy Math
Associates, Inc.)*

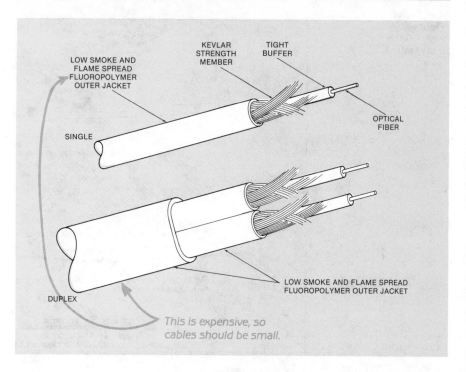

KEVLAR
STRENGTH
MEMBER

TIGHT
BUFFER

LOW SMOKE AND
FLAME SPREAD
FLUOROPOLYMER
OUTER JACKET

SINGLE

OPTICAL
FIBER

LOW SMOKE AND FLAME SPREAD
FLUOROPOLYMER OUTER JACKET

DUPLEX

*This is expensive, so
cables should be small.*

Aerial cables can run between overhead poles and can also be used in underground ducts.

- Aerial cables are made to be strung outdoors from poles, and typically can also be installed in underground ducts. They normally contain multiple fibers, and the internal arrangements of fibers can become elaborate. There are two types of aerial installations shown in *Figure 5-4*, which typically use different types of cables. One is the classical suspension of the cable between poles. The other requires attaching or lashing the cable to a messenger wire that runs between poles. Lashing supports the cable at more frequent intervals and reduces stress applied along its length, which can be large if the only support points are at the poles. Many fiber cables are designed only for lashing, not to withstand the high stress applied by suspension between poles. All such cables have strength members and structures that isolate the fibers from stress. The outer plastic jacket is a material such as polyethylene, which can withstand temperature extremes. The cable's internal structure is designed to keep moisture out.
- Ducted cables are similar to aerial types, but some may include armor to prevent rodent damage. In most cases, steel strength members, rather than the more expensive all-dielectric construction, are used.

**Figure 5-4.
Aerial Cable
Installations**

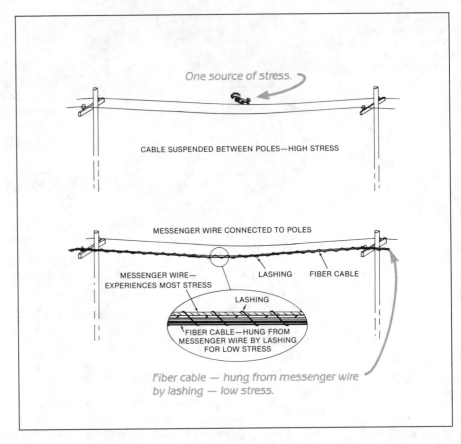

One source of stress.

CABLE SUSPENDED BETWEEN POLES—HIGH STRESS

MESSENGER WIRE CONNECTED TO POLES

MESSENGER WIRE—
EXPERIENCES MOST STRESS

LASHING   FIBER CABLE

LASHING

FIBER CABLE—HUNG FROM
MESSENGER WIRE BY LASHING
FOR LOW STRESS

*Fiber cable — hung from messenger wire
by lashing — low stress.*

- Direct-burial cables are similar to aerial cables except that they must have an outer armor layer to protect against gnawing by gophers. Normally, the metal jacket is surrounded inside and out with polyethylene layers that protect it from corrosion and cushion the inside from bending damage.
- Submarine cables can operate while submerged in fresh or salt water. Those intended to operate over relatively short distances—say no more than a few kilometers—are essentially rugged and waterproof versions of direct-burial cables. Cables for long-distance submarine use are much more elaborate, as will be described in Chapter 15. Some submarine cables are buried under the floor of the river, lake, or ocean, largely to protect them from damage by fishermen and boat anchors. The transatlantic submarine cable being installed by AT&T, shown in *Figure 5-5*, is designed in three layers, an inner core that can contain up to 12 fibers (although the cable being laid contains only 6), a deep-sea cable to withstand high pressure, and a double armoring layer for shallow waters where the cable could be damaged by fishing or shipping.

**Figure 5-5.
Fiber-Optic Cable
Design for Transatlantic
Cable** *(From Ali Adl, Ta-
Mu Chien, and Tek-Che
Chu, ''Design and Testing
of the SL cable,'' IEEE
Journal Selected Areas in
Communication, Nov.
1984, pp. 864-872, ©1984
IEEE. Used with
permission.)*

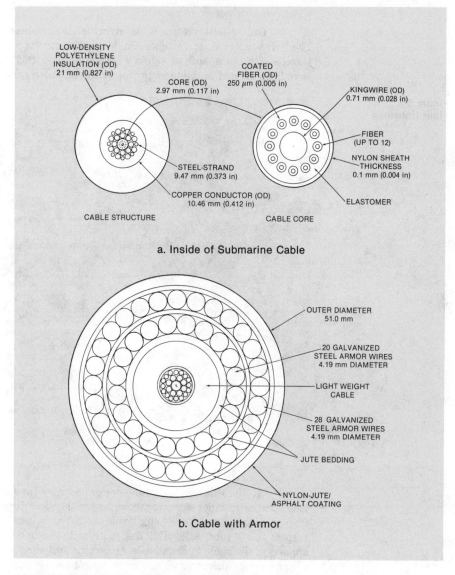

a. Inside of Submarine Cable

b. Cable with Armor

ELEMENTS OF CABLE STRUCTURE

*All fiber-optic cables are
made up of common
elements.*

As indicated above, fiber-optic cables are diverse in nature,
reflecting the diverse environments cables encounter. However, the same
basic elements are used in those different cables. Cut open a heavily
armored segment of submarine cable and you will find structures that
resemble those found in cables placed in much less demanding
environments.

Fiber Housing

One critical concern is the structure that houses individual fibers. There are two basic approaches, the loose-tube structure and the tightly jacketed structure, both shown in *Figure 5-6*. Their relative advantages have been argued in the industry for years, but both remain in use.

**Figure 5-6.
Fiber Housings**

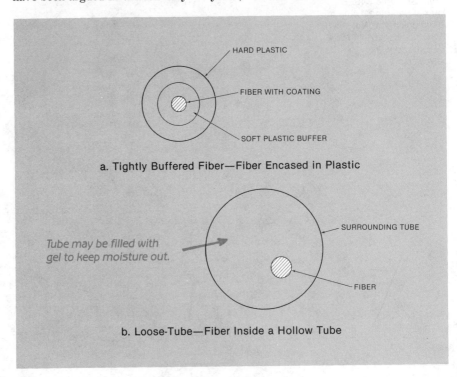

a. Tightly Buffered Fiber—Fiber Encased in Plastic

*Tube may be filled with
gel to keep moisture out.*

b. Loose-Tube—Fiber Inside a Hollow Tube

Loose-Tube Cable

In the simplest loose-tube design, a single fiber is contained in a long tube, with inner diameter much larger than the fiber diameter. The fiber is installed in a loose helix inside the tube, so it can move freely with respect to the tube walls. This design protects the fiber from stresses applied to the cable in installation or service, including effects of changing temperature. Such stresses can cause microbending losses as well as damage the fiber.

There are several variations on the loose-tube approach. Multiple fibers can run through the same tube. The tube does not have to be a physically distinct cylinder running the length of the cable. It can be formed by running grooves along the length of a solid cylindrical structure encased in a larger tube, as shown in *Figure 5-7*, or by pressing corrugated structures together and running fibers through the interstices. The end result is the same, the fiber is isolated from stresses applied to the surrounding cable structure.

Figure 5-7.
Variation on the Loose-Tube Cable

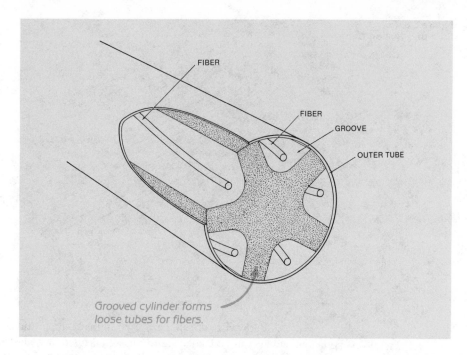

FIBER

FIBER

GROOVE

OUTER TUBE

Grooved cylinder forms
loose tubes for fibers.

Loose-tube cables are
filled with a gel for use
outdoors.

Loose tubes are used without any filling for indoor cables. However, if they are to be used outdoors, they normally are filled with a jelly-like buffer material. The gel acts as a lubricant so the fibers can move in the tube and helps keep out moisture. The fill also improves the cable's crush-resistance.

Tightly Buffered Fiber

Tightly buffered fibers
can be used in ribbon
cables.

A tightly buffered fiber is encased in one or more plastic layers. In many cases, the inner layer surrounding the (coated) fiber is a soft plastic that allows deformation and reduces forces applied to the fiber. This is surrounded by an outer layer of harder plastic to provide physical protection. Tightly buffered fibers may be stranded in conventional cables or in a ribbon structure such as the one shown in *Figure 5-8*.

Advantages and Disadvantages

Each structural approach has its advantages and advocates. Tight jacketing allows denser cables and better crush-resistance. It assures that the fibers are in precisely predictable positions, so splicing can be automated easily. The tight jacketing process also is easier to implement, so cables can be made less expensively. Loose-tube designs, on the other hand, offer better physical protection for the fibers and isolate them better from forces that could cause microbending losses.

**Figure 5-8.
Core of a Ribbon Cable**

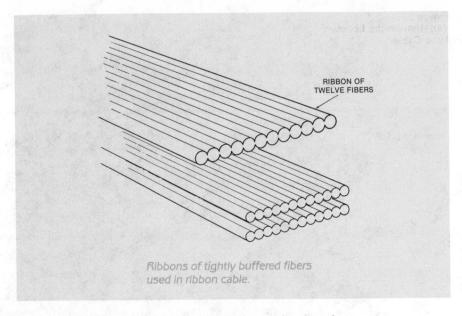

RIBBON OF
TWELVE FIBERS

*Ribbons of tightly buffered fibers
used in ribbon cable.*

Both types of cable are used in many applications, but some
distinctions can be made. Tightly jacketed fibers are most likely to be used
to satisfy demands for small cable size (e.g., crowded existing cable ducts),
extreme crush-resistance (e.g., cables running under carpets), or for low-
cost cables running over short distances where cable losses are
unimportant. Loose-tube cables are more likely in applications where
stresses will be applied in installation or service (e.g., aerial cables or cables
that must be pulled through twisted ductwork in buildings) or where stable
long-term, low-loss performance is critical (e.g., long-distance
telecommunications).

Fiber Arrangements in Cable

Duplex cables contain a
pair of fibers.

Fibers can be arranged in a cable in many different ways. The
simplest cables are round with a single fiber at their center. Duplex (two-
fiber) cables may either be circular or oval in cross section or be made like
electrical zip cord, with two single-fiber structures bonded together along
their length, as in *Figure 5-1.*

The more fibers that are included in the cable, the more complex
the structure. One common cable structure has six buffered fibers wound
loosely around a central strength member. The buffered fibers are wound
so they do not experience torsion in the cable. In loose-tube cables, the
fiber count can be raised by putting multiple fibers in each tube. Groups of
8 or 12 fibers also may be wound around strength members.

Cable structures can house up to several hundred fibers.

Cables with more fibers are likely to be built up of modular structures. For example, a 48-fiber cable could be made from eight loose-tube modules containing six fibers each, four 12-fiber ribbons, or from six subunits containing eight fibers each. (The last approach is shown in *Figure 5-9.*) Design details depend on the manufacturer. Cables containing several hundred fibers have been made but are not in routine use.

**Figure 5-9.
Modular Cable**

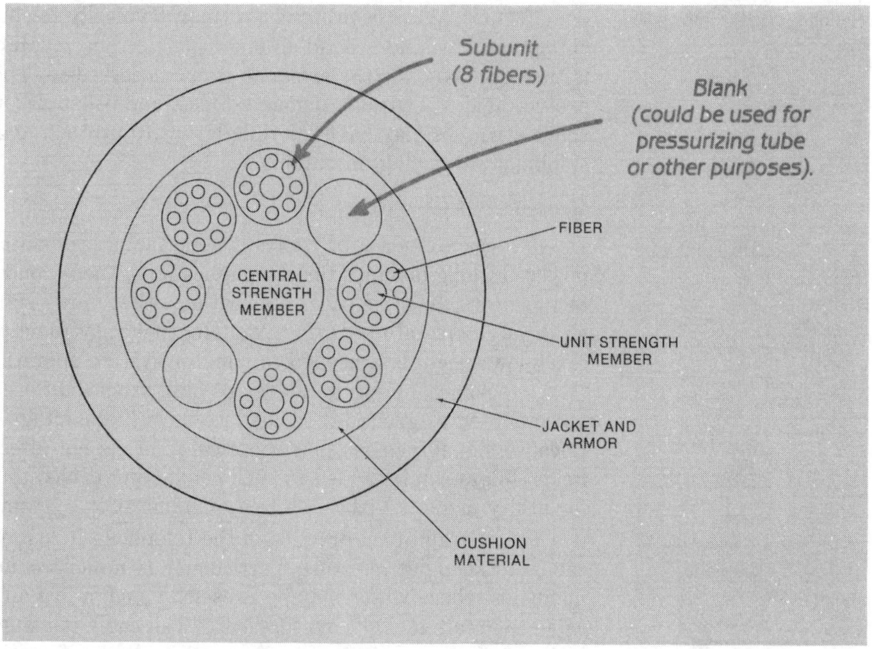

Subunit (8 fibers)

Blank (could be used for pressurizing tube or other purposes).

FIBER

CENTRAL STRENGTH MEMBER

UNIT STRENGTH MEMBER

JACKET AND ARMOR

CUSHION MATERIAL

Other Structural Elements

Cables have steel or Kevlar strength members to withstand stress.

Fibers and their buffers are not the only structural elements of cables. Many—but not all—fiber-optic cables include strength members. The usual strength members are steel or Kevlar, a high-strength polymer, which often are at the center of the cable structure. In some cables, strands of Kevlar are wound around the bundle containing the fibers, in addition to or instead of a central strength member. Cables apply tension along their length to them, rather than to the more vulnerable fiber. When a cable is pulled into a duct, the tension can be applied directly to the strength member. The strength member may be overcoated with a plastic or other material to make its size match cable requirements and to prevent friction between it and other parts of the cable.

The structure containing the fibers normally surrounds the strength member and, in turn, is surrounded by one or more outer jacketing layers. For light-duty cables, the typical jacketing material is polyvinyl chloride (PVC) or polyurethane; polyethylene is often used in heavier-duty and outdoor cables because of its better durability. Some cables may include layers of different plastics, such as PVC in an inner sheath and polyethylene in an outer sheath. Plenum cables, which must meet special fire-code requirements, have sheathing made of fluoropolymers, which produce little smoke and retard the spread of flames. Other materials may be used in special-purpose cables, such as those designed to run under carpets.

Buried and underwater cables require armor.

Underwater and buried cables are among the types that require one or more layers of protecting armor. Typically for buried cables, steel or aluminum is wound around an inner plastic sheath. An outer plastic sheath is then applied over the armor to prevent corrosion. The metal armor helps protect against crushing damage, rocks, and rodents. Underwater cables in shallow waters may have multiple layers to protect against damage from shipping and fishing operations.

Cable Auxiliaries

Fibers and wires may be combined in a single cable.

Some fiber-optic cables include metal conductors to carry electrical power. In long-distance telecommunications, these conductors supply power to repeaters. In some other applications, they carry signals that need not go through optical fibers; for example, they may function as order wires to carry low-speed data needed to monitor system operation.

Some fiber-optic cables are built from a different perspective—they are intended as ground-wire conductors in high-voltage power systems. Because signal transmission in optical fibers is not affected by high-voltage fields, fibers can be added to such conducting cables to carry signals either for utility use or as part of a telecommunication system.

Some cables are pressurized to keep out moisture.

A traditional practice in the telephone industry is to pressurize cables to keep out moisture, particularly in underground ducts and manholes where water may be present. Enough dry air is pumped into the cable to create a slight overpressure. The cable is made air-tight, so the internal pressure keeps any external liquid from flowing in. Such pressurization was effective in keeping water out of electrical cables, and many telephone companies continue to use it in optical cables, although some specialists do not believe that pressurization is needed.

CABLE INSTALLATION

Special techniques are used to install different types of cable.

Cable installation is not as simple as it sounds, but reliable methods developed for conventional metal cables have been successfully adapted for optical cables. The methods chosen depend on the type of installation.

- Submarine cables are laid from special ships built to lay cables.
- Buried cables normally are installed by digging a deep, narrow trench with a cable plow, laying the cable in the trench, and filling the trench in with dirt.

- Cables are installed in ducts by threading a stiff wire through the duct, clamping it to the cable, then pulling the cable through the ductwork. Manholes or other access points normally are available along the duct route, so the cable need not be pulled all at once through a long route.
- Aerial cables may be suspended directly from overhead poles, or from messenger wires, strong wires of steel or other metal strung between poles. If a messenger wire is used, the cable is strapped to it with a tape running around both the cable and the wire. This is a common installation for many overhead fiber cables because it minimizes strength requirements.
- Plenum cables are strung through interior air ducts.
- Interior cables may be installed within walls, above suspended ceilings, or elsewhere in buildings. Only special cables designed for installation under carpets should be laid on the floor where people walk.
- Temporary light-duty cables are laid by people carrying mobile equipment that requires a broad-band (typically video) connection to a fixed installation.
- Temporary military cables may be laid by helicopters from the air or by soldiers on the ground, during field exercises, in preparation for engagements, or in actual battle. Typically they would be unreeled from cable spools.

CHANGES IN CABLED FIBER

Microbending can cause fiber loss to change after cabling.

Ideally, characteristics of optical fibers should not change when they are cabled, but in practice some changes can occur, particularly in attenuation. A major cause of these changes is microbending, which depends on the fiber's local environment and the stresses applied to it. In some cases, comparison of fiber loss on the reel and in the cable shows that attenuation actually decreases upon cabling, apparently because the fiber had suffered from microbending on the reel. Cabling generally does not increase loss of single-mode fibers significantly, and it raises attenuation of multimode fibers only 0.1 or 0.2 dB/km.

Cabled fibers rarely suffer physical damage unless the entire cable is damaged. Most fiber manufacturers apply a stress test to fibers before they leave the factory, and cablers normally perform similar tests before the fiber is put into a cable. The test is a simple one in which a series of pulleys and wheels applies a given stress to the fiber. For the fiber to fail the test, it breaks and, thus, cannot be installed. These proof tests assure levels of fiber strength that meet normal cable requirements.

CAUSES OF CABLE FAILURE

Most cable failures are a result of physical abuse.

Telephone companies were very cautious before beginning their massive switchover to fiber-optic cables and conducted extensive tests and field trials to evaluate the reliability of optical cables. These studies have shown that optical fibers have excellent reliability. Proof testing of fibers before they are cabled assures that mechanical strength is high, and proper splicing produces splices about as strong as unspliced fiber.

Most cable failures are due to physical abuse. The archetypical problem is a backhoe digging up and breaking a buried cable. Aerial cables can be broken by falling branches or errant cranes. Fibers in light-duty indoor cables could be broken by slamming doors or windows on the cable, although the cable might not show serious damage. Applying a sharp stress along a short indoor cable (e.g., tripping over it) is not likely to break the cable. However, it could jerk the cable out of a connector at one end. In general, connector junctions are the physical weak points of short-distance cables.

If the cable itself breaks, the fibers in it also will break. Because fibers tend to break at weak points, they may not break at precisely the point of the cable break but should break close to it.

WHAT HAVE WE LEARNED?

1. Unlike copper wires, fibers can only stretch about 5% in length before breaking, so they must be protected from stretching forces.
2. Cabling packages fibers for protection and easier handling.
3. Cables must resist crushing as well as absorb tension along the fiber length.
4. Design requirements depend on the environment where a cable will be used.
5. Elements of cable structures include housing for the fiber, strength members, jacketing, and armor.
6. Fibers can be enclosed in a loose tube or a tight plastic buffer.
7. Physical arrangement of fibers in the cable depends on the number of fibers.
8. Fiber experiences only minimal changes in its optical properties when it is installed in a cable.
9. Most damage to cables is by application of sudden stresses, such as digging up buried cable.

WHAT'S NEXT?

In Chapter 6, we will examine the light sources used with fiber-optic cables.

Quiz for Chapter 5

1. What happens if an optical fiber is pulled along its length?
   a. It stretches out and does not return to its original length.
   b. It can stretch by about 5% before breaking.
   c. Its length is unchanged until it breaks.
   d. It breaks if any force is applied to it.

2. Cables cannot protect fibers effectively against:
   a. gnawing rodents.
   b. stresses during cable installation.
   c. careless excavation.
   d. static stresses.
   e. crushing.

3. Light-duty cables are intended for use:
   a. within office buildings.
   b. in underground ducts.
   c. underground where safe from contractors.
   d. on aerial poles where temperatures are not extreme.

4. The special advantages of plenum cables are:
   a. They are small enough to fit in air ducts.
   b. They meet fire codes for running through air ducts.
   c. They are crush-resistant and can run under carpets.
   d. They have special armor to keep rodents from damaging them.

5. Aerial cables are not used in which of the following situations?
   a. Suspended overhead between telephone poles.
   b. Tied to a separate messenger wire suspended between overhead poles.
   c. Buried directly in the ground.
   d. Pulled through underground ducts.

6. A loose-tube cable is:
   a. a cable in which fibers are housed in hollow tubes in the cable structure.
   b. a cable for installation in hollow tubes (ducts) underground.
   c. a hollow plastic tube containing several fibers for use indoors.
   d. none of the above.

7. Which of the following are present in direct-burial cables but not in aerial cables?
   a. Strength members.
   b. Outer jacket.
   c. Armor.
   d. Fiber housing.

8. Which type of cable installation requires pulling the cable into place?
   a. Direct burial.
   b. Underground duct.
   c. Aerial cable.
   d. Submarine cable.

9. The main cause of differences in properties of a fiber before and after cabling is:
   a. microbending.
   b. temperature within the cable.
   c. application of forces to the fiber.
   d. damage during cabling.

10. The major reason for failure of
cabled fiber is:
  a. hydrogen-induced increases in
  attenuation.
  b. corrosion of the fiber by
  moisture trapped within the
  cable.
  c. severe microbending losses.
  d. physical damage to the cable.

Transmitters

ABOUT THIS CHAPTER

Fiber-optic transmitters come in many types—from cheap LEDs directly driven by signal sources to sophisticated transmitters based on costly semiconductor lasers. Some operate at telephone-like speeds and bandwidths over several meters; others send hundreds of thousands of megabits per second through tens of kilometers of fiber.

This chapter examines transmitters and how they and their key optical components work. It begins by describing the major functional considerations for transmitters, then concentrates on the specifically optical elements of transmitters—the light sources. (Although the electronic functions of transmitters are important, we will not explore them in detail.)

A FEW WORDS ABOUT TERMINOLOGY

Fiber-optic transmitters may be packaged with receivers as systems. Short digital systems are called data links.

Fiber-optic transmitters often are packaged with receivers and cables and sold as systems. For short-distance digital transmission between two points, these systems are labelled data links. Local-area networks interconnect multiple terminals spread over relatively small distances (typically no more than a few kilometers); only a small fraction use fiber optics. Equipment for long-distance transmission normally is called a system rather than any specific term. Industry terminology is vague, and more than one short analog system has been called a data link. Details of how transmitter, receiver, cable, and other components come together to make fiber-optic systems will be described in later chapters on system design and specific applications.

TRANSMITTER PERFORMANCE

Type of signal, speed, operating wavelength, and light source impact the performance of a fiber-optic transmitter.

Several factors enter into the performance of a fiber-optic transmitter, including the type of signal being sent, the speed, the operating wavelength, the type of light source, and the cost. Each of these deserves a brief explanation.

Analog versus Digital Transmission

Fiber-optic systems can transmit analog or digital signals.

Most of the communications world is shifting from analog to digital transmission, but everyone isn't there yet. Some applications require continuous analog waveforms; others operate best with discrete digital pulses. In theory, a simple LED source could be modulated by either an analog or digital signal, but in practice transmitters are designed for one or the other type of modulation.

As we saw in Chapter 3, digital signals can withstand distortion better than analog signals. *Figure 6-1* shows how the inherently analog process of signal transmission can distort signals by not precisely reproducing the input waveform. Distortion presents a serious problem for analog signals because the output should be a linear reproduction of the input. That is, if the input signal is F(t), the output should be cF(t), where c is a constant. Digital systems can tolerate such non-linear distortion because they only have to detect presence or absence of a pulse—not its shape.

Figure 6-1.
Effects of Distortion on Analog and Digital Signals

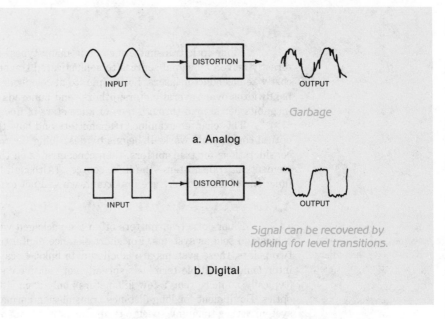

a. Analog

b. Digital

Digital transmission, on the other hand, demands faster response than analog, to follow the rapid rise of signals. Breaking a digital signal down into its component frequencies shows that the sharp edge of a digital pulse is made up of high frequencies. However, the system need only detect the difference between off and on.

Thus analog systems must accurately reproduce inputs, but speed is not as crucial as in digital systems, which must be fast but need not be accurate. These difference mean that analog and digital transmitters should use different designs. They may be able to use the same light source but with different electronics.

Bandwidth and Data Rate

Bandwidth usually defines capacity of an analog system. Digital capacity is measured as data rate.

The operating speed of a fiber-optic transmitter is measured in two ways: bandwidth for analog signals and data rate for digital signals. Both refer to the amplitude modulation of light from the source, which is shown in *Figure 6-2*. (Light waves are actually much smaller than shown. If each digital pulse was 1 ns [10^{-9} s] long, it would contain 300,000 waves of 1-μm light.) For analog modulation, bandwidth normally is defined as the point where signal amplitude drops 3 dB below the normal level (equivalent to a reduction of 50% in power). For digital transmission, the data rate is the maximum number of bits per second that can be transmitted with an error rate below a specified level (typically one error in 10^9).

**Figure 6-2.
Light Modulation by
Digital and Analog
Signals**

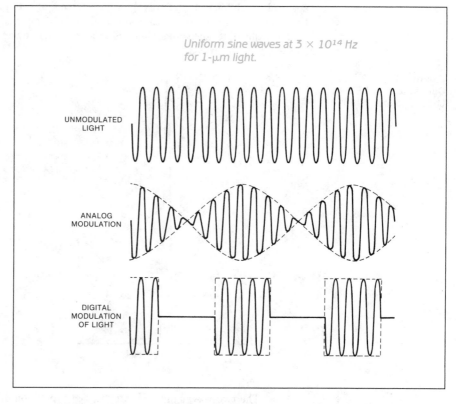

Uniform sine waves at 3×10^{14} Hz for 1-μm light.

UNMODULATED
LIGHT

ANALOG
MODULATION

DIGITAL
MODULATION
OF LIGHT

Neither of these quantities actually measures the limiting characteristic of light sources: their rise time or response time. Rise time normally is the time it takes light output to rise from 10 to 90% of the steady-state level for a given drive current. This can be related roughly to bandwidth by the equation:

$$BW = \frac{0.35}{\text{RISE TIME}}$$

where BW is in megahertz if rise time is in microseconds. The precise relationship between bandwidth and rise time differs among light sources and transmitters.

Rise time is an important variable in light source selection. Many semiconductor lasers have rise times of a fraction of a nanosecond. LED rise times range from a few nanoseconds to a few hundred nanoseconds, depending on design.

Operating Wavelength

Source wavelength affects performance of fiber-optic systems in two ways. The choice of wavelength determines the attenuation (and to some extent the pulse dispersion) the signal experiences in the fiber. The spectral width, or range of wavelengths emitted, affects pulse dispersion in the fiber, which increases with the wavelength range. Spectral width is one of the major differences between LEDs and the narrower-line laser sources, as shown in *Figure 6-3*.

**Figure 6-3.
Comparison of LED and
Laser Spectral Widths**

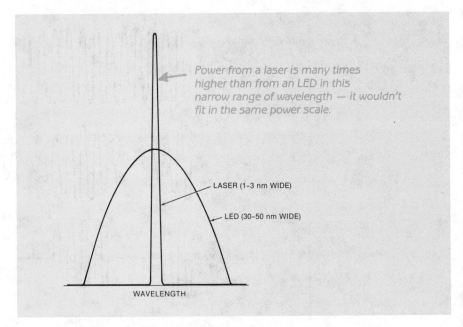

Power from a laser is many times higher than from an LED in this narrow range of wavelength — it wouldn't fit in the same power scale.

LASER (1–3 nm WIDE)

LED (30–50 nm WIDE)

WAVELENGTH

The two wavelength characteristics depend on different factors. The emitted wavelength depends on the semiconductor material from which the light source is made, as described below. The range of wavelengths depends on device structure. A laser and LED made of the same material would have the same center wavelength, but the LED would emit a much broader range of wavelengths. Two lasers with the same structure made of different materials would have comparable spectral widths but different center wavelengths.

Output Power

Power delivered by the light source can range from tens of milliwatts for semiconductor lasers to tens of microwatts for LEDs. Not all of that power is useful. For fiber-optic system applications, the relevant value is the power that can be coupled into an optical fiber. That power depends on the angle over which light is emitted, the size of the light-emitting area, the alignment of the source and fiber, and the coupling characteristics of the fiber, as shown in *Figure 6-4*. For all practical light sources, the light intensity is not uniform over the entire angle at which light is emitted but rather falls off with distance from the center. Typical semiconductor lasers emit light that spreads at an angle of 10–20°; the light from LEDs spreads out at larger angles.

Figure 6-4.
Light Coupling from an
Emitter into a Fiber

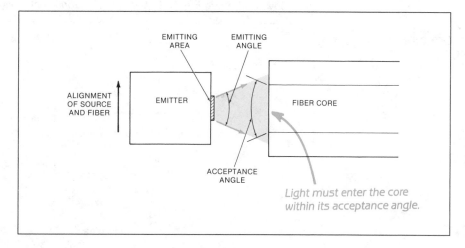

Losses of several decibels can easily occur in coupling light from an emitter into a fiber, especially with LED sources that have broad emitting areas and wide emitting angles. Some specification sheets do not clearly indicate if output power is that emerging from the source or collected by the fiber (e.g., by a fiber pigtail attached directly to the emitter). Be sure you know which is meant.

Cost/Performance Trade-Offs

As any student of engineering reality would expect, light sources with the most desirable characteristics cost the most. The cheapest light sources are LEDs with slow rise times, large emitting areas, and relatively low output power. Diode lasers that emit the 1300- and 1550-nm wavelengths where optical fibers have their lowest losses are the most expensive. The higher-power and narrower-line emission of lasers come at a marked price premium. The only real performance advantage of LEDs is generally longer lifetime than lasers, which are driven harder during operation.

LED SOURCES

An LED emits light when a current flows through it.

The most common fiber-optic light sources are LEDs that emit invisible near-infrared light. The basic idea behind operation of a light-emitting diode is shown in *Figure 6-5*. A small voltage is applied across a semiconductor diode, causing a current to flow across the junction. The diode is made up of two regions, each doped with impurities to give it the desired electrical characteristics. The p region is doped with impurities having fewer electrons than the semiconductor compound, which creates "holes" where there is room for electrons in the crystalline lattice. The n region is doped with impurities that donate electrons, so extra electrons are left floating in the crystalline matrix. Applying a positive voltage to the p region and a negative voltage to the n region causes the electrons and holes to flow toward the junction of the two regions, where they combine (the process is actually called recombination). As long as the voltage is applied, electrons keep flowing through the diode and recombination continues at the junction.

Figure 6-5.
LED Operation

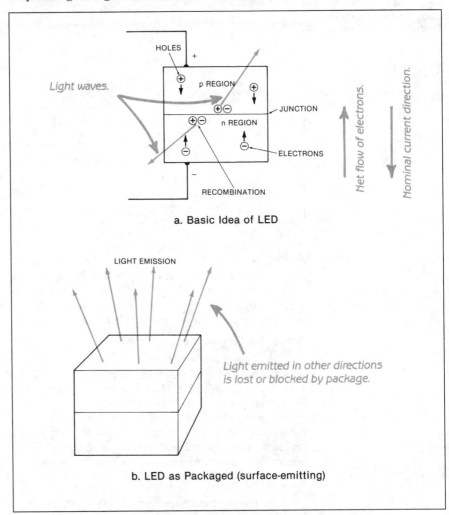

a. Basic Idea of LED

b. LED as Packaged (surface-emitting)

In many semiconductors, such as silicon, the released energy is simply dissipated as heat (vibrations of the crystalline lattice). However, in other materials, usable in LEDs, the recombination energy is released as a photon of light, which can emerge from the semiconductor material. The most important of these semiconductors, gallium arsenide and related materials, are made up of elements from the IIIa and Va columns of the periodic table:

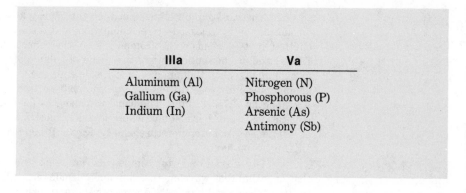

IIIa	Va
Aluminum (Al)	Nitrogen (N)
Gallium (Ga)	Phosphorous (P)
Indium (In)	Arsenic (As)
	Antimony (Sb)

The wavelength emitted depends on the semiconductor's internal energy levels. In a pure semiconductor at low temperature, all the electrons are bonded within the crystalline lattice. As temperature rises, some electrons in this valence band jump to a higher-energy conduction level, where they are free to move about in the crystal. The valence and conductor bands are separated by a void where no energy levels exist—the band gap that gives semiconductors many of their special properties.

Conduction-band electrons leave behind a hole in the valence band, which is considered to have a positive charge. This hole can move about, as electrons from other spots in the crystalline lattice move to fill in the hole and leave behind their own hole (i.e., the hole moves from where the electron came to where the electron was). Impurity doping of semiconductors also can generate free electrons and holes. When an electron drops from the conduction level to the valence level (i.e., when it recombines with a hole), it releases the difference in energy between the two levels, as shown in *Figure 6-6.*

**Figure 6-6.
Semiconductor Energy
Levels**

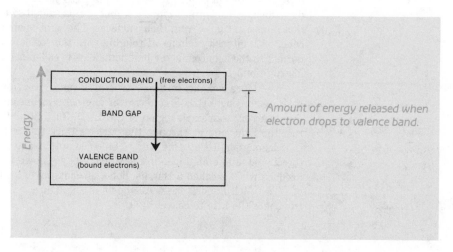

The usual LED wavelengths for fiber optics are 820 or 850 nm.

The band-gap difference between the energy levels—and hence the amount of energy released by recombination, and the wavelength of emitted light—depends on the composition of the semiconductor. The usual LEDs used in fiber-optic systems are made of gallium aluminum arsenide or gallium arsenide. Gallium arsenide LEDs emit near 930 nm, but adding aluminum can increase the energy gap and shift emission to shorter wavelengths of 750–900 nm. The usual wavelengths for fiber-optic applications are 820 or 850 nm, where losses of most fibers are lower than at 930 nm. At room temperature, the typical 3-dB bandwidth of an 820-nm LED is about 40 nm.

Visible LEDs are used with plastic fibers, which transmit poorly in the infrared.

Other semiconductor compounds can be used to make LEDs that emit different wavelengths. Gallium-arsenide-phosphide (GaAsP) LEDs emitting visible red light at 665 nm are used with plastic fibers, which are most transparent in the red, and transmit poorly at GaAlAs wavelengths. GaAsP LEDs cost less than GaAlAs LEDs but are lower in performance.

InGaAsP LEDs emit at 1300 and 1550 nm.

The most important compound for high-performance fiber optics is InGaAsP, made of indium, gallium, arsenic, and phosphorous mixed so the number of indium plus gallium atoms equals the number of arsenic plus phosphorous atoms. The resulting compound is written as $In_xGa_{1-x}As_yP_{1-y}$, where x is the fraction of indium and y the fraction of arsenic. These so-called "quaternary" (four-element) compounds are harder to make than "ternary" (three-element) compounds such as GaAlAs but are needed to produce output at the fibers' 1300- and 1550-nm windows. In practice, LEDs are often used at 1300 nm, where conventional fibers have low chromatic dispersion, but are rarely used at 1550 nm, where dispersion is much higher.

Other LED characteristics depend on device geometry and internal structure. The description of LEDs so far hasn't indicated in which direction they emit light. In fact, simple LED junctions emit light in all directions, as shown in *Figure 6-6*, and are packaged so that the bulk of that emission comes from their surfaces. The light is emitted in a broad cone, with intensity falling off roughly with the cosine of the angle from the normal to the semiconductor junction. (This is called a Lambertian distribution.)

The Burrus diode emits light from a hole etched in its surface.

More complex internal structures can concentrate the emission of surface-emitting LEDs in a narrower angle, by means such as confining drive current to a small region of the LED. Such designs typically require that the light emerge through the device's thick substrate, which can lead to transmission losses. One way to enhance output and make emission more directional is to etch a hole in the substrate, as shown in *Figure 6-7*, to produce what is called a Burrus diode, after its inventor Charles A. Burrus of AT&T Bell Laboratories. As shown in the figure, a fiber can be inserted right into the hole to collect light.

Figure 6-7.
A Burrus Diode

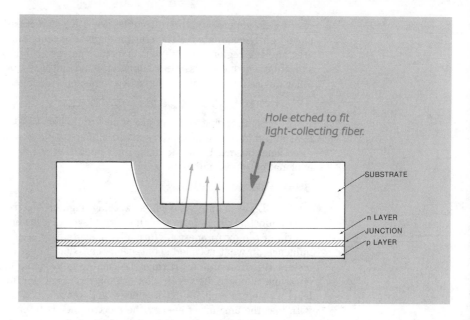

Hole etched to fit light-collecting fiber.

SUBSTRATE

n LAYER
JUNCTION
p LAYER

An edge-emitting diode emits light from its ends.

A fundamentally different configuration is the edge-emitting diode, shown in *Figure 6-8*. Electrical contacts cover the top and bottom of an edge emitter, so light cannot emerge there. The LED confines light in a thin, narrow stripe in the plane of the pn junction. This is done by surrounding that stripe with regions of lower refractive index, creating a waveguide, much as an optical fiber functions as a waveguide. The waveguide channels light out both ends, and one of these outputs can be coupled into a fiber. A disadvantage of this approach is that it increases the amount of heat the LED must dissipate and often requires cooling.

Figure 6-8.
An Edge-Emitting LED

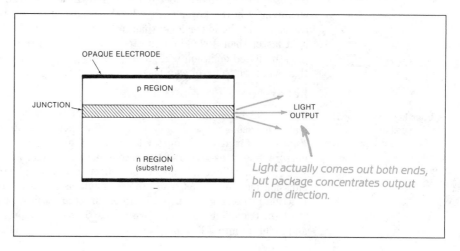

OPAQUE ELECTRODE
+
p REGION

JUNCTION

LIGHT OUTPUT

n REGION
(substrate)

−

Light actually comes out both ends, but package concentrates output in one direction.

In general, the more complex the LED structure, the brighter and more tightly collimated is the emitted light. Concentration of the emitting area and the region through which current passes also decreases rise time and, thus, enhances possible modulation bandwidth. Of course, as with other devices, the greater complexity comes only at higher cost.

SEMICONDUCTOR LASER SOURCES

Semiconductor lasers are superficially like LEDs, but they produce light in a different way that results in higher output powers and more directional beams. Understanding their operation requires a quick explanation of laser physics.

Stimulated Emission

Laser emission is stimulated.

Light is emitted when something (e.g., an electron in a semiconductor) drops from a high energy level to a lower one and needs to get rid of the extra energy. Normally, it emits light without outside influence, in what is called spontaneous emission, but it does not do so the instant it is first able to. It takes its time to get around to spontaneous emission. Suppose, however, that the electron is sitting in the upper energy level waiting to emit its extra energy, and another photon comes along, with just the amount of energy the electron needs to emit. That external photon can stimulate the electron in the upper energy level to drop to the lower one and emit its energy as light of the same wavelength. The result is a second identical photon. The process is called Light Amplification by the Stimulated Emission of Radiation, and the acronym spells "Laser."

Population Inversion

Special conditions are needed for laser emission. One requirement is that there be more electrons (or atoms or molecules) in the upper energy level than in the lower one. (Specialists call this a population inversion because normally more electrons are in lower levels.) The reason is that whatever is in the lower energy level can absorb the emitted light. If more things are in the lower level than in the upper level, they would absorb the light faster than it could be emitted. However, if only that condition is met, the stimulated emission will go off in every direction, more like a light bulb than what we think of as a laser.

Laser Beam Formation

A laser beam is formed by a resonator that reflects light back and forth through the recombination region.

The laser beam is formed by a resonator, which confines light and makes it pass again and again through the excited medium. As shown in *Figure 6-9* for a semiconductor laser, this resonator can be a pair of mirrors, one at each end of the recombination region. Light emitted straight toward one mirror will be reflected back and forth, stimulating emission from electrons ready to recombine as it passes through the junction plane. Light emitted in other directions will leak away. Thus only the light travelling back and forth along the laser stripe will be amplified and build up into a beam.

Figure 6-9.
Basic Operation of a
Semiconductor Laser

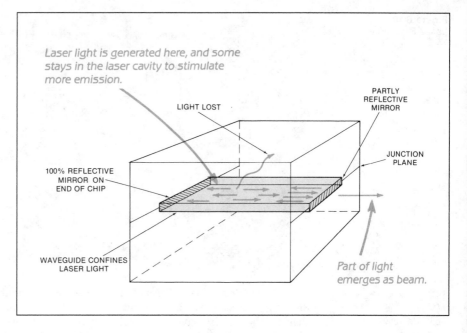

Laser light is generated here, and some stays in the laser cavity to stimulate more emission.

LIGHT LOST

PARTLY
REFLECTIVE
MIRROR

JUNCTION
PLANE

100% REFLECTIVE
MIRROR ON
END OF CHIP

WAVEGUIDE CONFINES
LASER LIGHT

Part of light emerges as beam.

A closer look shows that the process is a little more complex. The mirrors are really the ends of the semiconductor crystal, called facets. The rear facet is coated with a reflective layer, so all light that strikes it is reflected. In practice, the front facet is left uncoated so most light escapes but some is reflected back into the semiconductor. Semiconductors are such strong stimulated emitters that this is all that is needed to produce laser action.

Some LED materials are not suitable for use in semiconductor lasers. Much to the frustration of people who want visible semiconductor lasers, this includes the materials used in red, yellow, green, and blue LEDs. However, all semiconductor laser materials can be operated as LEDs.

Functional Differences

There are two important functional differences between LEDs and diode lasers. One is that LEDs lack reflective facets and, in fact, may be designed to minimize reflection back into the semiconductor. The other is that lasers must operate at higher drive currents to get the high density of ready-to-recombine electrons needed at the pn junction.

The output of a semiconductor laser depends on the drive current passing through it, as long as the bias voltage is above the minimum required (the band-gap energy). At low currents, the laser emits feeble spontaneous emission, operating as an inefficient LED. However, as drive current passes a threshold value, the device shifts over to laser emission, and output rises steeply, as shown in *Figure 6-10*. For most diode lasers, LED emission is so weak it can be ignored.

**Figure 6-10.
Laser Emission**

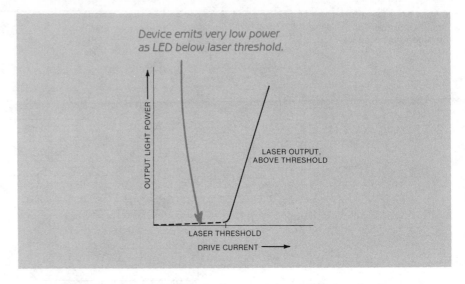

Device emits very low power
as LED below laser threshold.

OUTPUT LIGHT POWER

LASER OUTPUT,
ABOVE THRESHOLD

LASER THRESHOLD

DRIVE CURRENT

Lasers are much more
powerful than LEDs and
emit a narrower range of
wavelengths.

These internal differences lead to some important functional differences. Lasers convert electrical input power to light more efficiently than LEDs and also have higher drive currents, so lasers are much more powerful than LEDs. The concentration of stimulated emission leads to a beam much narrower than from an LED (although semiconductor laser beams are broad by laser standards). The higher drive currents and optical power levels make laser lifetimes shorter than LEDs. The amplification inherent in laser action tends to concentrate emission in a much narrower range of wavelengths than LED output, normally about a couple of nanometers. In essence, the center of the emission curve is amplified much more than the fringes, making the laser curve much more steeply peaked, as shown in *Figure 6-3*.

Structural Differences

Semiconductor lasers are more complex in structure than LEDs, reflecting the more demanding conditions for successful laser operation. The first semiconductor lasers were made in 1962 using simple structures, but they required liquid-nitrogen cooling and were severely limited in lifetime and duty cycle. Developers soon found ways to concentrate drive current and light generation so the laser would work better. The first step was the single-heterojunction laser in which a heterojunction between semiconductor materials with slightly different refractive indexes helps confine light on one side of the junction. The next step was the double-heterojunction laser, where a pair of heterojunctions formed a sandwich that confined the emitted light to the central junction region (the active layer). Further refinements confine light horizontally as well as vertically by concentrating the drive current and optical power in a narrow stripe running the length of the chip.

Lasers emit light from a
small region similar in
size to the cores of mul-
timode fibers.

Most refined designs confine laser emission to a small region. State-of-the-art lasers have active stripes just a few micrometers across and a fraction of a micrometer high. Like an optical fiber, the stripe functions as a waveguide. Because of its narrow dimensions, it is a single-mode waveguide. Light is emitted from a region at the end of the chip the same size as the cross section of the active area. In state-of-the-art lasers, that region is nicely matched to the cores of single-mode fibers. The beam spreading or divergence is comparable to the acceptance angle of fibers.

Wavelengths

Output Spectrum

Single-frequency lasers
are needed to overcome
high chromatic dispersion
at 1550 nm.

Although diode lasers emit a much narrower range of wavelengths than LEDs, that range is large enough that laser output suffers serious chromatic dispersion at 1550 nm in step-index single-mode fibers. Overcoming that problem will require true single-frequency lasers that emit an even narrower range of wavelengths.

The output spectrum of a conventional narrow-stripe semiconductor laser is shown in *Figure 6-11*. Although much of the power is concentrated at one wavelength, other wavelengths also are present. This happens because of the way the light oscillates within the laser resonator. The only wavelengths that are amplified in the laser are those for which the round-trip distance between the mirrors is an integral number of wavelengths λ:

$$2D = N\lambda$$

where 2D is the round-trip distance and N is an integer. In addition, the wavelength λ must be within the laser's gain curve, the range of wavelengths where the laser light can be amplified by stimulated emission, which is much broader than the individual peaks shown in *Figure 6-11*. The result is a series of narrow-wavelength spikes, called longitudinal (for along the length of the laser) modes. Unfortunately, the range of wavelengths covered by those multiple longitudinal modes is large enough for significant dispersion to accumulate at 1550 nm in step-index single-mode fibers, seriously limiting transmitted bandwidth.

Figure 6-11.
Wavelengths in Multiple
Longitudinal Modes

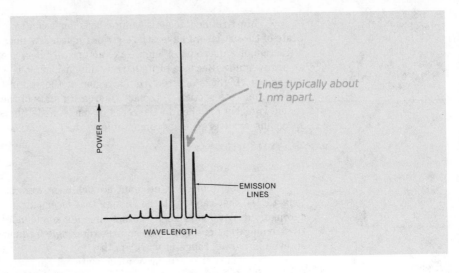

Lines typically about
1 nm apart.

POWER →

EMISSION
LINES

WAVELENGTH

Single-Frequency Lasers

Distributed feedback, C-
cubed, and external cavity
lasers emit only a single
frequency.

To restrict oscillation to a single longitudinal mode, researchers have developed laser resonators more elaborate than simply facets on the ends of a semiconductor chip. Three leading approaches are shown in *Figure 6-12*. One is the distributed-feedback laser, in which a series of corrugated ridges on the semiconductor substrate (replacing the mirrored end facets) reflect only certain wavelengths of light back into the laser, and light at only one resonant wavelength is amplified. In the cleaved-coupled-cavity (C-cubed or C³) laser, two separate laser chips are coupled optically but isolated electronically in a way that limits emission to just one wavelength. The third is the external-cavity laser, in which external optics form a much longer resonator than ordinary diode lasers, limiting oscillation to a single wavelength.

External modulation of lasers is needed to achieve
the best possible high-
speed modulation.

Even further refinements are needed to achieve the utmost in fiber-optic transmission. The problem comes from directly modulating the laser by changing the drive current passing through it. Although that makes diode lasers much easier to modulate than other types, it also subtly shifts their wavelength. As the density of electrons in the semiconductor changes, so does the refractive index of the material, effectively changing the optical length of the laser cavity (D = nL, where n is refractive index and L is the physical separation of the cavity mirrors). From the earlier equation for the wavelength at which the laser resonates, you can see that this means the wavelength λ changes by an amount Δλ:

$$\Delta\lambda = \frac{\Delta n \times L}{N}$$

where Δn is the change in refractive index and N is an integer, the number of wavelengths in the cavity. Although that change is small, it can limit performance.

**Figure 6-12.
Three Single-Frequency
Lasers**

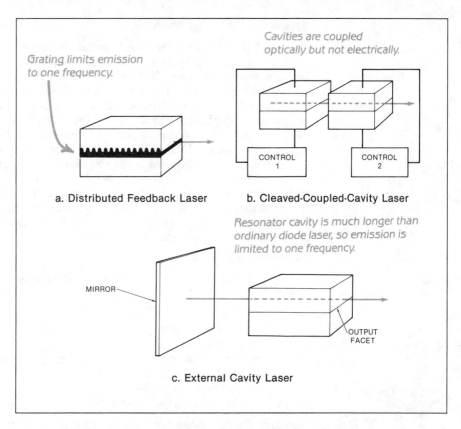

*Grating limits emission
to one frequency.*

*Cavities are coupled
optically but not electrically.*

CONTROL 1

CONTROL 2

a. Distributed Feedback Laser

b. Cleaved-Coupled-Cavity Laser

*Resonator cavity is much longer than
ordinary diode laser, so emission is
limited to one frequency.*

MIRROR

OUTPUT
FACET

c. External Cavity Laser

Performance records in fiber-optic transmission are defined by the product of bit rate times distance. The record at the time this book was written was set by the "hero experiments" group at AT&T Bell Laboratories, who sent 8 Gbit/s through 68.3 km of repeaterless fiber. They drove a cleaved-coupled-cavity laser with a constant bias current to keep its wavelength steady. An external optical modulator rapidly changed its transparency to modulate the light signal. Such systems now are too cumbersome for much practical use, but they show how far the limits of fiber-optic performance can be pushed.

Lifetimes

Output power of semiconductor lasers drops with age.

The output power of semiconductor lasers declines with age. Although lifetimes have improved greatly, aging and increasing temperature can cause laser output at a steady drive current to drop slowly. Eventually, power drops to a point where the laser no longer meets system needs and is pronounced "dead," although it still may emit some light. To deal with this problem, sophisticated laser transmitters include circuits to stabilize light output and temperature, as we will see later in this chapter.

TRANSMITTER DESIGN

A transmitter includes housing, drive circuitry, and monitoring equipment as well as a light source.

A light source is the most important component of a transmitter, but it is not sufficient by itself. A housing is required to mount and protect the light source and to interface with the electronic signal source and the transmitting optical fiber. Internal components may be needed to optimize coupling to the fiber. Drive circuitry is often needed, and temperature control and output monitoring can be crucial for sophisticated lasers.

The practical boundaries between transmitters and light sources can be vague. Simple LED sources can be mounted in a case with optical and electronic connections and little or no drive circuitry. On the other hand, a high-performance laser may be packaged as a transmitter in a case that also houses an output monitor and thermoelectric cooler. Then that whole package may be incorporated into a larger transmission system that performs electronic functions such as multiplexing. Thus, you can say that some transmitters can contain transmitters.

Elements of Transmitters

Several elements make up fiber-optic transmitters.

The basic elements that may be found in transmitters, as shown in *Figure 6-13* are:

- Housing
- Electronic interfaces
- Optical interfaces
- Drive circuitry
- Temperature sensing and control
- Optical output sensing
- Electronic signal preprocessing.

Note that the figure does not show an actual physical arrangement, only the relationships of the elements. If the transmitters are part of a transceiver or repeater, they may include optical inputs. Often transmitters do not contain all of these elements.

Housing

The housing for a fiber-optic transmitter, in its simplest form, is just a box designed to be mounted conveniently. Screws or other mounting equipment attach it mechanically to printed circuit boards or other electrical components. Some transmitters are built to fit within electronic connector housings, so the user sees just a cable with special connectors on each end. Telephone transmitters often are packaged to mount on standard equipment racks.

**Figure 6-13.
Elements Used in Fiber-
Optic Transmitters**

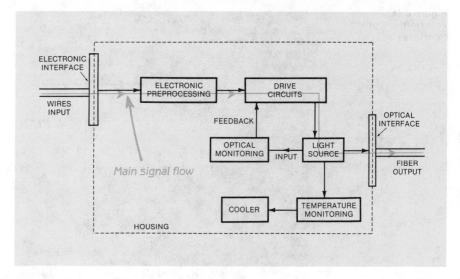

Electronic Interfaces

Electronic interfaces may be wires, standard electronic connectors, or pins emerging from packages. Some simple transmitters can be driven directly by electronic input signals. More complex transmitters may require power and may accept multiple electronic inputs (and provide one or more electronic outputs as well).

Optical Interfaces

Common optical interfaces are connectors and fiber pigtails.

The optical interface between light source and fiber can take various forms, as shown in *Figure 6-14*. One of the most common forms is integration of a fiber-optic connector in the housing. Light is delivered to that connector by internal optics, including a collimating lens and sometimes a short fiber segment. Another approach is a short fiber pigtail that collects light from the emitting area and delivers it outside the case, where it can be spliced to an external fiber. The choice between these types depends on several factors, including cost, whether connections are to be temporary or permanent, type of fiber used, importance of minimizing interconnection losses, and environmental requirements.

Drive Circuitry

Drive circuitry depends on application requirements, data format, and the light source.

The type of drive circuitry needed, if any, depends on application requirements, data format, and the light source. LEDs can be directly driven by a suitable current source (although most signals are in the form of voltage and must be converted to current). However, semiconductor lasers must be biased to a current level near laser threshold. Some LEDs work better with drive circuitry to tailor electrical input. For example, the proper drive waveform can effectively reduce the rise time of an inexpensive LED and allow its use at higher-than-specified bandwidths.

**Figure 6-14.
Transmitter Optical
Interfaces**

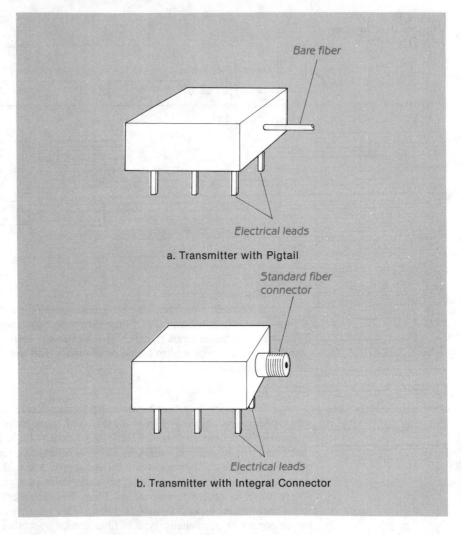

Bare fiber

Electrical leads

a. Transmitter with Pigtail

Standard fiber
connector

Electrical leads

b. Transmitter with Integral Connector

Electronic Preprocessing

The drive circuitry of some transmitters electronically
preprocesses the input electrical signals to put them into a form suitable for
driving the light sources. A simple example is the conversion of signals
from the voltage variations that drive electronic circuitry to the current
variations that modulate lasers and LEDs. Other preprocessing may change
signal formats to types better suited for fiber transmission.

Output Sensing and Stabilization

Some laser transmitters include output-stabilization circuits.
Lasers used in such transmitters emit a small amount of light from their
rear facets, which is sensed by a photodetector. Changes in the rear-facet

output, which indicate changes in transmitter output, trigger a feedback circuit which adjusts drive current so total output power remains stable, avoiding age-induced drops in laser output.

Temperature Sensing and Control

Characteristics of a semiconductor laser change with temperature.

Operating characteristics of a semiconductor laser—notably threshold current, output power, and wavelength—change with temperature. The threshold current increases roughly exponentially with the relative change in temperature, $\Delta T = (T_1 - T_2)/T_1$, where T_1 is initial temperature and T_2 is final temperature. Because optical output is proportional to how much the drive current exceeds the threshold current, output power decreases as temperature rises. Also, more electrical power must be dissipated within the laser, so laser lifetime decreases with operating temperature. The change in wavelength is more subtle. It is caused by temperature-induced changes in the semiconductor's refractive index and thus in the effective length of the laser cavity.

Thermoelectric coolers in transmitters can assure operation at a stable temperature. Temperature control of the laser can reduce temperature-induced degradation of output power. Operation at stable output power is important to assure proper receiver operation. As detected power decreases, bit-error rate increases in digital systems and signal-to-noise ratio decreases in analog systems reflecting degraded performance. The more subtle wavelength changes are not important in most present systems, although they could pose problems in future wavelength-division multiplexed systems that require transmission of specific wavelengths.

TYPES OF TRANSMITTERS

As indicated earlier, there are many different types of fiber-optic transmitters, each designed for particular uses. The same basic elements go into them all. The same LEDs and semiconductor lasers may be used in different transmitters. The same fiber-optic connector housings may be mounted on the outside. The major differences lie in the electronics.

Functional Differences

Transmitters differ in modulation scheme and speed.

Most major functional differences are in the modulation scheme and speed. Analog modulation typically is for audio, video, or radar signals, at progressively higher bandwidths. Audio signals require bandwidths measured in tens of kilohertz at most, while each standard video signal requires 6 MHz. Higher bandwidths are needed for multichannel video systems or for transmitting radar signals.

Digital transmitter circuits must have fast response to produce the fast rise-time pulses needed for digital transmission. Again, the speed depends on the application. Some simple computer data links may require no more than tens of thousands of bits per second, but long-haul telephone systems can operate at speeds above 1 Gbit/s (1000 Mbit/s).

The choice among light sources depends primarily upon speed, transmission type, and operating distance. All other things being equal (which they rarely are), LEDs may be preferred for analog transmitters because of their more linear response. (However, lasers also are used for analog transmission, especially over long distances or at high speed.) Both LEDs and lasers can be used in digital transmitters. Speed may dictate the choice of fast LEDs or of even faster lasers. As mentioned earlier, the wavelength dependence of fiber attenuation makes long-wavelength sources desirable for long-distance transmission.

Important differences lie in the transmitter electronics. Suitable circuitry can increase the effective speed of transmitters, for example. Logic circuits can change digitally encoded signals from one format to another. Drive circuits can compensate for nonlinearities in light source analog response.

The design of transmitter and receiver circuitry is a specialized realm of electronics, which we will not explore deeply because this book is concentrating on optics. Sophisticated transmitters and receivers are typically manufactured and sold to users, who see them as complete, functional black boxes, or as fiber-optic modems that mate on each end with electronic devices and are linked by a fiber-optic cable. Most fiber-optic users don't need to know the details of circuit design.

Coherent Transmission

We mentioned earlier that current digital and analog transmitters are based on intensity modulation of the light source. Other transmission schemes are in development. The most interesting, and probably most important, plan is called coherent or heterodyne transmission. As shown in *Figure 6-15*, it operates like a heterodyne radio system and requires two lasers, one at the transmitter and one (a local oscillator) at the receiver. The two lasers transmit at slightly different frequencies ν_1 and ν_2. At the receiver, the two laser beams are mixed together to obtain the difference frequency, $\nu_1 - \nu_2$, in the microwave region. That microwave signal—at about 1 GHz in experiments so far—is amplified and demodulated to give the desired signal.

**Figure 6-15.
Coherent Transmission
Scheme**

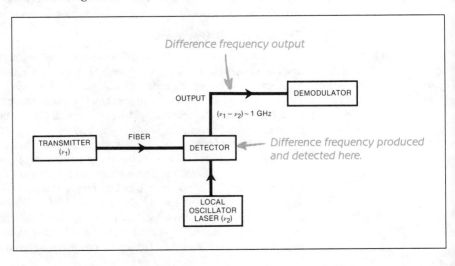

Why go to all that trouble? Because coherent receivers can pick out weaker signals from a noisy background than conventional direct-detection receivers, just as in radio sets. Impressive performance has been demonstrated in laboratory systems. However, much work remains to be done. Critical issues include stabilizing the frequencies of the semiconductor lasers and limiting emission to a narrow frequency range.

SAMPLE TRANSMITTERS

Although we won't go into detail on transmitter design, it is useful to have a general idea of what is inside the black boxes you're likely to encounter. The best way to get a feeling for the internal workings of transmitters is to look at some simple circuits and some examples of more complex types. As you would expect, the complexity of transmitters increases rapidly with the demands for high performance, but the basic concepts remain the same.

An LED emits light when it is forward biased at a voltage higher than the band-gap voltage, about 1.5 V for GaAlAs LEDs emitting at 800–900 nm and around 1 V for InGaAsP LEDs emitting near 1300 nm. As with other forward-biased semiconductor diodes, you can consider the voltage drop across the diode constant regardless of current. To modulate the light output, you modulate the drive current, not the voltage.

Modulation of a voltage fed into a transistor causes current passing through the LED—and its light output—to vary.

A very simple drive circuit for an LED is shown in *Figure 6-16*. A modulated voltage is fed into the base of a transistor, causing variations in the current passing through the LED and current-limiting resistor. The LED can be on either side of the transistor, but it is important that the transistor or some other circuit element be there to modulate the current. The transistor can be in parallel as well as in series with the LED. The resistor is needed to limit current passing through the LED. The auxiliary drive circuitry (not shown) becomes increasingly complex in more sophisticated transmitters as it is called upon to perform more functions (e.g., converting digital signal encoding format). However, the basic idea remains the same.

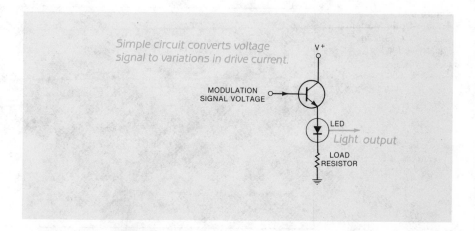

Simple circuit converts voltage signal to variations in drive current.

V+

MODULATION SIGNAL VOLTAGE

LED

Light output

LOAD RESISTOR

Simple commercial transmitters come in simple packages, such as the one shown in *Figure 6-17*. In this case, the whole assembly fits into a fiber-optic connector mount; the LED and any electronic circuitry are very small. In other cases, the transmitter elements are mounted in a small housing common to a range of transmitters and receivers manufactured by a company. A more complex package, in which a transmitter and receiver are mounted in a single package as a fiber-optic modem, is shown in *Figure 6-18*. A standard RS-232-C electrical connector is on one end, and a pair of fiber connectors is on the other.

Figure 6-17.
Cross Section of LED
Transmitter *(Courtesy Hewlett-Packard Co.)*

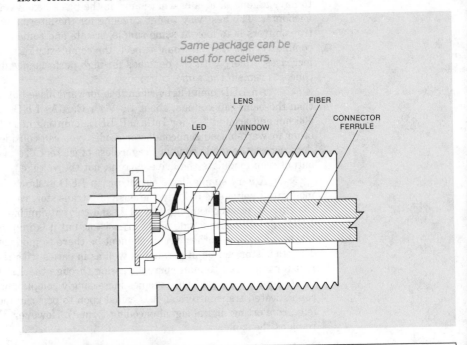

Same package can be used for receivers.

LENS   FIBER
LED   WINDOW   CONNECTOR FERRULE

Figure 6-18.
A Fiber-Optic Modem
(Courtesy American Photonics, Inc.)

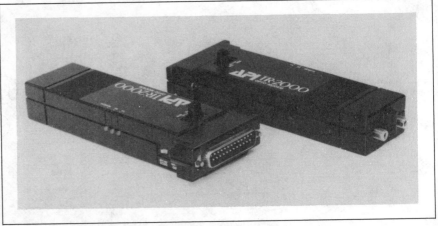

Laser Transmitters

The electrical characteristics of a semiconductor laser are similar
to those of an LED because both are semiconductor diodes. However, the
laser, unlike the LED, does not emit much light until drive current passes
a threshold value well above zero. Also, because lasers need much higher
drive currents than LEDs, their current-limiting resistors usually are
smaller.

Typically, the drive circuit must pre-bias a laser with a current
close to, but still below, the threshold value. Modulation should drive it
above the threshold. This approach enhances speed by avoiding the delay
needed to raise drive current above the laser threshold. At high speeds, it
may be useful to bias so it is emitting a little light in the off state, but a
much higher power when turned on. One concern with pre-biasing is that
the small amount of light emitted in the off state could confuse the receiver
if transmission losses are too low. However, careful design can minimize
the cost in receiver sensitivity.

The basic circuits used in LED drivers also can be used for laser
drivers. In practice, the major changes are addition of other circuit
elements (e.g., feedback monitoring of laser output to control bias current),
and changes in component values to match laser characteristics.

Transceivers

For some applications it is desirable to have a fiber-optic
transmitter and receiver in a single package called a transceiver. Like a
telephone, a transceiver both sends and receives signals. Normally, the
signals are sent over one fiber and received over a second fiber. Circuitry
for the transmitter and receiver are separate, as are the electrical signal
input and output. In fact, in most cases the transceiver is simply a way to
package a separate transmitter and receiver when two-directional
transmission is required, as, for example, between a remote computer
terminal and a mainframe.

Repeaters

Repeaters also contain both a receiver and a transmitter, but in
repeaters the two are connected in series. The receiver detects a signal
from a distant transmitter, amplifies and regenerates it, and produces an
electrical signal that drives the transmitter in the repeater. Repeaters are
used in long-distance telecommunication systems, when signals must be
transmitted over distances so long that they would otherwise fade away to
undetectable levels.

WHAT HAVE WE LEARNED?

1. Both digital and analog transmitters rely on direct amplitude modulation of
   light intensity. Changes in drive current change the light output of LEDs
   and semiconductor lasers.
2. The most common light sources for short fiber-optic systems are GaAlAs
   LEDs operating at 820 or 850 nm.

3. LEDs and semiconductor lasers are both semiconductor diodes that emit light when forward biasing causes current carriers to recombine at the junction between p- and n-doped materials.

4. Wavelength emitted by an LED or diode laser depends on the material from which the diode is made.

5. Laser light is produced by the amplification of stimulated emission, which can occur at a semiconductor diode junction when there is a population inversion. A resonant cavity creates a laser beam.

6. Semiconductor lasers are more complex in structure than LEDs and can emit higher powers. Unlike LEDs, they do not operate below a certain threshold current. Above that threshold, they are much more efficient.

7. LEDs are longer-lived and less expensive than semiconductor lasers and do not require as careful control of operating conditions.

8. Special types of lasers are needed to provide the single-frequency light needed to transmit signals through high-dispersion fibers with large bandwidth.

9. Light may be coupled out of transmitters through integral fiber-optic connectors or through fiber pigtails attached to the light source.

WHAT'S NEXT?

Now that we have discussed fiber-optic transmitters, Chapter 7 will examine the other end of the system, the receiver.

Quiz for Chapter 6

1. Digital transmission capacity is measured as:
   a. bandwidth in megahertz.
   b. rise time in microseconds.
   c. frequency of 3-dB point.
   d. number of bits transmitted per second.

2. Operating wavelengths of GaAlAs LEDs and lasers are:
   a. 820 and 850 nm.
   b. 665 nm.
   c. 1300 nm.
   d. 1550 nm.
   e. none of the above.

3. Light emission from an LED is modulated by:
   a. voltage applied across the diode.
   b. current passing through the diode.
   c. illumination of the diode.
   d. all of the above.

4. Which of the following statements about the difference between semiconductor lasers and LEDs are true?
   a. Lasers emit higher power at the same drive current.
   b. Lasers emit light only if drive current is above a threshold value.
   c. Output from LEDs spreads out over a broader angle.
   d. LEDs do not have reflective end facets.
   e. All of the above.

5. Laser light is produced by:
   a. stimulated emission.
   b. spontaneous emission.
   c. black magic.
   d. electricity.

6. The spectral width of a semiconductor laser is about:
   a. 2 nm.
   b. 30 nm.
   c. 40 nm.
   d. 850 nm.
   e. 1300 nm.

7. A distributed-feedback laser is:
   a. a laser that emits multiple longitudinal modes from a narrow stripe.
   b. a laser with a corrugated substrate that oscillates on a single longitudinal mode.
   c. a laser made of two segments that are optically coupled but electrically separated.
   d. a laser that requires liquid-nitrogen cooling to operate.

8. Which of the following is not needed in some present fiber-optic transmitters?
   a. Output monitoring.
   b. Temperature control.
   c. Wavelength control.
   d. Drive circuitry.

9. The principal advantage of coherent transmission is that it:
   a. requires two lasers.
   b. cannot be tapped.
   c. can pick weaker signals from a noisy background.
   d. is inexpensive.

10. What circuit element receives the drive voltage in a fiber-optic transmitter?
    a. A filter capacitor.
    b. A load-limiting resistor.
    c. A temperature sensor.
    d. A transistor.

Receivers

ABOUT THIS CHAPTER

The receiver is as essential an element of any fiber-optic system as the optical fiber or the light source. The receiver's job is to convert the optical signal transmitted through the fiber into electronic form, which can serve as input for other devices or communication systems. In the case of repeaters, that additional communication system may merely be another fiber-optic transmission system.

This chapter discusses the basic types of receivers, important performance considerations, and how they work. It will focus particularly on the types of detectors and to a lesser extent on the electronics.

BASIC ELEMENTS OF RECEIVERS

Fiber-optic receivers come in many varieties, from simple types that are little more than packaged photodetectors to sophisticated systems that process a weak signal to allow accurate, high-speed transmission. The basic elements of digital and analog fiber-optic receivers are shown in the block diagram of *Figure 7-1*.

The basic functional elements of a receiver are:

Receiver functions include detection, amplification, and demodulation.

1. The detector (to convert the received optical signal into electrical form)
2. Amplification stages (to amplify the signal and convert it into a form ready for processing)
3. Demodulation or decision circuits (to reproduce the original electronic signal).

In practice, the functional distinctions can be hazy, because some detectors have internal amplification stages. Some receivers do not have separate demodulation or decision circuitry because the electrical signal from the amplification stages is good enough for use by other electronic equipment. The entire receiver is packaged in a housing designed to meet user requirements, which includes interfaces with the optical fiber and whatever is to receive the signal.

Figure 7-1.
Basic Elements of
Analog and Digital
Receivers

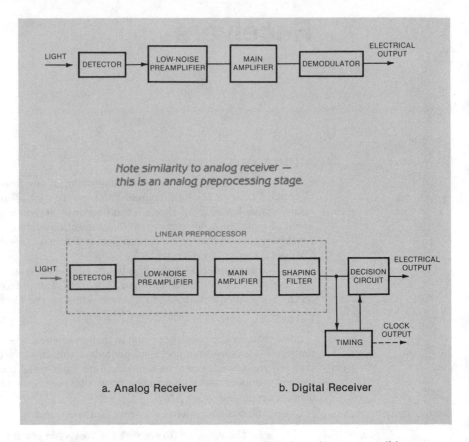

Note similarity to analog receiver —
this is an analog preprocessing stage.

a. Analog Receiver   b. Digital Receiver

Analog and digital receivers use the same type of detectors and amplifiers; the differences are in the demodulation or decision circuits.

The similarities between analog and digital receivers are striking. The initial stages are the same in the two types of receivers; the differences come where the signal is converted into final form for output to other equipment. Why is there an analog receiver at the front end of a digital receiver? Because the real world is analog. The signal reaching the detector may have started in digital form, but by the time it reaches the receiver it varies continuously in level like an analog signal. It must be converted to electrical form and amplified as an analog signal before electronic decision circuits can convert it back to digital form.

Then why bother with digital transmission and receivers? Because analog transmission requires precise replication of the original waveform. Any changes are distortion. Once distortion is added, the electronics don't know what is distortion and what is signal. Digital transmission, on the other hand, does not require precise replication of the waveform. Instead, it requires only the ability to decide whether the signal level is off or on, which can be done even in the presence of distortion (although severe distortion can lead to bit-interpreting errors).

DETECTOR BASICS

The detectors used in fiber-optic communications are semiconductor photodiodes or photodetectors, which get their name from their ability to detect light. The simplest semiconductor detectors are solar cells, where incident light raises conduction electrons to the valence band, generating an electric voltage. Unfortunately, such photovoltaic detectors are slow and insensitive.

Semiconductor photodi-
odes are reverse-biased to
detect light; they produce
a current proportional to
the level of illumination.

Detectors are much faster and more sensitive if electrically reverse-biased as shown *in Figure 7-2*. (This is the opposite of LEDs, which are forward-biased.) When a detector operates in this mode, the bias voltage draws current-carrying electrons and holes out of the junction region. This effort creates a depletion region, which essentially blocks current from passing through the diode. Illuminating this region with light of a suitable wavelength creates electron-hole pairs by raising an electron from the normal valence band to the conduction band, leaving a hole behind. The bias voltage causes these current carriers to drift quickly away from the junction region, and a current flows. The current is proportional to the light illuminating the detector. Several types of detectors can be used in fiber-optic systems, as described below.

Wavelength sensitivity of
photodetectors depends
on the materials from
which they are made.

Photodetectors can be made of silicon, gallium arsenide, germanium, indium phosphide, or other semiconductors. The composition determines the wavelengths at which the detectors are sensitive. To produce a photocurrent, photons must have enough energy to raise an electron across the band gap from the valence band to the conduction band. This phenomenon gives most photodetectors a fairly sharp cut-off at long wavelengths; in silicon, for example, the sensitivity drops sharply between 1000 and 1100 nm. Other effects limit short-wavelength cut-off, which is more gradual and rarely a factor in design of fiber-optic receivers.

Integration of detector and electronics is possible and economical because silicon is a usable photodetector material. So are many other semiconductors usable for electronic components, including gallium arsenide and germanium. We'll go back and look at the details later.

PERFORMANCE CONSIDERATIONS

Many interrelated factors
affect receiver
performance.

The factors influencing receiver performance are complex and often interrelated. At first glance, they might seem to be sensitivity, speed, and cost. In practice, these factors are controlled by other factors, including operating wavelength, choice of fiber and transmitter, dark current, noise-equivalent power, and nature of transmission coding. They depend upon both the response of the detector itself and the processing performed by the electronics.

This section reviews the most important factors affecting receiver performance and their relationship to inherent characteristics of detectors and performance of electronics. Most parameters are both receiver and detector characteristics.

**Figure 7-2.
Photodetector
Operation**

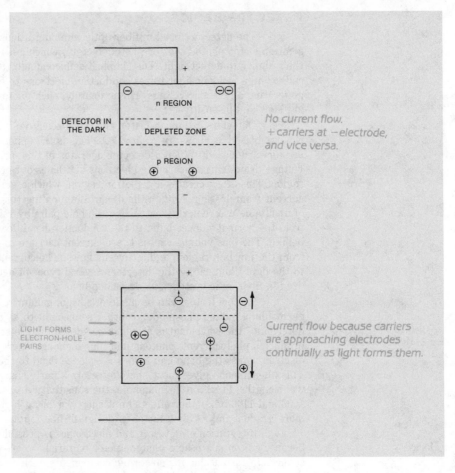

Sensitivity

Sensitivity measures how well a receiver responds to a signal as a function of its intensity.

Although sensitivity sounds like a simple concept, it actually is a conceptual umbrella covering how detectors and receivers respond to signal intensity. To understand the nature of the problem, we should stop briefly and look at how a receiver handles a weak signal.

Signal Quality

Signal strength and noise affect how well a receiver can reproduce a signal.

The role of a receiver is to reproduce accurately the signal it receives through a length of optical fiber. Two fundamental characteristics affect how well this can be done. One is the signal strength. The other is noise, which tends to obscure or degrade the signal. For analog systems, the signal-to-noise (S/N) ratio (the signal power divided by noise power, normally expressed in decibels) measures quality. The higher the signal-to-noise ratio, the better the received signal. Think of background noise in recorded music. In a high-quality stereo system, you can't hear the noise. However, the noise of a scratchy record or a poor tape recording can overwhelm the signal. The practical definition of a good S/N ratio depends on the application. In many fiber-optic systems, 40–50 dB is considered good to excellent, while S/N ratios in the 30-dB range are acceptable for many applications.

In digital systems, where the received information is either a one or zero, quality is measured as the probability of incorrect transmission, the bit-error rate. That probability depends on how effectively the receiver can tell zeros from ones, which in turn depends on factors such as received power, sensitivity, noise, and transmission speed. For a given receiver, the dependence on received power is striking, as shown in *Figure 7-3*. In certain ranges, a 5-dB decrease in received power can make the error rate soar from 10^{-12} to 10^{-3}. The bit-error rate also depends on data rate; the slower the data rate, the lower the error rate. Typical bit-error rates in telecommunication systems are 10^{-9}, or one error per billion bits, but even lower rates (10^{-12}) are needed for computer data transmission.

Signal-to-noise ratio and bit-error rate are introduced here because they become important at the receiver. However, they really measure overall system performance because they depend on received power as well as receiver sensitivity. Receiver performance enters the picture because of the importance of sensitivity, but overall performance also depends on transmitter output and losses in transmission. The effect on overall system design will be described in Chapter 14.

**Figure 7-3.
Bit-Error Rate as a
Function of Power**
(Courtesy AMP Inc.)

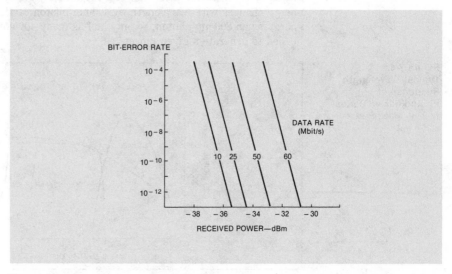

Wavelength

Sensitivity or a receiver depends on the wavelength of light reaching the detector.

Receiver sensitivity generally refers only to how well a receiver responds to a signal of a given amplitude—not to its time response or bandwidth. The sensitivity depends on the detector itself and the electronic circuits in the receiver that amplify and process the electrical signal from the detector. The detector parameters themselves depend on wavelength of the light and operating conditions.

Responsivity and Quantum Efficiency

One measure of photodetector sensitivity is responsivity, the ratio of electrical output from the detector to the input optical power. Most fiber-optic detectors generate signals as current, so this is normally measured as amperes per watt (A/W). (Because input optical powers are in the microwatt range, responsivity might be more properly given as microamperes per microwatt [μA/μW], and sometimes is, but the two measurements are equivalent.) If electrical output signals are voltages, response can be measured in volts per watt (V/W).

A closely related quantity for detectors is quantum efficiency. This measures the fraction of incoming photons which generate electrons in the output signal:

$$\text{QUANTUM EFFICIENCY} = \frac{\text{ELECTRONS GENERATED}}{\text{INPUT PHOTONS}}$$

It and responsivity both depend on wavelength, as can be seen in the plots of quantum efficiency in *Figure 7-4*. Like responsivity, quantum efficiency depends on the detector material and structure. Shape of a quantum efficiency curve differs from that of a responsivity curve because photon energy changes with wavelength. A 400-nm photon carries twice as much energy as an 800-nm photon, so only half as many 400-nm photons are needed to generate a given power.

**Figure 7-4.
Typical Wavelength
Response of
Photodiodes** (*Courtesy
Tran V. Muoi, Plesscor
Optronics*)

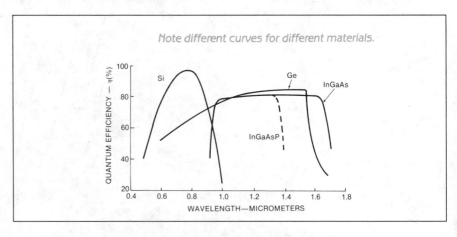

Quantum efficiency cannot be larger than one by that formal definition. However, values larger than one are possible if quantum efficiency is defined as the ratio of output electrons to incoming photons. That be greater than one in detectors (or receivers) with internal amplification because a single electron generated at the pn junction may produce more than one electron in the output current.

Different Materials

Silicon detectors are insensitive to wavelengths longer than 1000 nm, so germanium, InGaAs, or other materials must be used.

The dependence of detector response on wavelength means that different types must be used at different wavelengths. Silicon is fine for 800–900 nm, but it is insensitive at wavelengths longer than 1000 nm. Germanium, InGaAsP, and InGaAs photodiodes can be used at longer wavelengths because their smaller band-gap energies let them respond to less-energetic photons. While the shape of the responsivity curve depends mostly on the material, the value of the responsivity also depends on device structure. Typical values for simple silicon photodiodes at 800 nm are about 0.6 A/W.

Amplification

A final factor that enters into sensitivity is amplification. Adding internal amplification stages, as in a phototransistor or avalanche photodiode, can multiply responsivity. Electronic amplification can multiply receiver responsivity. However, amplification is indiscriminate; it multiplies noise as well as signal and can add some of its own noise.

Dark Current and Noise-Equivalent Power

Detector output includes noise as well as signal.

The electrical signal emerging from a detector includes noise as well as signal. Some noise comes along with the input optical signal (optical noise), and some is generated within the detector. Other noise is added later by the amplifier. Electromagnetic interference causes noise when stray electromagnetic fields induce currents in conductors in the receiver. Noise mechanisms are complex enough to be a field in themselves, and the details are beyond the scope of this book. However, it is important to understand two key concepts: dark current and noise-equivalent power.

Dark current is the noise current a detector produces when it is not illuminated.

Any detector will produce some current when it is operated in the normal manner but not exposed to light (i.e., kept in the dark). This dark current measures inherent electrical noise within the detector, which also will be present when the detector is exposed to light. It sets a floor on the minimum detectable signal because for a signal to be detected it must produce measurably more current than the dark current. Dark current depends on operating temperature, bias voltage, and the type of detector.

Noise-equivalent power is optical power needed to generate output current equal to root-mean-square noise.

Noise-equivalent power (NEP) is the input optical power needed to generate an electrical current equal to the root-mean-square noise of the detector (or receiver). This is a more direct measurement of the minimum detectable signal because it compares noise level directly to optical power. NEP depends on frequency of the modulated signal, the bandwidth over which noise was measured, area of the detector, and operating temperature. Its units are the peculiar ones of watts divided by the square root of frequency (in hertz) or $W/Hz^{0.5}$. The values found on specification sheets normally are measured with a 1-kHz modulation frequency and a 1-Hz bandwidth.

Operating Wavelength and Materials

Each photodiode material has a unique response as a function of wavelength.

As we saw in *Figure 7-4*, the sensitivity of a detector—and hence of the entire receiver—depends strongly on wavelength. Response drops sharply at wavelengths longer than a characteristic value for each material because longer-wavelength light lacks the energy needed to free electrons. At shorter wavelengths, response rises to a peak value, then drops because material absorption becomes so strong that much light is absorbed before it can reach the semiconductor junction where photocurrents are generated.

The approximate operating ranges for the most important detector materials are shown in *Table 7-1*.

Table 7-1. Detector Operating Ranges

Material	Wavelength (nm)
Silicon	400–1000
Germanium	600–1600
GaAs	800–1000
InGaAs	1000–1700
InGaAsP	1100–1600 (doping dependent)

Individual photodiodes often are designed to work best in part of the material's range. For fiber optics, silicon may be optimized for either the red region, where plastic fibers transmit best, or the 800- to 900-nm region of GaAlAs emitters. Germanium, InGaAs, and InGaAsP typically are designed for use at 1300–1550 nm.

Spectral response of InGaAs and InGaAsP changes with material composition.

For germanium, silicon, and gallium arsenide, the response depends on chip design because the material always has the same composition and, thus, the same response. However, the spectral response of InGaAs and InGaAsP depends on material composition, which affects the band gap. Thus specific compositions of the two compounds are best for use at specific wavelengths.

Photodetector sensitivity is temperature-dependent. The effect is not pronounced at the wavelengths used in present fiber-optic systems. However, the sensitivity of detectors that operate at wavelengths much longer than 2 or 3 μm is greatly improved by cooling.

The choice of material affects more than the wavelengths that can be detected. Electrical characteristics, such as speed and dark current, also differ among materials. Germanium detectors tend to be noisier and slower than those made of silicon or the III–V materials such as GaAs and InGaAs. Another relevant concern for some applications is how well electronic components can be made from the detector material. In the case of silicon, electronic components can be made readily enough to allow integration of the detector on a chip with other components. Integration techniques are not as well developed for other materials.

Speed and Bandwidth

Detectors take finite times to respond to changes in input. That is, there is a delay between input of an optical signal to a detector and its production of an electrical current. The delay depends on the material and the device design.

A second type of internal speed limit more directly affects operation of fiber-optic systems: the time it takes the electrical output signal to rise from low to high levels and the corresponding fall time. The rise time normally is defined as the time the output signal takes to rise from 10 to 90% of the final level after the input is turned on abruptly. Analogously, fall time is the time the output takes to drop from 90 to 10% after the input is turned off. Device geometry, material composition, electrical bias, and other factors all combine to determine rise and fall times, which may not be equal. Generally the longer of these two quantities is considered the device response time. (In some cases, the fall time is markedly slower than the rise time.)

While internal delay does not directly affect bandwidth or bit rate of a fiber-optic receiver, the rise and fall times do. Propagation delays shift the signal in time, but rise and fall times spread it out. A 10-ns delay, for example, means that a 10-ns pulse arrives at the output 10 ns late, but still only 10 ns long. However, a 10-ns response time doubles the pulse length to 20 ns, in effect halving the bit rate. Frequency response and limiting bandwidth of a detector are inversely proportional to response time, as is maximum bit rate.

For relatively slow devices, response time is proportional to the RC time constant—the photodiode capacitance multiplied by the sum of the load resistance and the diode series resistance. Speed of such devices can be increased by reducing equivalent capacitance. As speeds increase, two other factors can limit response time: diffusion of current carriers in the photodiode and time needed for carriers to cross the depletion region.

There are wide differences among detector response times. The slowest are photodarlingtons with response times measured in tens of microseconds. With avalanche photodiodes and fast pin photodiodes, response time can be under a nanosecond. We'll get into these characteristics when we examine the different types of detectors.

Signal Coding, Analog and Digital Modulation

Performance of a fiber-optic receiver also is influenced by the signal format. The simplest type is straightforward analog intensity modulation. In this case, it is the job of the receiver to reconstruct the transmitted waveform with as little distortion as possible. Accuracy of the reproduced waveform depends on intensity of the received signal, linearity and speed of the receiver, and noise levels in the input signal and the receiver.

Detectors do not respond instantly to changes in input.

Rise and fall times limit receiver bandwidth and bit rate.

Coding format is impor-
tant for digital signals.

The detection of digital signals is strongly influenced by the coding scheme. Bit rates in digital fiber-optic systems are so much slower than the frequency of the light waves that digital modulation schemes can be considered as being superimposed on a steady optical carrier. The modulation is strictly binary; the light is either off or on. (However, in some high-speed systems, off may actually be a very low-level light signal when a laser light source is prebiased.)

The two most common ways of optically encoding this kind of modulation are shown in *Figure 7-5*. One is return-to-zero (RZ) coding, where the signal level returns to a nominal zero level between bits. The other is non-return-to-zero (NRZ) coding, in which the signal does not return to zero but remains at one if two successive one bits are transmitted. The two differ in their effective speed, because RZ signals have twice as many pulses. Each modulation scheme has its own advantages.

**Figure 7-5.
Signal Levels for Digital
Coding**

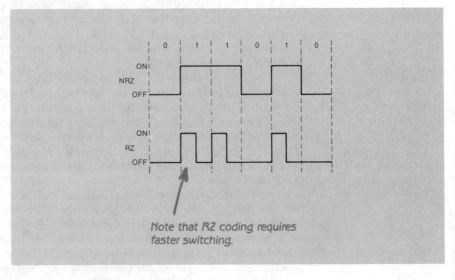

Note that RZ coding requires faster switching.

Device Geometry

As mentioned earlier, detector geometry can influence detector speed and sensitivity. You also should be aware of another geometrical factor: detector active area.

All light emerging from a fiber should fall onto a detector's light-sensitive area.

For a detector to operate efficiently, all light emerging from the fiber should fall onto its light-sensitive (active) area. This means that the detector's active area should be larger than the fiber core (how much larger depends on how light is coupled to the detector). If the output fiber has core diameter d and half-acceptance angle θ, and is a distance S from the detector; then it will project onto the detector a spot with diameter D:

$$D = d + 2S \tan\theta$$

If that spot size is larger than the active area of the detector, some light will be lost and sensitivity will be reduced.

Most detectors have active areas larger than the cores of multimode fibers, so normally little light misses them. However, losses can occur if the fiber is far from the detector, if the fiber core is large, or if the fiber end and detector are misaligned.

Dynamic Range

Input signals must be within a detector's dynamic range to avoid distortion.

Another concern in receivers is dynamic range—the range of input power over which they produce the desired output. At first glance, it might seem that the higher the input power, the better the response would be. However, that isn't the case because any receiver responds linearly to input over only a limited range. Once input power exceeds the upper limit of that range, signals are distorted, leading to high noise in analog systems and errors in digital transmission.

The basic problem can be seen in *Figure 7-6*, which shows output from a receiver as a function of input light power. In the lower part of the curve, the response is linear. An increase in input power by an amount Δp produces an increase in output current Δi, where $\Delta i/\Delta p$ is the slope of that part of the curve. Typically, the response is nearly linear at low levels. (The lower end of the dynamic range is where noise overwhelms the signal.) However, if the input power is too high, the receiver response saturates and output power falls short of the expected level. The result is distortion, much as when an audio speaker is driven with more power than it can handle.

**Figure 7-6.
Receiver Output as
Function of Input Light
Power**

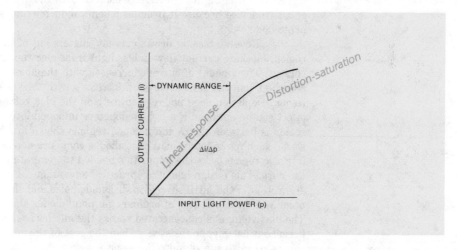

Attenuators may be needed to reduce power to within the dynamic range of a digital receiver.

The same basic principles apply for digital receivers. Exceeding the dynamic range causes an increase in the bit-error rate. However, the way the dynamic range might be exceeded differs. In analog transmission, the dynamic range is most likely to be exceeded briefly—analogous to a loud passage of music in an audio system. In digital systems, the peak power is consistent from pulse to pulse. If that peak exceeds the receiver's dynamic range, the reason is that the system attenuation is lower than expected between transmitter and receiver. Inserting attenuators can reduce average signal intensity to a level within the receiver's dynamic range.

The dynamic ranges of detectors vary widely, with those intended for short-distance communication often having the widest ranges because they may be used with different lengths of fiber. Some receivers used in long-distance systems have much more limited dynamic ranges and may be limited to input powers below 100 μW. Important limits on dynamic range come from the receiver circuitry; as load resistance rises, non-linear response begins to occur at lower input powers.

DETECTOR TYPES

The detectors used in fiber-optic receivers are more complex than the simple photodiode described earlier in this chapter. More sensitive detectors, with higher current outputs, normally are used. Some provide enough output current to serve as receivers by themselves (when packaged), while others require external electronics to amplify and process electrical signals. These detectors are described below.

Pn and Pin Photodiodes

Photodiode sensitivity is improved by sandwiching an undoped intrinsic region between p and n regions.

We saw earlier that fiber-optic photodiodes are reverse-biased so the electrical signal is a current passing through the diode. Such a detector is said to be operating in the photoconductive mode because it produces signals by changing its effective resistance. However, it is not strictly a resistive device because it includes a semiconductor junction, forming a pn photodiode.

Reverse biasing draws current carriers out of the central depleted region, blocking current flow unless light frees electrons and holes to carry current. The amount of current increases with the amount of light absorbed, and the light absorption increases with thickness of the depleted region. Depletion need not rely entirely on the bias voltage. The same effect can be obtained if a lightly doped or undoped intrinsic semiconductor region is between the p- and n-doped regions shown in *Figure 7-7*. In a sense, such pin (p-intrinsic-n) photodiodes come pre-depleted because the intrinsic region lacks the impurities needed to generate current carriers in the dark. This design has other practical advantages. By concentrating absorption in the intrinsic region, it avoids noise and slow response that occur when the p region of ordinary pn photodiodes absorbs some light. The bias voltage is concentrated across the intrinsic semiconducting region because it has higher resistivity than the rest of the semiconductor, and this helps raise speed and reduce noise.

Figure 7-7.
A Pin Photodiode

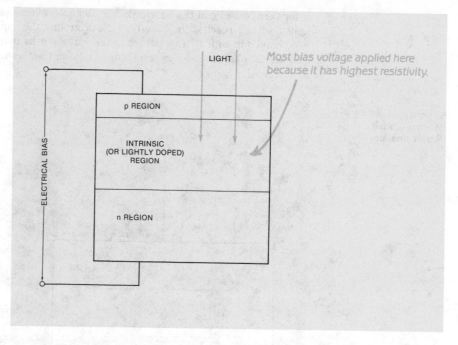

LIGHT

Most bias voltage applied here
because it has highest resistivity.

p REGION

INTRINSIC
(OR LIGHTLY DOPED)
REGION

ELECTRICAL BIAS

n REGION

Pin detectors can have response times under 1 ns
and dynamic ranges of
50 dB.

The speed of pin photodiodes is limited by variations in the time it takes electrons to pass through the device. This time spread can be reduced in two ways—by increasing the bias voltage and/or by decreasing thickness of the intrinsic layer. Changing intrinsic layer thickness must be traded off against detector sensitivity because this reduces the fraction of the incident light absorbed. Typical biases are 5–20 V, although some devices have specified maximum bias voltages above 100 V. Fastest response times are under 1 ns, but sensitivity in the 800-nm region is limited to about 0.7 A/W. An important attraction of pin photodiodes is a large dynamic range; their output-current characteristics can be linear over six decades (50 dB).

Pin photodiodes are
widely used because of
their high speed and good
sensitivity.

The speed and sensitivity of pin photodiodes are more than adequate for many fiber-optic applications, and they are widely used. Often they are attached to amplifiers to boost sensitivity for high-speed communications. However, they are not well-matched to the needs of low-cost, low-speed fiber systems.

Phototransistors

A phototransistor both
senses light and amplifies
light-generated current.

The simplest detector that includes internal amplification is the phototransistor shown schematically in *Figure 7-8*. It can be viewed from several perspectives. Perhaps the simplest for those familiar with transistors is to consider it as a transistor in which base current is generated by light rather than by being injected from an external source. An alternative is to see a phototransistor as a pn photodiode within a transistor. Often phototransistors have no base electrode and have the bias voltage applied between emitter and collector so the base-emitter junction is forward-biased and the base-collector junction reverse-biased. In practice, light normally reaches the base through or around the emitter, which has a wide band gap so it is transparent to the wavelengths detected. Most commercial phototransistors are made of silicon; their commonest uses are in inexpensive sensors.

**Figure 7-8.
Structure of a
Representative
Phototransistor**

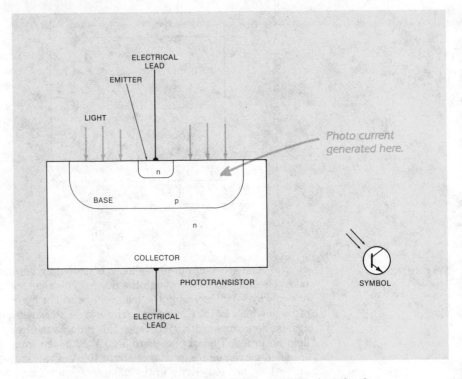

The photocurrent generated in the base-emitter junction is amplified, as is base current in a conventional transistor. The result is a much higher responsivity than a simple photodiode. However, this increase comes at a steep price in response time and linearity and at some cost in noise. You can see that by looking at *Table 7-2*, which compares typical characteristics of silicon photodetectors. In practice, there is a large range in possible values, particularly for detectors that include amplification stages.

The slow response of phototransistors limits their use to systems below the megahertz range. However, their high responsivity lets them detect faint signals and directly drive some external devices, such as simple logic circuits. The main uses of phototransistors (and photodarlingtons) are in inexpensive systems for short-distance, low-speed transmission.

*The main uses of photo-
transistors are in low-
cost, low-speed systems.*

**Table 7-2.
Typical Characteristics
of Important Silicon
Detectors**

Device	Responsivity	Rise Time	Dark Current
Phototransistor	18 A/W	2.5 μs	25 nA
Photodarlington	500 A/W	40 μs	100 nA
Pin photodiode	0.5 A/W	1 ns	10 nA
Avalanche photodiode	75 A/W	1 ns	Voltage-dependent
Pin-FET (detector-preamplifier)	15,000 V/W	10 ns	NA (output is V)

Recent research has shown much faster operation is possible in phototransistors containing heterojunctions of two materials, such as GaAs and GaAlAs, similar to the heterojunctions in lasers described in Chapter 6. That structure reduces the transit times and device capacitance that slow response of silicon phototransistors. (Heterojunctions cannot be made in silicon or germanium, only in III–V semiconductors, such as gallium arsenide and indium phosphide.) Laboratory devices have shown bandwidths well over 100 MHz, but the technology has yet to find practical use.

Photodarlingtons

The photodarlington is a simple integrated darlington amplifier in which the emitter output of a phototransistor is fed to the base of a second transistor for amplification. Addition of the second transistor increases responsivity but at the cost of lower speed and higher noise. Thus, the photodarlington's uses are even more narrowly constrained than the phototransistor's, but it offers higher responsivity for low-cost, slow-speed applications.

Avalanche Photodiodes

The avalanche photodiode gets high sensitivity from internal multiplication of light-generated electrons.

A different approach to providing internal amplification within a detector is the avalanche photodiode (APD). It relies on the phenomenon of avalanche multiplication. Here, a strong electric field accelerates current carriers so much that they can knock valence electrons out of the semiconductor lattice. The result at high enough bias voltages is a veritable avalanche of carriers—thus the name.

Multiplication factors—the degree to which an initial electron is multiplied—typically range from 30 to about 100. The multiplication factor M is:

$$M = \frac{1}{1 - (V/V_B)^n}$$

where V is the operating voltage, V_B is the voltage at which the diode would break down electrically, and n is a number between 3 and 6 dependent on device characteristics. Care must be taken in operation because exceeding the breakdown voltage cold fatally damage the device. A representative plot of this characteristic as bias voltage approaches breakdown is shown in *Figure 7-9*. (Note that to reach multiplication factors of 100, bias must approach within a few percent of breakdown.) Breakdown voltages normally are well over 100 V and in some devices can range up to a few hundred volts.

**Figure 7-9.
Increase of
Multiplication Factor**

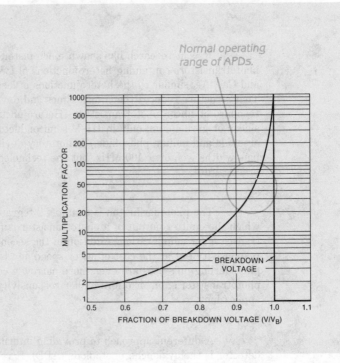

The high operating voltage can make an APD faster than a pin photodiode, with response below 1 ns. However, the uneven nature of multiplication also introduces noise. Avalanche gain is an average; not all photons may be multiplied by the same factor. Signal power increases roughly with the square of the multiplication factor M; for moderate values of M, it increases faster than noise. However, as M increases to high levels, noise increases faster than the square of M (approximately as $M^{2.1}$). As a result, avalanche photodiodes have an optimum multiplication value, typically between 30 and 100.

Note that the voltages required to drive avalanche photodiodes are
much higher than the few volts normally used in semiconductor electronics.
They range from tens of volts to over 100 V, depending on the device and
material. Special circuits are needed to provide the bias voltage needed for
desired multiplication factor and to compensate for changes in temperature
that affect APD parameters. These provide automatic gain control.

Pin-FET and Integrated Receivers

As mentioned earlier, external electronics can amplify the
electrical signal from a photodetector. Some receivers integrate the
functions of detector and amplifier (or preamplifier) in a single circuit that
serves as a detector–(pre)amplifier. These devices are called integrated
detector–amplifiers, detector—amplifiers, or pin-FETs (the last because
field-effect transistor or FET circuitry is used in the preamplifier).

Figure 7-10 shows the type of circuit used in a hybrid receiver
with pin photodiode and low-noise FET preamplifier. This circuit amplifies
the electrical signal before it encounters the noise associated with the load
resistor, increasing S/N ratio and output power.

**Figure 7-10.
Circuit for a Pin-FET
Receiver**

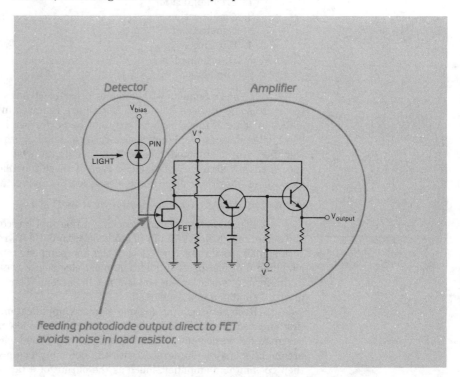

The pin-FET output is a voltage rather than current, with a
typical responsivity around 15 mV/μW (15,000 V/W). Because responsivity
depends on details of the electronics, there is a wide range in commercial
devices.

Integrated detector–preamplifiers have become popular for many moderate-speed fiber-optic applications because of their simplicity and reasonable cost. Unlike avalanche photodiodes, they do not require voltages above the 5-V levels normally needed by semiconductor electronics. Their rise times of 10 ns and up are markedly slower than the 1-ns range for discrete pin photodiodes or avalanche photodiodes, but they are much faster than phototransistors.

ELECTRONIC FUNCTIONS

Converting an optical signal into electrical form is only the first part of a receiver's job. The raw electrical signal generally requires some further processing before it can serve as input to a terminal device at the receiver end. Typically, photodiode signals are weak currents that require amplification and conversion to voltage. In addition, they may require such cleaning up as squaring off digital pulses, regenerating clock signals for digital transmission, or filtering out noise introduced in transmission. The major electronic functions are:

1. Preamplification
2. Amplification
3. Equalization
4. Filtering
5. Discrimination
6. Timing.

If you're familiar with audio or other electronics, you will recognize some of these functions. Not all are required in every receiver, and even some of those included may not be performed by separate, identifiable devices. A phototransistor, for example, both detects and amplifies. And many moderate-performance digital systems don't need special timing circuits. Nonetheless, each of these functions may appear on block diagrams. Their operation is described briefly below.

Preamplification and Amplification

Typical optical signals reaching a fiber-optic receiver are 1–10 μW and sometimes lower. If a pin photodiode with 0.6 A/W responsivity detects such signals, its output current is in the microampere range and must be amplified for most uses. In addition, most electronics require input signals in the form of variations in voltage, not in current. Thus, detector output must be amplified and converted.

Receivers may include one or more amplification stages. Often the first is called preamplification because it is a special low-noise amplifier designed for weak input. In some cases, as mentioned earlier, the preamplifier may be integrated with the detector. The preamplifier output then goes into an amplifier, much as the output of a tape-deck preamplifier goes to a stereo amplifier that can produce the power needed to drive speakers.

Equalization

Detection and amplification can introduce nonlinear distortions into the received signal. For example, high and low frequencies may not be amplified by the same factor. The equalization circuit evens out these differences, so the amplified signal is closer to the original. Much the same is done in high-fidelity equipment, where standard equalization circuits process signals from tape heads and phonograph cartridges so they more accurately represent the original music.

Filtering

Filtering blocks noise while transmitting the signal.

Filtering helps increase the S/N ratio by selectively attenuating noise. This can be important when noise is at particular frequencies (e.g., a high-frequency hiss on audio tapes). It is most likely to be used in fiber optics to remove undesired frequencies within or close to the desired signal, such as second harmonics.

Discrimination

Discrimination circuits generate digital pulses from an analog input.

So far, the functions we've looked at are needed to regenerate the original waveform for both analog and digital receivers. However, a further stage is needed to turn the received analog signal back into a series of digital pulses—decoding and discrimination. What started as rectangular pulses with sharp turn-on and turn-off edges have been degraded into unboxy humps, as shown in *Figure 7-11*. Dispersion may have blurred the boundaries between pulses.

**Figure 7-11.
Results of Transmission Through Fiber-Optic System**

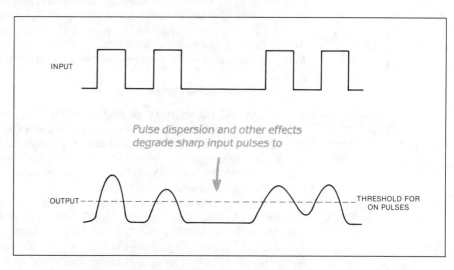

From a theoretical standpoint, this rounding of square pulses represents loss of high frequencies in the signal. Mathematically, a square-wave pulse is the sum of many sine waves of different frequency, the first with frequency equal to the square-wave frequency and others with frequencies that are integral multiples (harmonics) of the square-wave frequency. The highest-frequency signals (the highest-order harmonics) are the ones that make up the sharp rising and falling edges of the pulse. This means that if the high-frequency components are lost, the pulses lose their square edges.

The remaining low-frequency components contain most of the information needed, but they are not clean enough to serve as input to other electronic devices. Regeneration of clean pulses requires circuitry that decides whether or not the input is in the on or off state by comparing it to an intermediate threshold level. Power above the threshold is an on pulse; power below the threshold is off. Care must be taken in selecting this threshold level to avoid misinterpreting input; too low a threshold, for example, could interpret noise spikes in the off state as on pulses.

Timing

Another essential task in many receivers is resynchronization of the signal. Conventional digital signals are generated at a characteristic clock rate, such as once every microsecond. As a signal is transmitted, random errors in timing gradually build up and can eventually reach a level where they are comparable to the duration of a pulse. Then the receiver can confuse successive pulses, causing transmission errors.

The random timing errors that can accumulate to cause such problems are called jitter. They can pose serious practical problems. Suppose, for instance, that a digital signal passed through ten regenerators, each of which could reconstruct the series of pulses with accuracy of ± 0.1 μs. If all of those errors were negative, they would add to 1 μs; if that is the clock rate, the electronics at the receiving end might assume that the received bit was the bit supposed to come in the next time slot. Accumulated jitter can have unpredictable effects on external circuits. Retiming the pulses when they are regenerated can help combat jitter problems.

PACKAGING CONSIDERATIONS

As with transmitters, packaging can be an important concern with receivers. The basic requirements are simple and easy-to-use mechanical, electronic, and optical interfaces. The main mechanical issues are mounts. The electronic interfaces must allow both for input of bias voltage and amplifier power (where needed) and for output of signals in the required format. The details can differ significantly.

Optical interface requirements are simpler than those for transmitters because coupling light from fibers to detectors is simpler. Detector active areas are larger than the emitting areas of most fibers. Thus, simple butt-coupling of the fiber is adequate for most purposes. Most short-distance receivers using multimode fiber come with integral connectors that mate with fiber connectors. The connector case either delivers light directly to the detector or couples light to a fiber that carries it to the detector. Connector mounts also are provided on long-distance single-mode fiber receivers.

SAMPLE RECEIVER CIRCUITS

Details of receiver circuitry vary widely with the type of detector used and with the purpose of the receiver. For purposes of this book, we will cover only a few simple circuits for important devices and avoid detailed circuit diagrams.

Photoconductive Photodiodes

Photoconductive photodiodes have a load resistor in series with the bias voltage.

The typical pin or pn photodiode used in a fiber-optic receiver is used in a circuit with a reverse-bias voltage applied across the photodiode and a series load resistor, such as that shown in *Figure 7-12*. In this mode, the photodiode is photoconductive because the photocurrent flowing is proportional to the nominal resistance of the illuminated photodiode.

**Figure 7-12.
Basic Circuit for
Photoconductive Pin or
Pn Photodiode**

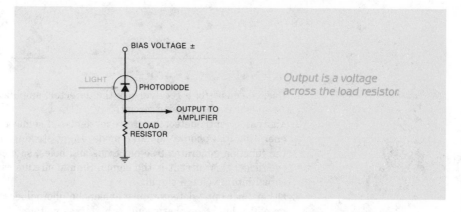

Output is a voltage across the load resistor.

The division of the bias voltage between the photodiode and the fixed resistor depends on illumination level. The higher the illumination of the photodiode, the more current it will conduct and, thus, the larger the voltage drop across the load resistor. In the simple circuit shown, the signal voltage is the drop across the load resistor. Most circuits are more complex with amplification stages beyond the load resistor, as in pin-FET and detector–preamplifier circuits.

Avalanche Photodiode Circuits

The circuits used for avalanche photodiodes are conceptually similar to those used for photoconductive pin photodiodes. However, because of the high bias voltages required and the sensitivity of the photodiode to bias voltage, care must be taken to assure stable bias voltage. This adds to circuit complexity, as shown in the block diagram of *Figure 7-13*.

Phototransistor and Photodarlington Circuits

The types of circuits used for phototransistors also are similar to photoconductive photodiode circuits because they, too, require bias voltages and current-limiting resistors. Amplification stages can be added to the voltage output line. Phototransistors with base connections also can be used as photodiodes in circuits, although it generally would be simpler to start with a photodiode. Photodarlington operation is similar to that of phototransistors.

**Figure 7-13.
Basic Receiver Circuit
for Avalanche
Photodiode**

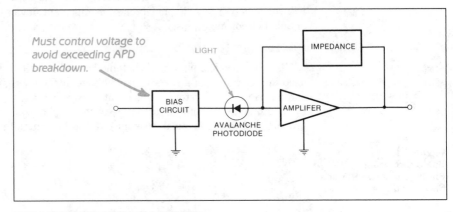

WHAT HAVE WE LEARNED?

1. Basic elements of a receiver are the detector, amplification stages, and demodulation or decision circuits.
2. Many fiber-optic detectors are reverse-biased semiconductor photodiodes operating in a photoconductive mode. Light striking the depleted region near the junction generates free electrons and holes, so a current can flow through the diode; this current is the signal. Simple circuits can convert the current signal into a voltage signal.
3. Other fiber-optic detectors use avalanche photodiodes, with an internal amplification stage that requires high bias voltage.
4. Different detectors are needed for different wavelengths. Silicon is used at 800–950 nm. Germanium and InGaAs are the usual choices at longer wavelengths.
5. Phototransistors and photodarlingtons have internal amplification that gives them high responsivity, but they are too slow for most fiber-optic applications.
6. Detectors operate best over a limited dynamic range. At higher powers, they distort received signals, while at lower powers the signal can be lost in the noise.
7. Pin photodiodes often are packaged with FET preamplifiers in pin-FETs that are among the most common fiber-optic detectors. That packaging avoids load-resistor noise but slows response time.
8. The input stages of analog and digital receivers are similar because by the time the signals reach the receivers they are weak and rounded. The difference in those receivers is in the electronic processing after amplification.

WHAT'S NEXT?

In Chapter 8, we move on from transmitters and receivers to the connectors that hold cables, transmitters, and receivers together.

Quiz for Chapter 7

1. What is the main difference between an analog and a digital receiver?
   a. Special amplification circuitry.
   b. The presence of decision circuitry to distinguish between on and off signal levels.
   c. The two are completely different.
   d. Digital receivers are free from distortion.

2. Photodiodes used as fiber-optic detectors normally are:
   a. reverse-biased.
   b. thermoelectrically cooled.
   c. forward-biased.
   d. unbiased to generate a voltage like a solar cell.

3. What bit-error rate is typically specified for digital telecommunication systems?
   a. 40 dB.
   b. 10^{-4}.
   c. 10^{-6}.
   d. 10^{-9}.
   e. 10^{-12}.

4. Silicon detectors are usable at wavelengths of:
   a. 800–900 nm.
   b. 1300 nm.
   c. 1550 nm.
   d. All of the above.

5. The fastest photodetectors have response times of:
   a. a microsecond.
   b. hundreds of nanoseconds.
   c. tens of nanoseconds.
   d. a few nanoseconds.
   e. under a nanosecond.

6. A pin photodiode is:
   a. a point-contact diode detector.
   b. a detector with an undoped intrinsic region between p and n materials.
   c. a circuit element used in receiver amplification.
   d. a photovoltaic detector.

7. Match the characteristics listed below with the type of detector.
   a. pin photodiode.
   b. avalanche photodiode.
   c. pin-FET receiver.
   d. photodarlington.
   e. phototransistor.
   A. Response time can be less than 1 ns; responsivity under 1 A/W.
   B. Rise time a few microseconds; responsivity about 20 A/W.
   C. Rise time about 10 ns; responsivity thousands of volts per watt.
   D. Response time 1 ns or less; responsivity tens of amperes per watt.
   E. Rise time tens of microseconds; responsivity hundreds of amperes per watt.

8. Which of the following does not include an internal amplification stage?
   a. Photodarlington.
   b. Phototransistor.
   c. Avalanche photodiode.
   d. Pin-FET.
   e. Pin photodiode.

9. A receiver's bandwidth can be limited by:
   a. dynamic range.
   b. rise time.
   c. responsivity.
   d. quantum efficiency.
   e. bias voltage.

10. A discrimination or decision circuit:
   a. filters out noise in analog receivers.
   b. tells off from on states.
   c. decides which pulses to amplify.
   d. controls input level to avoid exceeding a receiver's dynamic range.

Connectors

In the world of fiber optics, connectors are not the only way to make connections. The term "connector" has a specific meaning: a device that makes a temporary connection between two fiber ends or between a fiber end and a transmitter or receiver. The connector typically is mounted on the end of a cable or on a device package, where it can be mated to a matching connector. There are other ways to make connections, but they do not use connectors. A permanent junction between two fibers is called a splice, which we'll learn about in Chapter 9. A device to interconnect three or more fiber ends or devices is called a coupler, which will be described in Chapter 10.

This chapter first explains connectors and how they work, starting with the basic concepts behind fiber-optic connectors and the mechanisms causing their inherent loss or attenuation. Then it discusses various types of connectors and how they're used. There are many types of connectors, but only a few are compatible with each other. We'll close the chapter by looking at some important types of connectors.

WHY CONNECTORS ARE NEEDED

Connectors make temporary connections among equipment that may need to be rearranged.

Electrical connectors are commonplace in modular electronic, audio, or telephone equipment, although you may think of them as plugs and jacks. Their purpose is to connect electrically and mechanically two devices, such as a cable and an amplifier. A plug on the end of the cable goes into a socket in the back of the amplifier, making electrical contact and holding the cable in place. Both the electrical and mechanical junctions are important. If the cable falls out, it can't carry signals; if the electrical connection is bad, the mechanical connection doesn't do any good. (You'll understand the problem all too well if you've ever tried to find an intermittent fault in electronic connectors.)

Fiber-optic connectors are intended to do the same job, but the signal being transmitted is light through an optical fiber, not electricity through a wire. That's an important difference because, as we learned in Chapter 4, the way light is guided through a fiber is fundamentally different from the way current travels in a wire. Electrons can follow a convoluted path through electrical conductors (wires) as long as all the wires are touching. However, the cores of the input and output fibers in a connector must be precisely aligned with respect to each other. Just how precisely we'll see later.

Electrical connectors are used for audio equipment and telephones because the connections are not supposed to be permanent. You use fiber connectors for the same reason. For permanent connections, you splice or solder wires, and you splice optical fibers. Permanent connections have some advantages, including better mechanical stability and—especially for fiber optics—lower signal loss. However, those advantages come at a cost in flexibility; you don't want to unmake a splice each time you move a computer terminal or telephone.

Splices and connectors are used in different places.

In the world of fiber optics, connectors and splices are far from interchangeable. Connectors normally are used at the ends of systems to join cables to transmitters and receivers. A connector may be put at the junction between a long-distance system and a local one, such as the point where a long-distance cable enters a telephone switching office. Connectors are used on patch boxes, where cables come together to route signals through a building. They are used where configurations are likely to be changed, such as with terminals that may be moved from place. Examples include:

- Interfaces between devices and local-area networks
- Connections with short intrabuilding data links
- Patch panels where signals are routed in a building
- The point where a telecommunication system enters a building
- Connections between terminal equipment and other computer devices
- Temporary connections between remote mobile video cameras and recording equipment or temporary studios
- Portable military systems.

Splices are used where junctions are permanent or where the lower loss of splices is critical. For example, splices are made in long cable runs because there is no need to disconnect the cable segments and because connector losses would reduce maximum transmission distance. (Splices offer better mechanical characteristics in outdoor cables.) Splices also are easier to make in the field.

Distinctions between splices and connectors are not always sharp.

The distinctions between connectors and splices are not always as sharp as they would seem from the above guidelines. The most common fiber connectors, which superficially resemble those for metal coaxial cables, differ obviously from the most common splices—the welding or fusion of two fiber ends together. However, between these extremes are such hybrids as demountable splices, which nominally bond fibers together permanently but can be removed. There also are connectors that are attached to the end of cables to ease installation but are not designed for repeated mating and unmating. Some of these approaches will be covered in Chapter 9.

CONNECTOR ATTENUATION

The key functional parameter of fiber connectors is attenuation—the fraction of the signal lost within the connector. This loss is measured in decibels. Attenuation of the best connectors is well below 1 dB, but it may be a few decibels for inexpensive connectors. The type of fiber is an important variable in connector performance because of the different ways light enters and leaves different fibers. Most connectors are made for specific types of fibers.

To prevent any misunderstanding, be sure to remember that when fiber-optic specialists talk about connector loss, they mean the loss of a mated connector pair, that is, the loss in going from one fiber (or other device) to the other. When light goes between two fibers, it actually passes through two connectors. However, the loss of a single connector is not meaningful since the signal isn't going anywhere.

Note also that in most of the discussion that follows we assume that a connector is joining the ends of two fibers. Connectors can be mounted on transmitters and receivers as well, and although details differ, the principles are the same.

Connector attenuation is the sum of losses caused by several factors, which are easier to isolate in theory than in practice. All these factors stem from the way light is guided in fibers. The major ones are:

- Overlap of fiber cores
- Alignment of fiber axes
- Fiber numerical aperture
- Fiber spacing
- Reflection at fiber ends.

These factors interact to some degree. One of them—overlap of fiber cores—really is the sum of many different effects, including variation in core diameter, concentricity of the core within the cladding, eccentricity of the core, and lateral alignment of the two fibers.

Overlap of Fiber Cores

To see how the degree of overlap of fiber cores affects loss, look at *Figure 8-1*, where the end of one fiber is offset from the end of the other. For simplicity, assume that light is distributed uniformly in the cores of identical fibers and the two fiber ends are next to each other and are otherwise well aligned. The loss then equals the fraction of the area of one core that does not overlap with that of the output fiber. If the offset is 10% of the core diameter, the excess loss is about 0.6 dB, just from the effects of the lack of coverage. In other words, for a 50-μm core multimode fiber, this loss would occur if the cores were offset just 5 μm with respect to each other. In a single-mode fiber with a 10 μm-core, such a loss would occur if the fibers were offset by just 1 μm!

**Figure 8-1.
Offset Fibers Can
Cause Loss**

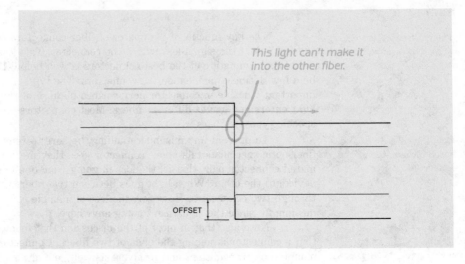

This light can't make it into the other fiber.

OFFSET

Variations in core diameter within normal manufacturing tolerance can cause 0.5-dB loss.

Similar mismatches of emitting and collecting area can occur if there are small differences in core diameter, either within the same fiber or between different batches of fiber. Suppose that the fibers were perfectly aligned but that the 50-μm nominal fiber core diameter varied within tolerance of ±3 μm, as specified on a typical commercial graded-index fiber. If one fiber core was 53 μm and the other 47 μm, the relative difference in area would be:

$$\frac{0.5r_1^2 - 0.5r_2^2}{0.5r_1^2}$$

or a factor of 0.21, about 0.5 dB.

Mismatches in area can arise from other factors, too, as shown in *Figure 8-2*. The fiber core might be slightly elliptical, or the core might be slightly off center in the fiber. (These problems have been exaggerated in *Figure 8-2*.) Variations in cladding dimensions can throw off alignment in some connectors which hold the fiber in position by gripping its outside.

Alignment of Fiber Axes

Angular misalignment of fiber ends can cause significant losses.

The importance of aligning fiber axes is shown in *Figure 8-3*. As the fibers tilt out of alignment and the angle θ increases, the light enters the second fiber at increasingly steeper angles, and some rays emerging from one fiber are not confined in the other. The severity of this loss decreases as numerical aperture increases because the larger the NA, the larger the collection angle.

**Figure 8-2.
Losses Arise when
Cores Are Elliptical or
Off-Center**

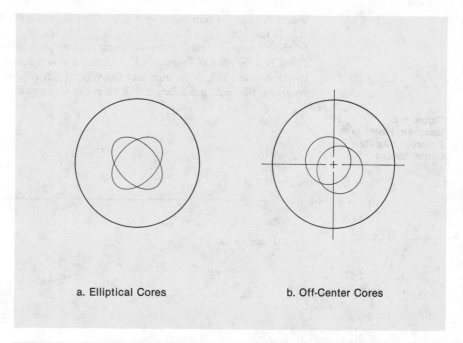

a. Elliptical Cores   b. Off-Center Cores

**Figure 8-3.
Misaligned Fiber Axes
Cause Losses**

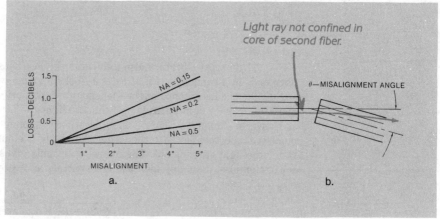

a.   b.

Fiber Numerical Aperture

Differences in NA can
contribute to connector
losses.

Differences in NA between fibers also can contribute to connector losses. If the fiber receiving the light has a smaller NA than the one delivering the light, some light will enter it in modes that are not confined in the core. That light will quickly leak out of the fiber, as shown in *Figure 8-4*. In this case, the loss can be defined with a simple formula:

$$\text{LOSS (dB)} = 10 \log_{10}\left(\frac{NA_2}{NA_1}\right)^2$$

where NA_2 is the numerical aperture of the fiber receiving the signal and NA_1 is the NA of the fiber from which light is transmitted. The NA must be the measured value for the segment of fiber used (which for multimode fibers is a function of length, light sources, and other factors), rather than the theoretical NA. Note also that there is no NA-related loss if the fiber receiving the light has a larger NA than the transmitting fiber.

**Figure 8-4.
Matching Fibers with
Different NAs Can
Cause Losses**

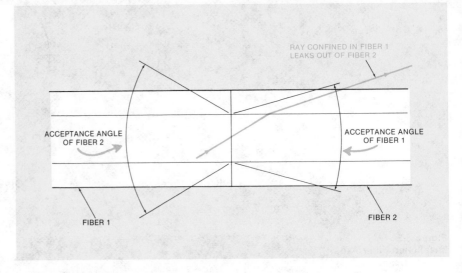

Spacing Between Fibers

Numerical aperture also influences the loss caused by separation of fiber ends in a connector. Light exits a fiber in a cone, with the spreading angle—like the acceptance angle—dependent on numerical aperture. The more the cone of light spreads out, the less light the other fiber can collect, as shown in *Figure 8-5*. This is one case where coupling losses increase with numerical aperture because the larger the NA of the output fiber, the faster the light spreads out. The following formula gives values for end-separation loss, assuming identical transmitting and receiving fibers:

$$LOSS(dB) = 10 \log_{10} \left[\frac{d/2}{(d/2) + (S \times \tan[\arcsin(NA/n_0)])} \right]^2$$

where d is core diameter, S is the distance between them, and n_0 is the refractive index of the material between them.

**Figure 8-5.
End-Separation Loss**

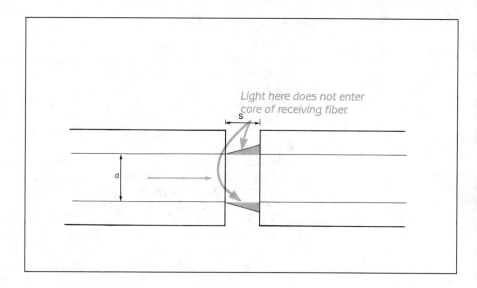

*Light here does not enter
core of receiving fiber.*

*Fresnel reflection losses
occur when light enters
material of a different refractive index.*

Even if the light going through the gap between fibers did not spread out, it would suffer another type of loss from Fresnel reflection. This is a reflection of light that occurs whenever light passes between two materials with different refractive indexes. It occurs for all transparent optical materials, even ordinary window glass. (It causes the reflections you see on windows when in a lighted room looking out into darkness.) Fresnel loss depends on the difference in refractive index between the fiber core and the material in the gap. For uncoated fiber ends in air, it is about 0.32 dB. This loss can be reduced by antireflective coatings on optics, which have a refractive index between that of glass and air, or by filling the gap with a transparent material with a refractive index closer to that of glass. Special materials called index-matching fluids are made for that purpose.

Other End Losses

Other effects can add to connector loss. So far, we have assumed that the fiber ends are cut cleanly and perfectly perpendicular to the fiber axis. However, the ends may be cut at a slight angle, causing a loss that depends on the size of the angle and how the fiber ends are aligned relative to each other. Other losses can arise if the ends are not smooth or if dirt gets into the connector.

With all these loss mechanisms, it is no wonder early fiber-optic developers were very worried about connectors. Concern about achieving the tight tolerances needed with small-core fibers led to interest in large-core fibers. Tremendous progress has been made, but connector losses still are high enough that they must be considered explicitly in designing fiber-optic systems, as we will see in Chapter 14. However, the problems are not as severe as had been feared, allowing production of inexpensive connectors with losses of a few decibels and of high-performance connectors with attenuation below 1 dB.

INTERNAL REFLECTIONS

Losses are not the only potentially harmful things that happen within connectors. Recall that earlier we discussed reflections at fiber ends. These reflections are weak enough that under normal circumstances they should not interfere with signal transmission through a fiber. However, they could cause problems with a laser light source.

As mentioned in Chapter 6, the operation of a semiconductor laser relies on optical feedback, the reflection of light from front and rear facets of the semiconductor cavity. Reflection from the ends of a fiber in a connector can add to that feedback, but the addition is not always constructive. Light waves from a laser are nominally coherent (in phase with one another). The light reflected from the connector is likely to be out of phase with the light reflected from the laser facet. The resulting interaction can effectively add a spurious modulation signal to the signal modulated on the drive current, adding noise to analog systems or increasing the digital error rate.

Because internal reflections depend on the difference in refractive index between the fiber and surrounding materials, there are ways to suppress them. One is butting the fiber ends together, so they appear (to the light) to be a single continuous piece of glass. Another is to fill the inside of the connector with a fluid or gel (index-matching fluid) having a refractive index close to that of the glass. If the index matches perfectly, in theory there would be no reflective losses. If it comes closer to the index of the glass than air (which is the normal case), it can reduce reflective losses greatly.

Wet connectors, which have index-matching fluids or gels, do have some practical problems. Dust or dirt can contaminate the fluid, or some of it can leak out, creating messy junctions or increasing losses. In practice, index-matching fluids are useless unless they are viscous enough to stay inside the connector, a consideration that depends both on the choice of fluid and on the design of the connector.

SINGLE AND MULTIMODE CONNECTORS

Now that you have a feeling for the mechanisms that cause connector loss, you can understand why different types of connectors are used with different fibers. The small core and small NA of single-mode fibers demand tight tolerances to achieve moderate losses. Indeed, the problems of splicing and connecting single-mode fibers appeared so formidable in the 1970s that there was a several-year hiatus in work on single-mode fiber and a large investment in developing graded-index fibers. Good results with multimode fibers encouraged renewed interest in single-mode connectors, leading to types with losses below 1 dB. Nonetheless, single-mode connectors tend to cost more and have higher losses than multimode connectors.

Because the fiber generally must fit into a hole, ferrule, or slot, its outer diameter is a critical dimension and different connector models (even from the same family) are needed for fibers with different outer diameters.

In general, the outer diameter is defined relative to the fiber cladding (any outer plastic coating layers are removed). Small-seeming differences are important for fiber-optic connectors; fibers with claddings 125 and 140 μm in diameter take different connector components.

Multimode fiber connectors generally can be used with fibers that have identical cladding diameters and similar core diameters and NAs. For example, 50/125 and 62.5/125 fibers may work adequately with the same multimode fiber connector. However, to maintain loss at a reasonable level, only single-mode fiber connectors should be used with single-mode fiber.

Although the same connector can work with different fibers, connecting two fibers with different core diameters can lead to large losses. For example, if light from a 62.5/125 fiber is coupled into a 50/125 fiber, there will be a loss of about 2 dB because the core of the 50/125 fiber has only 64% as large an area as the 62.5/125 fiber. Losses caused only by the mismatch in core areas are shown in *Table 8-1*.

**Table 8-1.
Losses Caused by
Mismatches in Core
Area**

Input Fiber	Output Fiber	Extra Loss, dB
50/125	10/125 single-mode	14
62.5/125	10/125 single-mode	16
62.5/125	50/125	1.9
85/125	50/125	4.6
85/125	62.5/125	2.7
100/140	50/125	6.0
100/140	62.5/125	4.1
100/140	85/125	1.4

MOUNTING CONNECTORS

Connectors are mounted on cables, not on bare fibers.

As discussed in Chapter 5, bare fibers aren't practical for most communications. Because fibers are contained in cables, connectors must be mounted on cables, not just on fibers. In many ways, that's good news because it would be awkward to mount connectors on bare fibers. However, it also adds another dimension of complexity beyond the wide variation in connector types—the connectors must work with cables of various types and with different numbers of fibers.

Gripping Points

The cable gives the connector multiple points to grip, including:

- The fiber itself (either its cladding or the plasic coating)
- The plastic buffer in tightly buffered fibers

- The tube or other structure creating a loose buffer for fibers in loose-tube structures
- Cable strength members
- Cable armor (where present)
- Cable jackets.

However, not all of these gripping points are effective ones. Plastic jackets can slip off fibers, much as plastic insulation slips readily off wires.

A typical connector, such as the one shown in exploded form in *Figure 8-6*, holds onto more than one of these entities. Normally, the ferrule grips the fiber firmly enough to hold it in place so the cores are properly aligned, but not tightly enough to keep the connector from pulling off the cable if stressed. Mechanical strength comes from attachments to either strength members (such as Kevlar fibers within the cable structure) or to the cable jacket.

Most connectors hold onto more than one element of the cable.

**Figure 8-6.
Cutaway of a Single-Fiber Connector**
(Courtesy NEC)

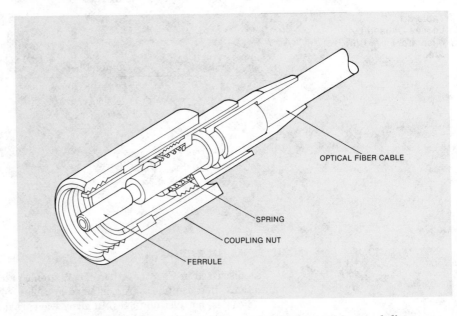

OPTICAL FIBER CABLE

SPRING

COUPLING NUT

FERRULE

Most connectors are sealed in some way to prevent dust and dirt from entering. The mating surfaces perform this function at the fiber ends. The cable-to-connector junction may be covered with a rubber or plastic shield so that a weak point is not exposed. Often a heat-shrink tubing is mounted on the cable and slipped over the end of the connector and then heated to form a tight seal.

Connecting Single- and Multifiber Cables

Connectors for multifiber cables are more complex than those for single-fiber cables. Often multifiber cables may be broken out at each end into many single-fiber connectors.

The connector shown in *Figure 8-6* is a simple one, called a simplex connector because it contains only one fiber. Things get more complicated for multifiber cables. The next step up is a duplex cable. It contains two fibers, and typically uses a duplex connector, often made by mounting the internal elements of two single-fiber connectors side by side in a single housing, as shown in *Figure 8-7*. However, some two-fiber cables —particularly those with two parallel tubes for the two fibers—may use a pair of simplex optical connectors. This is done by splitting the cable apart into two segments at the ends and mounting a connector on each piece.

**Figure 8-7.
A Duplex Connector**
(Courtesy AMP Inc.)

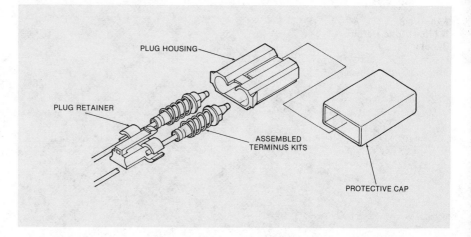

Single- and dual-fiber cables are common in indoor applications where connectors are most likely to be used. Multifiber cables are more difficult to join using single connectors because of the tight tolerances and the need to align all the fibers properly with respect to each other. Many connector designs are not usable for multifiber cables because they require that components be installed separately on each fiber and that fiber pairs be individually joined. Often the problem is solved by breaking individual fibers out of the cable at each end, and putting one simplex connector on each fiber.

Multifiber connectors are used, particularly in military field applications, where concerns of physical and environmental integrity mitigate against using many separate simplex connectors. They also may be used in some intrabuilding networks, where the ease of making connections justifies the extra expense. However, such connectors must be custom-designed for a particular cable and may not be available for non-standard cables. In many cases, simplex and duplex connectors, perhaps installed on separate fibers broken out from multifiber cables, may be more cost-effective.

Multifiber cables may terminate in patch boxes with simplex or duplex connectors that connect with intrabuilding cables.

In some cases, connectors may not be mounted directly on the end of a cable but rather on a box in which the cable terminates, similar to the patch panels used in telephone networks. In *Figure 8-8,* a multifiber cable entering a building from the outside is routed directly into a patch box. Individual fibers are routed to points where they are attached to connector assemblies mounted on the box. (Only one cable to the inside of the building is shown for simplicity.) Alternatively, the cable may end in a multifiber connector that mates with one installed in the box. In the factory, fibers are run from that multifiber connector to simplex or duplex connectors on the box. This simplifies cable installation at the cost of requiring more complex connectors mounted on the cable.

**Figure 8-8.
A Fiber-Optic Patch
Panel**

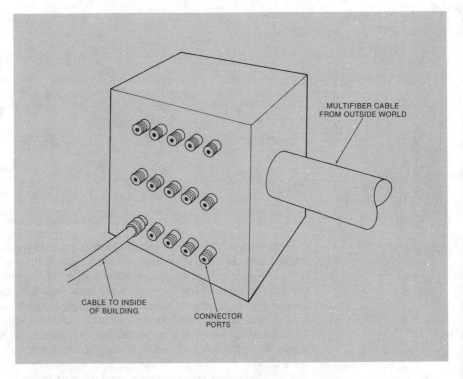

In both cases, cables inside the building mate with the connectors on the box. The intrabuilding cables may be either single- or double-fiber cables running directly to terminal equipment or multifiber cables running to other patch boxes that further distribute the signals. Such an approach lets the system designer take advantage of both the ease of running multifiber cables through ducts and walls and the simplicity and low cost of simplex and duplex connectors.

Connectors also are mounted directly on transmitters and receivers. In some cases, the connectors interface with a fiber inside the package, which delivers light to or from the detector or source. However, in many cases, the light source or detector is so near the connector interface that no fiber is needed, although optics typically are used to focus the light for more efficient coupling.

Factory and Field Installation

Many connectors are factory-mounted and sold on preterminated cables.

Because of the tight tolerances, many fiber-optic connectors are designed for installation in the factory rather than in the field. This has led to sales of preterminated or connectorized cables that are cut to user-specified lengths with connectors attached in the factory. However, other connectors are made for field as well as factory installation.

Special tools are needed to cut cables and install connectors.

Special tools normally are needed to:

- Open the cable
- Remove cable jacketing and fiber buffer layers to expose the fiber
- Strip plastic coating away to expose fiber cladding (mechanically or with a solvent)
- Cut or break the fiber (a fiber scribe or cleaver)
- Polish or smooth the fiber end
- Insert the fiber into the connector and hold it there (typically by crimping or gluing)
- Attach connector to cable structure (typically by crimping or gluing)
- Inspect fiber ends and completed connector (a microscope)
- Seal the junction between connector and cable.

Major makers of fiber-optic connectors supply field installation kits, and some make connectors specifically for field installation on cables. However, fiber-optic connectors are more complex to install than electrical connectors, and the job is not one that can be done with tools as simple and familiar as a soldering iron, wire strippers, and pliers. A sampling of special tools and supplies used in installing fiber connectors is listed in *Table 8-2.*

Environmental Considerations

Fiber ends must be kept free of contaminants to avoid excess losses.

Most fiber-optic connectors are designed for use indoors, protected from environmental extremes. Keeping them free from contaminants is even more important than it is for electrical connectors. Dirt or dust on fiber ends or within the connector can scatter or absorb light, causing excessive connector loss and poor system performance. This makes it unwise to leave fiber-optic connectors open to the air, even inside. Most patch panels and other types of connectors that are not always mated come with covers to prevent the entry of dust. These small covers should be mounted whenever the connector is to be left unconnected for any period of time.

Military interest in fiber-optic communication on the battlefield has led to the development of rugged, hermetically sealed connectors for use outdoors. These special-purpose connectors are expensive and not just because of military gold-plating procurement. The connector must survive rough handling in the field and must resist the influx of water and mud when laid on the ground—even when unconnected. This requires special shutters, seals, and mounting techniques.

**Table 8-2.
Sampling of Tools and
Supplies Used in
Installing Fiber-Optic
Connectors**

Tool or Supply	Purpose
Crimping tool	To bond connector to fiber, strength member, or cable
Crimp dies	To crimp connectors to specific cable assemblies
Polishing disk or film and fixtures	To polish fiber ends
Cable stripper	To remove cable jacket, etc.
Epoxy and applicators	To bond fiber and cable to connector
Insertion tools	To fit plastic sleeves on connectors, etc.
Scribing tool	To scribe fiber so it can be broken at desired point
Magnifier	To examine fiber ends and connector
Microscope	To examine fiber ends and connector
Scissors	To cut strength members in cable
Utility knife	To cut cable components
Mechanical fiber coating strippers	To remove plastic coating from fiber
Solvents for stripping coatings	To remove plastic coating from fiber
Recoating materials	To replace protective plastic coatings on fibers, after connector assembly, and to recoat spliced regions of fiber
Index-matching fluid or gel	To place inside "wet" connectors to minimize internal losses and reflections
Heat gun	To mount heat-shrink seals between connector and cable
Special wrenches	To assemble/disassemble connectors

Durability

Fiber connectors are specified for lifetimes of a few hundred matings. Most can be torn from cable ends by a sharp tug.

Durability can be a problem with standard fiber-optic connectors. Repeated mating and unmating can wear mechanical components, introduce dirt into the optics, strain the fiber and other cable components, and even damage exposed fiber ends. Typical indoor fiber connectors are specified for no more than a few hundred matings, which in practice should be sufficient for most applications. A few connectors serve splice-like functions like permanent junctions and are not intended for extensive mating and unmating.

Because sharp bends can increase losses and damage fibers, care should be taken to avoid sharp kinks in cables at the connector (e.g., when a cable mates with a connector on a patch box). Fibers are particularly vulnerable if they have been nicked during connector installation. Care also should be taken to be certain that fiber ends do not protrude from the ends of connectors. If fiber ends hit each other or other objects they can easily be damaged, increasing attenuation.

Connectors are attached to fibers and cables by crimping and/or epoxies. That physical connection is adequate for normal wear and tear but not for sudden sharp forces, such as those produced when someone trips over a cable. A vigorous tug can break the cable away from a mounted connector. The problem becomes more severe if the connector is attached only to fiber and jacket and is particularly severe for plastic-clad silica fiber since the soft plastic cladding does not adhere well to the glass core of the fiber.

TYPES OF CONNECTORS

A wide variety of fiber-optic connectors has been developed, tested, and marketed. Formal and de facto standards are emerging slowly, but the variety of connectors remains large and is sometimes bewildering. Many bear at least a superficial resemblance to electronic connectors. Indeed, some designs started out as adaptations of standard electronic connector types, and look very much like their electronic counterparts.

In the rest of this chapter, we'll look at a few generic fiber-optic connectors to see how they work. There are so many connector types available that we can't hope to cover them all.

Resilient Ferrule Connectors

One of the largest families of connectors uses resilient (i.e., flexible) ferrules to center and align the fiber. The fiber is inserted into a hole in the plastic ferrule, as shown in *Figure 8-9*. The ferrule, in turn, is inserted into a metal housing that compresses it in a way that centers the fiber in the housing. The metal housing is threaded so two connectors can be held together by a threaded bushing.

**Figure 8-9.
Fiber Mounting in a
Resilient Ferrule**

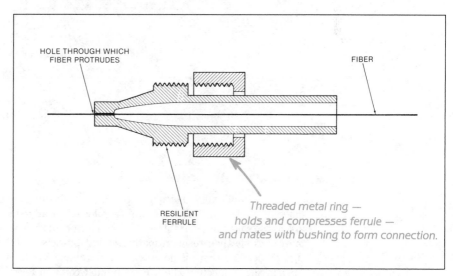

HOLE THROUGH WHICH
FIBER PROTRUDES

FIBER

RESILIENT
FERRULE

*Threaded metal ring —
holds and compresses ferrule —
and mates with bushing to form connection.*

There are many variations on this basic theme. One large connector maker uses a common ferrule in both simplex and duplex connectors. In the simplex connector, the single-fiber cable passes through a metal retainer, and the stripped fiber at the end of the cable is inserted into the ferrule. The ferrule then is inserted into the metal retainer, and heat-shrink tubing seals the cable to the connector. The fiber end is cut and polished to match the mating surface of the ferrule. In the duplex version, two parallel cables are passed through springs and eyelets, then terminated with ferrules. The eyelets are pushed over the ferrules, and the springs are put over them. Then the two units are inserted into a plug housing to form the complete connector.

Details of assembly can vary widely among manufacturers. *Figure 8-10* shows a biconical connector from another manufacturer, which comes in models that allow fiber ends to be butted together or kept separate. The contacting connectors are used in systems where it is essential to avoid laser noise; the non-contact connectors are used where the trade-off falls on the side of simplicity and avoiding damage to fiber tips.

**Figure 8-10.
Cross Section of
Biconical Connector
Plug** *(Courtesy Dorran
Photonics)*

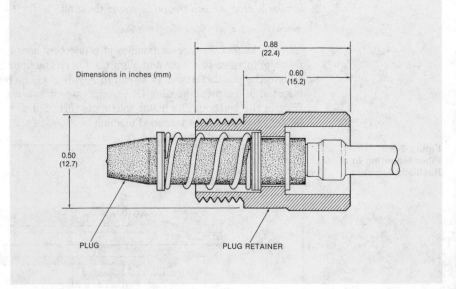

Other Ferrule Connectors

The ferrules that align optical fibers inside connectors need not be made of resilient materials. Metal ferrules—typically stainless steel—also are used, with holes that precisely match fiber diameter. Stainless steel is more durable than resilient materials, but fabrication tolerances are much tighter because there is no automatic alignment mechanism to center the fiber in the connector. The precision must come in ferrule manufacture, which leads to higher costs. Except for the choice of ferrule material, such connectors are similar in outside appearance to resilient ferrule connectors. The two types can meet the same interface requirements and can be mated.

Assembling Ferruled Connectors

To mount a ferruled connector on a single-fiber cable, the cable is first stripped, and the fiber is isolated. The fiber's plastic cladding then is removed, mechanically or with a solvent, to expose the bare cladding. Then the outer elements of the connector, including a protective rubber sleeve, and the outer metal parts of the connector are mounted on the stripped cable. Adhesive is applied to the ferrule and/or to the fiber, as well as to the parts of the cable to which the connector body is bonded. Then the fiber is pushed into the ferrule hole, with an end left protruding and the rest of the connector body partly or totally assembled (depending on design). After the epoxy has dried, the end of the fiber protruding from the ferrule is cut, and the end is polished with a special tool.

Not all ferruled connectors are assembled in precisely this manner. Some do not require epoxy. Instead, they are crimped onto the fiber and/or cable. This saves time by avoiding the need for adhesive drying time, but it risks a lower-strength fiber-cable-connector bond and damage to the fiber. If rigid ferrules are used, the fiber diameter must be precisely matched to ferrule hole size. Typically, this is done by inserting the fiber end in a series of ferrules with holes different in size (for example, $130 + 1, -0$ μm and so on down 1 μm each step) until the right one is found.

Procedures for mounting connectors on multifiber cables are more involved, with details depending on the cable structure. Duplex cables are the simplest because many are designed for use of side-by-side simplex connectors or for duplex connectors that contain two ferrule assemblies. Mounting also is reasonably straightforward for breakout cables, where single-fiber subcables can be isolated for connector mounting.

Other Fiber Alignment Mechanisms

The ferrule approach is the most popular one to mechanical alignment of fibers, but some connector designers have used different techniques. The primary requirement is that the end of the fibers be physically aligned at multiple points. Simply laying the fiber into a groove in a flat plate provides some alignment because the round fiber touches the plate at two points. Laying a flat, ungrooved plate on the fiber provides a third alignment point. Such three-point confinement is sufficient to confine the fiber end to a given location, but it leaves the connector vulnerable to losses caused by differences in the core and fiber diameter. To control such losses, the fiber should be confined at four points (e.g., in grooves in both top and bottom plates).

Such grooved structures are not as common in standard connectors as in splices and other devices that are in the hazy area between pure splice and pure connector. One example is the connector developed for multifiber ribbon cables in which fibers are laid in precision-etched grooves in a plate, and a matching plate is laid down on top of them. Laying fibers from separate cables in such grooves simplifies field connection of ribbon cables. Devices that use this alignment mechanism will be described in Chapter 9.

Expanded-Beam Connectors

All the connectors described so far require tight tolerances on fiber position because of the small size of the light-carrying cores being aligned. There is, however, an alternative that can loosen tolerances: the expanded-beam connector.

Expanded-beam connectors take advantage of the fact that optical fibers carry light signals. What they do is put a lens at the end of each fiber. The lens expands the beam emerging from the fiber end, then another lens collimates it to make a beam of parallel rays much larger in diameter than the fiber core. Then the equivalent lens assembly in the fiber connector receiving the signal focuses the beam of light down to the fiber end, as shown in *Figure 8-11*. Identical lens assemblies can be used (essentially, light is going forward through one assembly and backward through the other).

Figure 8-11. Expanded Beam Connector

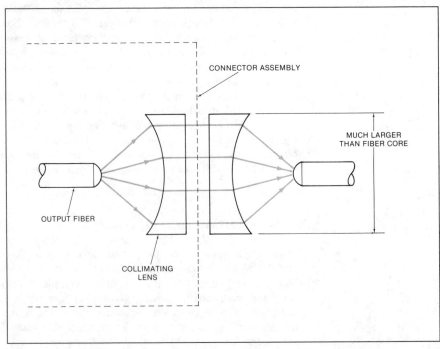

CONNECTOR ASSEMBLY

MUCH LARGER THAN FIBER CORE

OUTPUT FIBER

COLLIMATING LENS

The big attraction of lensed fiber connectors avoiding the tight mechanical tolerances needed to align fiber cores precisely—particularly for single-mode connectors. Suppose the lenses expand the beam a factor of 10 —from 10 μm for a single-mode fiber to 100 μm. Then a 10-μm tolerance would be equivalent to the 1-μm tolerance needed to get the same attenuation without beam expansion.

Expanded-beam designs have been used both in low- and moderate-cost connectors for multimode fibers and in high-performance connectors for single-mode fibers. The latter have losses in the range of about 0.6 dB. Although the use of optical components avoids some problems with mechanical alignment, it introduces others, including other loss mechanisms and the need to mass produce optics inexpensively with a sometimes-complex design. Such connectors remain less popular than ferruled designs.

Rotary Connectors

Special connectors are needed at rotary joints.

One special type of fiber connector is designed for use at rotating joints, where signals have to be transmitted through a pivoting junction. A few rotary connectors have been developed, but the problem is not common and they are not widely used. The biggest problems with rotary connectors are connecting cables with more than one fiber because it becomes hard to arrange the multiple signal-transmission paths.

OTHER CONNECTOR CONSIDERATIONS

As mentioned at the start of the chapter, the most important connector specification is the attenuation of the optical signal. However, there are other factors that are important in connector selection and use.

Compatibility

Always be sure all connectors used in a system can mate with each other.

Compatibility, the ability to mate different connectors, is a key consideration. The commercial development of fiber-optic connectors has come a long way from the early days, when each manufacturer had its own design, and the one thing you could be sure about was that no two types would mate properly. Now some connector designs are de facto standards, and such connectors made by one company can be mated with those from other suppliers. The most common types are fiber-optic equivalents of the electrical SMA connector, biconic connectors used in telecommunication systems, connector designs similar to AMP Inc.'s Optimate design, Amphenol-type connectors, and Japanese D3 and D4 format connectors. Military standards are evolving as well.

Fiber Size

Fiber size is also a concern because of the tight tolerances in coupling light between fibers. Most connectors require fibers of a particular diameter. Typically, a family of connector types includes many different models, each designed to accommodate a specific fiber size. The connectors may look alike from the outside and may even mate together physically, but that does not mean they are identical. In going from a smaller- to a larger-core fiber, losses may be small; however, losses will be higher than specified in going from a larger- to a smaller-core fiber. Using connectors with fibers of core sizes they are not designed to accommodate can cause excessive losses. Note that for most connector designs the difference between 125- and 140-μm cladding diameter is significant and that for some types a 1-μm difference in cladding diameter may affect performance.

Stress Tolerance

Stress tolerance, or the ability of a connector to stay on the cable when stressed, is another factor that must be considered. Some connectors are designed to release their hold on the cable at stresses slightly below the level at which they would break off the cable; this minimizes damage to the cable. From a practical standpoint, excessive stress caused by pulling or tugging on a cable is the most common cause of mechanical failure at the connector.

Durability

Two other considerations are the corrosion resistance of connector materials and the number of matings and dematings (often called cycles) that the connector is designed to withstand.

WHAT HAVE WE LEARNED?

1. Connectors make temporary connections between fiber ends.
2. The most important specification of connectors is attenuation, which is always given for a pair of connectors, measuring the loss in transferring a signal between two fibers.
3. Causes of connector loss include mismatch of fiber cores, misalignment of fiber axes, differences in numerical aperture, spacing between fibers, and reflection at fiber ends.
4. There are many different types of fiber-optic connectors. Some designs are widely accepted and are made by several companies. However, many connectors are not compatible, so care must be taken to be sure any different connectors used in a system can be mated with each other.
5. Typical losses range from a fraction of a decibel to several decibels, depending on the type of connector and fiber. Losses are typically higher the smaller the core of the fiber. Because of the smaller size of single-mode fiber cores, connectors for use with single-mode fibers require much tighter mechanical tolerances and generally are more expensive.
6. Connectors can be mounted on the ends of cables, on light sources and detectors, and on separate patch boxes or panels.
7. It is easier to install fiber-optic connectors in the factory than in the field, so cables often are supplied with connectors in pre-cut lengths, as is the case with many types of electronic cabling.

WHAT'S NEXT?

In Chapter 9, we will look at splices, the permanent connections between two fiber ends.

Quiz for Chapter 8

1. Connectors are used:
   a. to permanently join two fiber ends.
   b. to make temporary connections between two fiber ends or devices.
   c. to transmit light in only one direction.
   d. to merge signals coming from many devices.

2. Which of the following effects affect connector attenuation?
   a. Fiber core overlap.
   b. Alignment of fiber axes.
   c. Numerical apertures.
   d. End-to-end spacing of fibers.
   e. All of the above.

3. What is the role of index-matching fluid in a connector?
   a. Holds the fibers in place.
   b. Keeps dirt out of the space between fiber ends.
   c. Prevents reflections at fiber ends.
   d. Eliminates effects of numerical aperture mismatch.

4. What will be the excess loss caused by the mismatch in core diameters when a connector transmits light from a 62.5/125 multimode fiber into a 50/125 fiber?
   a. 0 dB.
   b. 0.1 dB.
   c. 1 dB.
   d. 1.9 dB.
   e. 12.5 dB.

5. Excess loss will occur in which of the following cases?
   a. Transfer of light from a single-mode to a multimode fiber.
   b. Transfer of light from a fiber with high NA to one with a low NA.
   c. Transfer of light from a fiber with a low NA to one with high NA.
   d. Transfer of light through a connector filled with index-matching gel.

6. Connectors for multifiber cables:
   a. are custom-made for the cable.
   b. are attached to individual fibers broken out of the cable.
   c. are more complex and costly than single-fiber connectors.
   d. are used in military field installations.
   e. all of the above.

7. Which of the following tools is not needed to install fiber-optic connectors?
   a. Fiber cleaver.
   b. End polisher.
   c. Soldering iron.
   d. Cable jacket stripper.
   e. Crimping tool.

8. What is the function of a patch panel?
   a. To repair broken cables.
   b. To provide an interface between exterior and interior cables.
   c. To connect fiber cables with metal wiring.
   d. To match connectors with the proper type of fiber.

9. What aligns fiber cores in resilient ferrule connectors?
   a. Pressure of the surrounding mount on the ferrule.
   b. Precision machining of the hole in the ferrule.
   c. Surface tension of the index-matching fluid or gel.
   d. Guiding of light in the fibers.
   e. Bonding of the resilient ferrules with epoxy.

10. Which of the following do connectors need not be protected against?
    a. Contamination of optical surfaces by dirt.
    b. Sharp tugs on the cable.
    c. Kinks in the cable near the connector.
    d. Power surges.

Splicing

ABOUT THIS CHAPTER

Splices, unlike the connectors discussed in the last chapter, are permanent connections between fibers. Splices weld, glue, or otherwise bond together the ends of two fibers, which may be cabled or uncabled. Like fiber-optic connectors, fiber-optic splices are functionally similar to their wire counterparts. However, as with connectors, there are important differences between splicing wires and optical fibers.

In this chapter you will see when and why optical fibers are spliced, the major considerations in fiber splicing, the types of splices, and the special equipment used in splicing.

APPLICATIONS OF FIBER SPLICES

Splices are low-loss, permanent connections between fiber ends.

As a simple rule of thumb, think of splices as being used in long-distance, high-capacity fiber-optic systems, while connectors are used in shorter-distance, lower-capacity systems. Alternatively, splices are used to join segments of cable that run over long distances, while connectors join short segments of cable and terminal devices. Still another viewpoint is to think of splices as being used outdoors and connectors, indoors.

All three ideas are useful, but the actual picture is more complex. Consider the list of advantages of splices and connectors in *Table 9-1*.

**Table 9-1.
Comparison of Splice
and Connector
Advantages**

Connectors	Splices
Non-permanent	Permanent
Factory installable on cables	Low loss easier to get in field
Some field installable	Lower attenuation
Allows easy reconfiguration	Spliced fiber can fit inside cable
Simple to use	Easier to seal hermetically
	Usually less expensive per splice
	Stronger fiber junction.

Permanent and non-permanent junctions are needed in different situations.

It might seem strange to see "Permanent" listed as an advantage of splices and "Non-permanent" as an advantage of connectors. However, in certain applications one characteristic or the other may be desirable. For example, if an underground cable has been broken, the repair should be permanent. However, permanent junctions are not desirable between a local-area network and terminals that may be moved about within a building.

The lower loss of splices is vital in long-haul systems.

The lower attenuation of splices is a crucial advantage in long-haul fiber-optic systems. Bare fiber normally comes on reels in standard lengths from 1 to 12 km. (Longer lengths can be drawn from large preforms, but there is little demand because such lengths are hard to handle and cable.) Cables are much bulkier than fibers—particularly the heavy-duty types intended for outdoor use—and normally come in lengths of no more than a few kilometers.

Long-distance fiber cables are spliced together as the cable is installed.

High-speed fiber-optic systems with repeaters tens of kilometers apart are made by splicing together shorter segments of cable. If the cables are installed in underground ducts, the splices are made and installed in manholes, with cable-segment length dependent on manhole spacing. Overhead cables are spliced in the field, from segments typically a kilometer long. Installers may bring the cable down to the ground so they can work in a truck or may work in a cherry-picker lift on the back of a truck. If a splice is needed every kilometer in a 30-km system, the total loss of the required 29 splices with an average 0.1-dB attenuation is only 2.9 dB. With connectors having a good (for connectors) loss of 0.7 dB each, total attenuation would be 20.3 dB—so large that the system could not operate properly.

Enclosures are needed to protect splices.

The physical characteristics of splices are important in many long-distance applications. If cables are to be spliced, the splice enclosures must be able to withstand the environment—including the bottom of the ocean for submarine cables. Although many splice enclosures are designed to be re-opened if repairs or changes are needed, they can be hermetically sealed. Spliced fibers can even be included in cables without requiring an expansion of the cable diameter, although this is not done in many types of cable.

From a practical standpoint, it is the low loss and physical strength of splices that make them preferred for joining lengths of fiber in long-haul telecommunication systems. These considerations are less important in shorter systems and systems in more controlled environments.

SPLICING ISSUES

Loss, durability, and ease of operation are major concerns in splicing.

There are three principal concerns in splicing: the attenuation of the finished splice, its physical durability, and ease of splicing.

Attenuation

Fiber-optic splices are subject to attenuation mechanisms similar to those described for connectors in Chapter 8. However, the methods used to splice fibers produce tighter tolerances and, hence, lower attenuation than in connectors. Some sources of connection loss are essentially eliminated in splices; others are greatly reduced.

In a splice, the two fiber ends are bonded together either by melting (fusing) them, gluing them, or mechanically holding them in a tightly confined structure. Bonding the two fiber ends together with no intervening air space reduces or eliminates losses caused by file spacing. Some interface and spreading losses can occur if the fiber ends are separated by a bonding agent (e.g., a transparent epoxy), but the losses are minimal if the material's refractive index is close to that of the glass and if the intermediate layer is not too thick.

Splice losses fall into two categories: intrinsic and extrinsic. Many are analogous to those encountered in connectors.

Intrinsic Losses

Intrinsic losses are those arising from differences in the two fibers being connected. These loss mechanisms include variations in fiber core and outer diameter, differences in index profile and in ellipticity and eccentricity of the core. They can occur even in fibers with nominally identical specifications, because of inevitable variations in the manufacturing process.

Extrinsic Losses

Extrinsic losses are those arising from the nature of the splice itself. These include alignment of fiber ends, end quality, contamination, refractive-index matching between ends, spacing between ends, waveguide imperfections at the junction, and angular misalignment of bonded fibers.

Total Splice Loss

Typically these two loss mechanisms are comparable in magnitude for well-made splices. Fortunately, the two types of loss combine to give a total splice loss that can be less than their sum. However, as will be described in Chapter 13, care must be taken in measuring multimode fibers because of modal distribution effects that can give misleading results.

Improper splicing can let dirt into the junction or form imperfect junctions that suffer from high loss. For example, a single 10-μm dust particle could almost totally block transmission through a single-mode fiber splice. However, if splices are properly made, losses can be low. *Table 9-2* shows representative splice losses measured in experiments at the Corning Glass Works where fibers were broken, then fused together at the break point. That technique avoids losses caused by mismatched fibers, showing only losses caused by the splicing process itself.

Fiber ends are fused, glued, or mechanically held together.

Extrinsic losses arise from the nature of the splice.

Loss can be very low in properly made splices.

Table 9-2.
**Average Attenuation
and Strength of Fusion
Splices of Matched
Fiber Ends** *(Data
reproduced with
permission of Corning
Glass Works)*

Fiber type	Samples	Splice Attenuation, dB		Strength, GN/m²	
		Mean @ nm	Std Dev	Mean	Std Dev
10/125 (Single-mode)	80	0.04 @ 1300	0.04	1.41	0.47
9/125 (Dispersion-shifted single-mode)	30	0.06 @ 1300	0.04	1.28	0.38
50/125	20	0.05 @ 850	0.02	1.54	0.31
85/125	20	0.02 @ 850	0.008	1.42	0.36
100/140	20	0.05 @ 850	0.01	1.71	0.53

As *Table 9-2* indicates, loss arising from the splice itself is in the
0.05-dB range for both single- and multimode fibers. The highest losses are
for dispersion-shifted single-mode fibers, which are not in widespread use.
As might be expected, the standard deviation of splice loss—the statistical
measure of variation—is highest for single-mode fibers that have the
smallest cores. Minor deviations in positioning such small-core fibers tend
to cause larger (and less-uniform) errors than similar deviations in
alignment of multimode fibers.

Mismatched Fibers

Average loss is only 0.1
dB for splicing fibers with
identical specifications.

If the fibers spliced together are not identical, other factors can
become important. For single-mode fibers, the most important contributions
come from mismatches in mode field diameter—the effective size of the
mode transmitted by the fiber—and lateral offset of the fiber cores. Even
when such effects are taken into account, the average loss of fusion splices
is usually under 0.1 dB when fibers with identical specifications are spliced
together. However, losses of individual splices can be higher, and
manufacturers of splicing equipment normally specify splice attenuation as
less than 0.2 or 0.25 dB to allow for normal variations. Losses are similar
or slightly lower for multimode fibers, where the larger core sizes ease
tolerances. Attenuations typically are higher for mechanical splices.

It is physically possible to splice dissimilar fibers, such as a single-
mode fiber to a multimode fiber, or even several small-diameter fibers to a
large-diameter fiber. However, sizable losses can be expected if the light
emerging from the output fiber is not completely collected within the core
of the input fiber. For example, losses would be minute in coupling light
from a 10-μm core single-mode fiber to a 62.5/125 multimode fiber, but they
would be very large—about 20 dB—in going the other way. The basic
mechanism involved, the difference in core areas, is the same as for
connectors.

Fusion splicing does have some effect on other optical characteristics of fibers, including their bandwidth and pulse dispersion. These effects are negligible for single-mode fibers in most cases, but not always for multimode fibers.

Strength

Carefully controlled fusion splices are about as strong as unspliced fiber.

If you pull hard on a spliced metal wire, you expect it to part at the splice long before the wire itself fails. This is not necessarily the case with optical fibers. Fibers spliced carefully in the factory with the fusion process can withstand roughly the same stress tests as unspliced fibers. However, this is not always true for other types of splices, including fusion splices made under conditions not stringently controlled.

Fusion splicing melts fiber ends so they weld together as they cool.

Fusion splices are analogous to welding together pieces of metal (and, indeed, people sometimes say they are welding fiber ends together). As with metal welds, contamination of the weld—even by such seemingly innocuous materials as fingerprints—can weaken the splice. When fusion splices fail, they typically do so within 0.5–1.0 mm of the splice interface. Examination of the failure zones indicates the presence of thermal stresses, which are thought to arise because contaminants present on the fiber surface during heating weakened the surface fatally. Thus, cleanliness and proper recoating of the area after the splice is completed are key requirements for strong splices.

Proper fusion splicing requires that the glass in the fiber be heated to the right temperature to melt, then fused together to form a sound joint. The temperatures required at the joint are about 2000°C. Temperatures too low can form the fiber-optic equivalent of a cold-solder joint—a splice with high loss and low mechanical strength. Excessive temperatures also can weaken the physical bond between fibers and cause excess loss. Problems can arise if the two fibers do not have the same melting point.

In general, mechanical splices and those in which fiber ends are bonded with epoxy lack the mechanical strength of fusion splices. The strength of such splices depends on how the fibers are bonded.

Ease of Splicing

Specialized equipment is used in fiber splicing.

Because splices often are installed in the field, the ease with which they can be made is an important concern. This has led to development of specialized fiber-splicing equipment.

Fusion splicing sounds complicated because it requires the alignment of two fiber ends with micrometer accuracy and the precisely controlled application of energy to fuse the ends together. Specialized instruments have been developed for the job, which are described in more detail later in this chapter. By automating the most sensitive operations in fusion splicing, they make the technique practical in the field. Trained technicians are most efficient at fusion splicing, but with proper instruction and a good instrument anyone with reasonable mechanical skill can make a fusion splice. However, the instruments are bulky, typically weighing 25–45 lb (10–20 kg), and expensive, costing many thousands of dollars.

Installation of mechanical splices requires special splice housings and tools. However, the housings are small and the tools are much less expensive than those needed for fusion splicing. This makes mechanical splicing practical on a smaller scale than fusion splicing. Like fusion splicers, mechanical splice kits are designed to let people without special training make adequate splices.

Mass Versus Individual Splicing

Cabled fibers can be spliced individually or—in some special cables—en masse.

So far we've implicitly assumed that one fiber is being spliced at a time. However, cables generally contain two or more fibers, all of which must be spliced. If fibers are spliced individually, by fusion or mechanical means, all the resulting splices normally are housed within a special splice enclosure as described later in this chapter. (An exception is the fiber that is spliced in the factory before cabling, so the spliced fiber is incorporated directly into the cable. In that case, the user sees an intact cable and is not aware of the splice.)

Sequentially splicing all fibers in a multifiber cable can be time-consuming and costly. Experienced operators need at least several minutes per fiber, and time requirements increase with cable complexity. To speed things up, a few methods have been developed for mass splicing all fibers in a cable. Many of these techniques require a special cable structure that pre-aligns the fibers. A good example is the ribbon cable designed for the mass splice shown in *Figure 9-1*. A dozen fibers are imbedded in a plastic ribbon. The ends of the fibers in the two cables being joined are stripped and inserted into a multigrooved plate. Then a second grooved plate is mounted on top of the first to confine the fibers tightly and align them with respect to each other. The result is a simple, easy mass splice that may require only about a minute per fiber. Note that, although specialized cables are helpful, they are not an absolute necessity because multiple fibers from a cable can be individually laid in a V-groove-type mass splice. However, that takes longer than mass splicing a cable designed for that purpose.

As with other mechanical splicing techniques, in mass splicing there generally is a trade-off between attenuation and ease of the splicing operation. Faster splicing operations generally have looser tolerances and lead to higher attenuation.

Figure 9-1.
Mass Splicing of 12-
Fiber Ribbon Elements

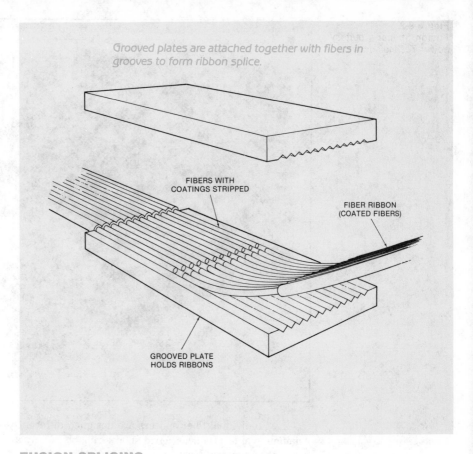

Grooved plates are attached together with fibers in grooves to form ribbon splice.

FIBERS WITH
COATINGS STRIPPED

FIBER RIBBON
(COATED FIBERS)

GROOVED PLATE
HOLDS RIBBONS

FUSION SPLICING

Fusion splicing is the
most common type for all-
glass fibers.

The most common type of splice is the fusion splice, formed by welding the ends of two optical fibers together. It is performed with a specialized instrument called a fusion splicer, shown in *Figure 9-2*, which includes a binocular microscope for viewing the junction and mounting stages and precision micrometers to handle the fiber. Most fusion splices have loss of 0.2 dB or less, often lower than 0.1 dB. As mentioned above, their mechanical strength can be similar to that of an unspliced fiber. Fusion splicing is intended only for all-glass fibers with their plastic coatings removed; it is not intended for plastic or plastic-clad fibers.

Figure 9-2.
Fusion Splicer *(Courtesy Power Technology Inc.)*

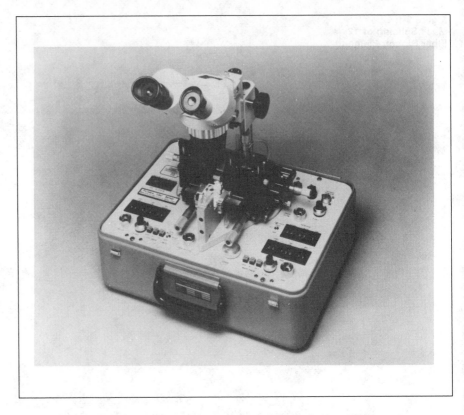

Fusion-splicing instruments usually are automated to help the operator.

Individual fiber splicers are designed differently, but all have the common goal of producing good splices reliably. Many are automated to assist the operator. Most models share the following key functions:

- A fusion welder, typically an electric arc, with spacing and timing of the arc adjustable by the user. Flames and the infrared beams from carbon-dioxide lasers also have been used, but virtually all commercial splicers use electric arcs. Portable versions are operated by batteries that carry enough charge for a few hundred splices before recharging. Factory versions operate from power lines or batteries.
- Mechanisms for aligning fibers with respect to the arc and each other.
- A microscope (generally a binocular model) with magnification of 50 power or more so the operator can see the fibers while aligning them.
- A way to check optical power transmitted through the fibers both before and after splicing.
- A cleaver to cut the fibers, producing smooth, flat ends perpendicular to within 1°–3° of the fiber axis.

Options include equipment to recoat the fiber after stripping. In some cases, some functions—such as measuring splice transmission—may be performed by devices separate from the splicer itself.

Before fusion splicing, plastic coatings must be removed from the fiber, and the end must be cleaved to have a face within a few degrees of perpendicular to the fiber axis.

The first step in fusion splicing is to prepare the ends by stripping the plastic protective coating from a few millimeters to a few centimeters of the fiber at the ends to be spliced, then cleaving them to produce faces that are within 1°–3° of being perpendicular to the fiber axis. Those ends must be kept clean until they are fused.

The fiber ends are not spliced immediately. First they are placed about 25 μm from the center of the arc and prefused for about a second to clean the ends and round their edges. Then the ends are pushed together just tightly enough to see some compression. At this stage, power transmission through the fiber junction can be measured to evaluate accuracy of alignment. After results are satisfactory, the arc is fired for a little over a second to weld the two fiber ends together. Care must be taken to ensure proper timing of the arc so the fiber ends are heated to the right temperature. After the joint cools, it can be recoated with a plastic material to protect against environmental degradation and to make size close to that of the original coated fiber.

Special splicers must be used with single-mode fibers.

Exacting tolerances are required to splice single-mode fibers because their cores are only about 10 μm in diameter. Thus, special single-mode splicers must be used with them. Single-mode splicers also can be used to splice multimode fibers because the tolerances required for the latter are looser. However, multimode splicers should not be used to splice single-mode fibers.

Fusion splicing can achieve the tightest tolerance of any present technique and in practice is the standard way to splice single-mode fibers.

MECHANICAL SPLICING

Mechanical splicing gives larger tolerances but requires simpler equipment than fusion splicing.

Mechanical splices join two fiber ends either by clamping them within a structure or by gluing them together. Because tolerances are looser than in fusion splicing, this approach is used more often with multimode than single-mode fiber. Although losses tend to be higher than those of fusion splices, mechanical splices are easier to perform and do not require expensive splicing equipment.

V-Groove Splice

One of the simplest types of splices relies on alignment of fiber ends in a V-shaped groove, as shown in *Figure 9-3*. The concept is similar to that used in some connectors. The fibers are confined between two plates, each containing a groove into which the fiber fits. This approach centers the fiber cores regardless of variations in the outer diameter of the fiber.

Figure 9-3.
V-Groove Mechanical
Splice

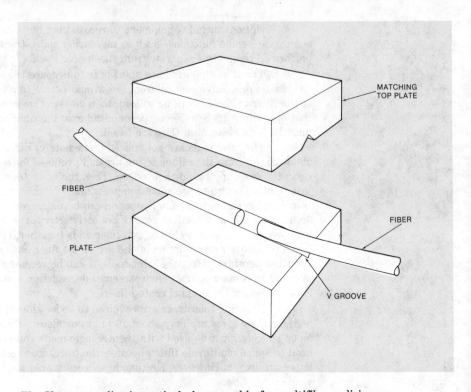

MATCHING
TOP PLATE

FIBER

FIBER

PLATE

V GROOVE

V-groove splice is valuable for multifiber splicing.

The V-groove splice is particularly amenable for multifiber splicing, as shown in *Figure 9-1* for ribbon cable. Each fiber in the cable slips into a separate groove, and mating the top and bottom plates automatically aligns all fibers with respect to one another. A practical limitation is the need for tight tolerances in the grooves. Field splicing of multifiber ribbon cables is aided by assembling end mounts on each cable in the factory and mating them in the field.

Elastomeric Splice

Elastomeric splices align fibers in a hole in a flexible plastic.

Another type of mechanical splice is the elastomeric splice shown in *Figure 9-4*. The internal structure is similar to the V-groove splice, but the plates are made of a flexible plastic material. An index-matching fluid or epoxy is first inserted into the hole through the splice. Then one fiber end is inserted until it reaches about halfway through the splice; the second fiber end is then inserted from the other end until it can be felt pushing against the first.

Proper preparation of fiber ends is crucial to avoid excess losses from reflection or the presence of contaminants. Reflection losses can be reduced by inserting a gel, index-matching fluid, or epoxy into the hole before inserting both fiber ends. Developers report that the maximum loss of this type is about 0.25 dB for multimode fibers.

Figure 9-4.
Elastomeric Splice
Developed for Field Use
(Courtesy GTE)

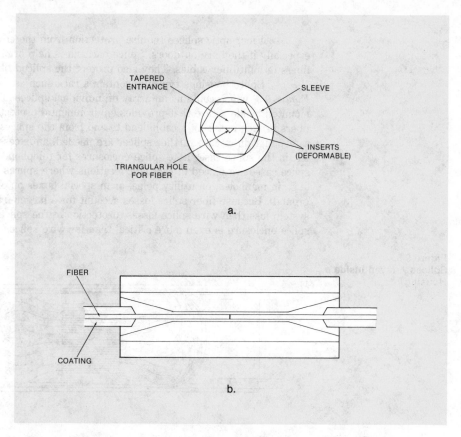

Fibers also can be spliced by inserting two ends into a hollow tube and gluing them in place with a transparent epoxy. The epoxy serves as an index-matching material to reduce reflective losses, allowing low overall loss.

The Connector-Splice Borderland

As we saw in the last chapter, the borderline between connectors and splices can be hazy. One example is the disconnectable splice made by attaching ferrules to the two fiber ends, joining the ferrules in a housing, and holding the assembly together with a spring clip. The assembly can be disconnected by removing the clip and is rated to survive 250 mating cycles. Although it is nominally a splice, it might be more proper to call it a special-purpose connector.

Some ribbon splices also fall within this hazy borderland. Parts of the splice may be factory mounted at the ends of a segment of cable, so the whole splice (or connector) can be assembled in the field by joining the two end-pieces with other mounting equipment.

SPLICE HOUSINGS

Fiber-optic splices require protection from the environment, especially if they are outdoors. Splice enclosures help organize spliced fibers in multifiber cables. They also protect the spliced fibers from strain.

Splice housings typically contain a rack such as the one shown in *Figure 9-5*, which contains an array of individual splices. This rack is mounted inside a case that provides environmental protection. Individual fibers broken out from a cable lead to and from the splices. To provide a safety margin in case further splices are needed, an excess length of fiber is left in the splice case. Like splice enclosures for telephone wires, fiber-optic splice cases are placed in strategic locations where splices are necessary (e.g., in manholes, on utility poles, or in special boxes on or below the ground). Because fiber splice losses account for a larger fraction of overall system loss than wire splice losses, protection of the splices within the splice enclosure is even more critical than for wire splices.

Figure 9-5.
Splices Arrayed Inside a
Housing

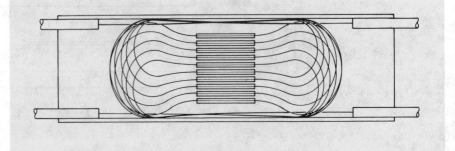

Fiber splice enclosures should be designed to:

- Hold cable strength members tightly
- Block entrance of water and retain gas if the cable is pressurized
- Provide redundant seals in case one level fails
- Electrically bond and ground any metal elements in the cable (e.g., strength members and armor)
- Be re-enterable if the splice must be changed or repaired
- Organize splices and fibers so they can be readily identified
- Provide room for initial splicing and future modifications
- Leave large enough bend radii for fibers and cables to avoid losses and physical damage.

SPLICE TESTING

Where high performance is crucial, attenuation of a splice can be measured as the splice is being made. As will be described in more detail in Chapter 13, precise measurement requires passing light through the splice from the remote end of one fiber being spliced to the remote end of the other. This is awkward in the field, where the remote ends may be several kilometers away.

A simpler approach is possible as long as the person making a splice needs only to know relative attenuation as the fibers are aligned. In practice, this is the needed information, because a peak in transmitted power indicates that the fiber alignment is best, and the fibers are ready to be spliced. Such relative measurements can be made by bending the fibers near the splice point so light can be coupled into (and out of) them. Although such measurements do not indicate actual splice loss, they can make field fiber splicing a one-person job.

SPLICE LIFETIME

Field experience indicates that if a splice is properly made and protected from environmental degradation, its physical and optical characteristics will not change appreciably. Thus, splice lifetime does not appear to be a significant problem under normal conditions.

WHAT HAVE WE LEARNED?

1. Splices are permanent connections between fibers. They have lower attenuation than connectors.
2. Major issues in splices are attenuation, physical durability, and ease of installation.
3. Splices normally are made in the field.
4. The most common way of making splices is by fusion, melting the ends of the fiber together with a device called a fusion splicer. Fusion splice loss can be under 0.1 dB.
5. Splices can also be made mechanically with devices that hold the fiber ends together either by using pressure or by gluing them.
6. The factors influencing attenuation of splices are very similar to those influencing connector attenuation.

WHAT'S NEXT?

In Chapter 10, we will look at the issues involved in coupling three or more fiber ends together, a job for fiber-optic couplers.

Quiz for Chapter 9

1. Which of the following is not an advantage of a splice?
   a. Permanent junction between two fibers.
   b. Ease of making changes.
   c. Low attenuation.
   d. Ease of installing in the field.
   e. Strength.

2. Where would a splice be used?
   a. In the middle of a long-distance cable.
   b. To connect a computer terminal with a local-area network.
   c. To couple light from an LED to a short-distance fiber system.
   d. To join an intrabuilding cable to a patch panel.

3. Splice loss typically is in the range of:
   a. 0.01 dB.
   b. 0.1 dB.
   c. 0.5 dB.
   d. 1.0 dB.
   e. 5.0 dB.

4. Which of the following mechanisms could not cause loss in a fusion splice?
   a. Differences in fiber core diameter.
   b. Dirt in the splice zone.
   c. Misalignment of fiber ends.
   d. Separation between fiber ends.

5. What would happen in a splice between fibers with identical outer diameters but different size cores?
   a. The splice would fail mechanically.
   b. Loss would be high in both directions.
   c. Loss would be high going from the large-core fiber to the small-core fibers, and low in the opposite direction.
   d. Loss would be high going from the small-core fiber to the large-core fibers, and low in the opposite direction.

6. The mechanical strength of a good-quality fusion splice is:
   a. much greater than that of unspliced fiber.
   b. about the same as unspliced fiber.
   c. slightly less than an unspliced fiber.
   d. much less than an unspliced fiber.

7. What is the major advantage of mechanical splices over fusion splices?
   a. Lower attenuation.
   b. Higher mechanical strength.
   c. Elaborate fusion-splicing system not required.
   d. Physically smaller.

8. Which of the following is not required in a fusion splicer?
   a. A welder to heat fiber ends.
   b. A microscope to view fiber ends.
   c. A mechanical alignment system.
   d. A device to cut fibers.
   e. A glue-delivery system.

9. Which of the following statements about splicing is false?
    a. A single-mode splicer can be used to splice multimode fibers.
    b. Mechanical splices are used more often for multimode than single-mode fibers.
    c. Losses of single-mode splices are much higher than those of multimode splices.
    d. Tighter tolerances are needed in splicing single-mode fiber.

10. Splice housings are important because they:
    a. reduce splice attenuation.
    b. protect splices from physical and environmental stresses.
    c. prevent hydrogen from escaping from splices.
    d. allow measurement of splice attenuation.

Couplers

The term "coupler" has a special meaning in fiber optics. A coupler connects three or more fiber ends (or optical devices such as detectors and transmitters). As such, it is distinct from connectors and splices, which join two entities. In fact, you splice or connect fibers to couplers. The distinction is far more important in fiber optics than in electrical signal transmission because the way in which optical fibers transmit light makes it hard to connect more than two points. To get around that problem, developers have turned to active couplers, which include a receiver and one or more transmitters.

In this chapter, you will see what fiber-optic couplers are and how they are used. You will also examine the various types of couplers and their strengths and weaknesses.

WHAT ARE COUPLERS?

Dividing Signals

Couplers connect three or more points.

Connectors and splices join two fiber ends together. That's fine for sending signals between two devices. However, many applications require connecting more than two devices. Connecting both a telephone and an answering machine on the same phone line, for example, requires a coupler (the other end of that phone line—the third point connected—is the rest of the world). That coupler must take the signal from one set of telephone wires and split it between two modular phone sockets, one for the phone and one for the answering machine.

That's not a problem with a telephone or other electrical equipment because of the way electricity flows in a circuit. If you hook one, two, or twenty identical resistors across an ideal voltage source, each will see the same voltage signal. Of course, that's only an approximation of what happens in a telecommunication system. Transmission line resistance and other effects can cause voltage across the load to drop somewhat as the load resistance drops (because more devices are put in parallel across the signal source). Nonetheless, with careful design, if many loads are in parallel, all will see voltage signals comparable in magnitude to what they would see individually. Likewise, if the signal is carried by variations in current from a current source, many devices in series will all see currents of magnitude comparable to what an individual device would see.

An optical signal must be divided among output ports, reducing its magnitude.

Optical signals are different from electrical signals and, thus, are transmitted and coupled differently. An optical signal is not a potential, like an electrical voltage, but a flow of signal carriers (photons), similar in some ways to an electric current. However, unlike a current, an optical signal does not flow through the receiver on its way to ground. It stops there, absorbed by the detector. That means multiple fiber-optic receivers cannot be put in series optically, because the first would absorb all the signal. If an optical signal is to be divided between two or more output ports, the ports must be in parallel. However, because the optical signal is not a potential, the whole signal cannot be delivered to all the ports. Instead, it must be divided among them in some way, reducing its magnitude.

Active and Passive Couplers

This puts an upper limit on the number of terminals that can be connected to a passive fiber-optic coupler that merely splits up the input signal. After some maximum number of output ports is exceeded, there is not enough signal to go around (i.e., to be detected reliably with low enough bit-error rate or high enough signal-to-noise ratio for the application). This limitation creates some serious problems for fiber-optic networks.

An active coupler includes a receiver and one or more transmitters to boost optical signals.

One way around this limitation is to generate stronger signals within the device that serves as a coupler. This can be done by directing the input signals into a receiver and using the electrical output from the receiver to drive one or more transmitters and (in many cases) provide electrical output to a terminal device. Such an active coupler is shown schematically in *Figure 10-1*. Active couplers aren't purely optical devices because they require electrical power input and can deliver electrical output signals. "Optikers" sometimes look down their noses at them, but they do solve the problem of running out of signal.

**Figure 10-1.
An Active Coupler**

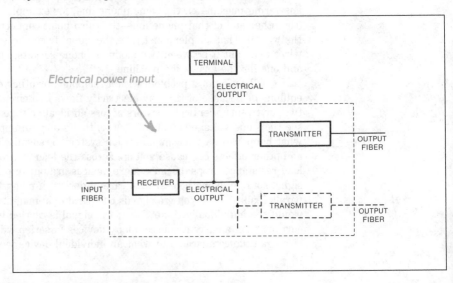

A further complication is that those simple-sounding passive couplers are hard to build. As we saw earlier, it is much harder to transfer light between fibers than current between wires because the light must be guided precisely. That makes design of passive couplers a tough problem, as we shall see later in this chapter. In practice, the more complex-sounding active couplers may be easier to build.

THE NEED FOR COUPLERS

Multiterminal Networks

Couplers are needed for the many systems that interconnect more than two points.

If fiber-optic couplers are so hard to make, why bother? The reason is that in many applications, such as the network shown in *Figure 10-2*, more than two devices must be interconnected to one fiber cable. Connectors and splices are fine for assembling cables to connect two points, whether they be on opposite sides of a room or on opposite sides of the Atlantic Ocean. Fiber optics are outstanding for such point-to-point communications, as long as one signal is going one way through each fiber. But if a third point must be added—like the answering machine mentioned earlier—a coupler is needed. That is very important in many situations, and it has been hard for fiber optics.

**Figure 10-2.
A Simple Local-Area
Network**

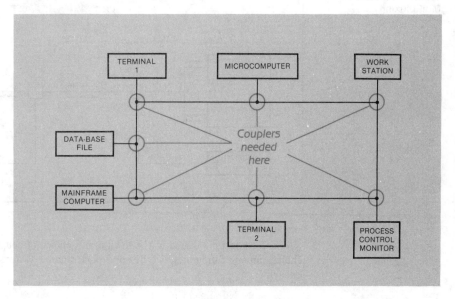

The standard example of an application requiring couplers is the local-area network shown in *Figure 10-2*. Like a telephone network, such a system interconnects many separate terminals, allowing each one to talk with any other. Local-area networks may be connected in various ways; the ring architecture shown is only one example. In other networks, all terminals are linked to a single data bus that carries all signals. In some cases, all signals may pass through a central mixer or coupler. The common feature is that all the architectures require couplers to expand connections beyond two terminals to multiple terminals.

Wavelength-Division Multiplexing

Couplers also are needed when two or more signals are sent at different wavelengths through the same fiber, as shown in *Figure 10-3*. The technique is called wavelength-division multiplexing (WDM). The idea is to save on fiber by having a single fiber carry multiple signals at different wavelengths. This is possible because light of different wavelengths travelling through the same fiber does not interact strongly enough to affect signal transmission. Couplers are needed to combine light signals at one end and separate them at the other. The separation (at least) requires couplers that are sensitive to wavelength, so light of different wavelengths can be directed along different paths. Note that all signals can be sent in the same direction, or they can be sent in different directions. In *Figure 10-3*, signals are sent to the right at λ_1 and λ_2 and to the left at λ_3.

Figure 10-3. Wavelength-Division Multiplexing

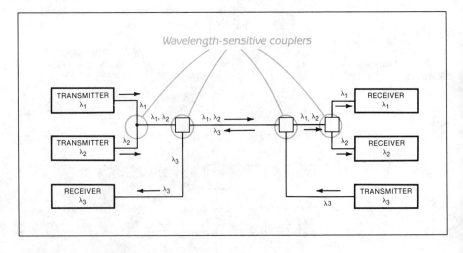

COUPLER ISSUES

Important issues in the design and choice of couplers include both choices among fundamental types and performance factors. The major ones are:

- Number of ports
- Sensitivity to direction in which light is transmitted
- Wavelength selectivity
- Passive or active coupling
- Type of fiber (single- or multimode)
- Power levels required and losses tolerable within the connector
- Cost.

These considerations have contributed to development of several different types of fiber coupler. The labels that follow all are attached to some types:

- Star coupler
- T (or tee) coupler
- Directional coupler
- Bidirectional coupler
- Splitters
- Combiners
- Wavelength-sensitive coupler
- Passive coupler
- Active coupler
- n×m coupler (where n and m are two integers, e.g., 2 and 3 or 4 and 4).

This proliferation of labels is confusing, particularly since the same coupler can wear two or more labels. For example, a T coupler can also be a directional coupler and a passive coupler (i.e., a passive, directional T coupler). Let's break this confusing mass into smaller chunks that should be more digestible.

Basics of Passive Couplers

Total output power from a passive coupler can be no more than the input power.

The fundamental fact of life for passive couplers that divide a light signal among two or more ports is that the total output power can be no more than the input power. From the viewpoint of each output device, the coupler has a characteristic loss, equal to the ratio of output to input power (in decibels). If the input power is equally divided between two output ports, that loss must be at least 3 dB because the input power is divided in half. The 3-dB loss is caused by the power division. Any additional loss above that theoretical minimum is called excess loss.

In the general case of a coupler with one input and i outputs, the total output, summed over all ports, equals input power minus excess loss. When power and losses are summed in decibels,

$$\text{INPUT} - \text{EXCESS LOSS} = \Sigma\text{OUTPUT}_i$$

Note that in general, power need not be divided equally among the output ports. In some cases, it may be desirable for 10% of input to emerge from one port and 90% from the other.

T and Star Couplers

A T coupler has three ports; a star coupler has more than three.

The two classic types of passive couplers are the star and the T types shown in *Figure 10-4*. They get their names from their geometry. The T coupler has three ports and is analogous to electrical taps that take a signal from a passing cable to a terminal. Thus, this type of coupler normally is shown as one fiber coming off a fiber passing by, in a T configuration, as in *Figure 10-4*. The input light need not be divided equally between the output ports, and T couplers are made with various power-splitting ratios.

**Figure 10-4.
T and Star Couplers**

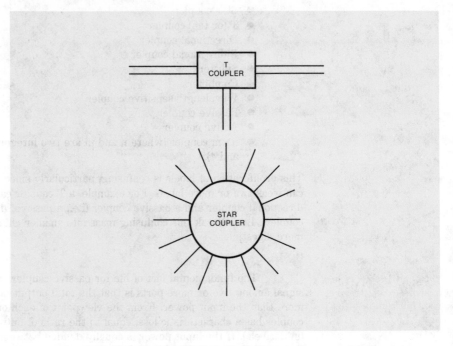

The star coupler has more than three ports—often many more, up to several dozen. In the classic concept of the star coupler, light is mixed in a central mixing element, and the output light emerges from that element into fibers. The actual geometry is not usually star-like, but such couplers are normally sketched in that way.

T and star couplers are used in different ways. The standard T coupler is used in a network such as shown in *Figure 10-2*, where individual devices are connected to a data bus or ring that carries signals. A star coupler is used in a network where all signals pass through a central point (where the coupler is inserted). In each case, the number of devices that can be connected is limited by the loss of passive couplers and receiver sensitivity. In a T-coupler system, the limit is reached when losses added along the ring or data bus reduce signal level below receiver requirements. With a star coupler, the limit is imposed by the number of output ports. For example, if a 1-mW signal is divided equally among 100 ports without excess loss, each receives 10 μW from the coupler, a 20-dB loss. On the other hand, seven T couplers that divided signals equally between two outputs (i.e., with 3-dB loss), would have a total loss of 21 dB. Even if the T coupler had throughput loss of 1 dB on the main fiber, a signal could pass through only 20 of them before experiencing 20-dB loss. Thus, star couplers generally can link more devices than T couplers, although an individual star coupler is more complex and costly than a single T coupler.

Directional Couplers

Some couplers work only if light is going in one direction; others work if light is going in either direction.

Another variable in coupler design is their operation on light going in different directions. Some types, called directional couplers, operate in only one direction. Others, bidirectional couplers, can operate in either direction. Examples of the two types for T couplers are shown in *Figure 10-5.*

**Figure 10-5.
Single- and
Bidirectional T Couplers**

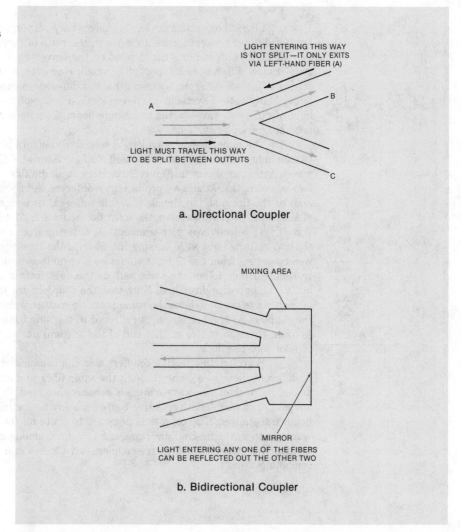

LIGHT ENTERING THIS WAY
IS NOT SPLIT—IT ONLY EXITS
VIA LEFT-HAND FIBER (A)

A

B

LIGHT MUST TRAVEL THIS WAY
TO BE SPLIT BETWEEN OUTPUTS

C

a. Directional Coupler

MIXING AREA

MIRROR
LIGHT ENTERING ANY ONE OF THE FIBERS
CAN BE REFLECTED OUT THE OTHER TWO

b. Bidirectional Coupler

In the directional coupler, light must be input via the left-hand fiber (A) to be split between the two output fibers (B and C). If light enters via B, all of it emerges through A and none goes to fiber C. Such a coupler also can take inputs from B and C and combine them as a single output in A, but it cannot combine A and B in C or A and C in B. In this example, directionality is caused by the way light is guided in fibers. The simple bidirectional coupler shown in *Figure 10-5* uses reflection in a mixing area to distribute signals input from any of three fibers to the other two. For purposes of signal splitting, it does not matter which is the input fiber and which are the outputs.

Wavelength Selectivity

Wavelength-selective couplers transmit light of different wavelengths in different ways. Applications are in wavelength-division multiplexing.

Like other optical devices, couplers may respond differently to light of different wavelengths. In general, the ratio of light coupled into two (or more) output ports can depend on the wavelength. In many couplers the differences are negligibly small. However, if the goal is to separate light of different wavelengths, the differences can be made very large. Many optical systems use devices called beamsplitters to separate light of different wavelengths in a single beam. The same concept can be used in fiber-optic systems.

The major use of wavelength-selective couplers is in WDM, where two (or more) signals are sent through a single fiber at different wavelengths, as shown in *Figure 10-6*. Here, a single fiber carries signals at two wavelengths, λ_1 and λ_2, produced by different light sources on different ends of the fiber, so the signals travel in different directions. At each end of the fiber, the two wavelengths must be separated. At top, the wavelength-selective coupler transmits λ_1 entering from the transmitter at top and reflects light at λ_2 leaving the fiber to the receiver. If light at λ_1 was emerging from the fiber, it would be transmitted back to the transmitter (i.e., follow the same path as the light from the λ_1 transmitter but in the opposite direction). Note that the couplers are bidirectional; light of each wavelength follows the same path, no matter which way it enters the coupler. At the bottom, λ_1 is reflected to the side to the receiver, and λ_2 (from the transmitter at that end of the system) passes through the coupler into the fiber.

Wavelength-selective couplers also can combine and separate light of different directions going through the same fiber and can discriminate among more than two wavelengths if suitably designed. One important advantage of wavelength-selective couplers is low loss at the wavelengths being transmitted. The coupler is designed to route all the light of each wavelength in the desired direction, not to split it among multiple outputs. In practice, wavelength-selective couplers have losses of 0.1–1.0 dB when combining signals.

Figure 10-6.
Use of Wavelength-
Selective Beamsplitting
Couplers

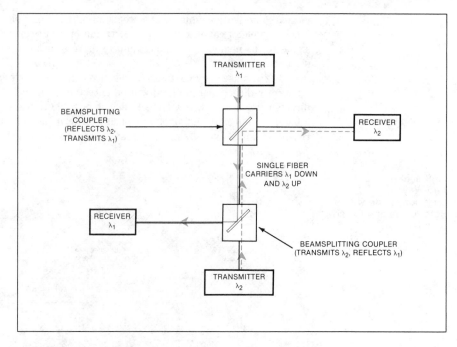

Splitters and Combiners

Couplers can split a single input into two or more outputs, or combine multiple inputs into a single output.

Couplers can serve two distinct functions in signal transmission. Sometimes they divide one input into two or more output signals to drive multiple devices. In other cases, they combine two or more inputs from separate devices to provide input to another device. These functions are inherently directional, depending on which way the signal is being transmitted. If signals are transmitted in both directions through the system, the same bidirectional coupler can serve both as a splitter and combiner. Sometimes these terms are used to describe couplers.

Passive and Active Coupling

Passive couplers can only divide the optical signal they receive; active couplers can boost the optical signal. Active couplers can drive more devices, but they require electrical power.

The basic distinction between passive and active couplers was described earlier, but it should be explored in more detail. Passive devices are purely optical, working by guiding waves and reflecting, refracting, and transmitting light. They require no input energy other than the light beam and don't have any active devices such as light emitters or modulators. They simply sit in place like a lens or any other optical component. They can be directional or bidirectional and sensitive or insensitive to wavelength.

Active couplers are not purely optical because they include a detector and at least one light source and require input of electrical power as well as the optical signal. This structure lets them boost a signal so the total output signal power can be greater than the input. Their signal outputs can be electrical as well as optical. One common configuration is a local-area-network coupler that detects an optical input signal and uses the receiver's electrical output both to drive an optical transmitter (to relay the signal to the next active coupler) and to drive a terminal device (with an electrical signal).

Because they contain a receiver at one port, active couplers are inherently directional (nothing will be detected if the optical input is directed to a transmitter). Note that it is possible to use both an active and a passive coupler in certain cases, such as a terminal that sends and receives signals over a single fiber at different wavelengths, as shown in *Figure 10-7*.

**Figure 10-7.
Two-Way
Communication with
Both Active and Passive
Couplers**

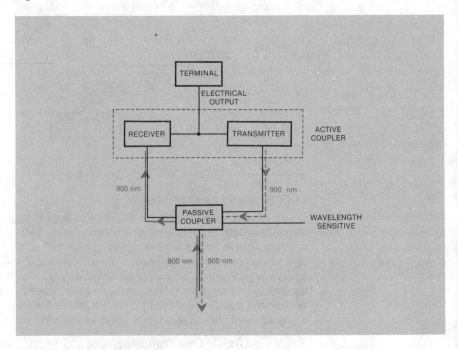

Number of Ports

An n×m coupler has n input ports and m output ports.

Each coupler has a characteristic number of input and output ports. Those numbers normally are identified by describing the coupler as an n×m type, where n is the number of inputs and m the number of outputs. For directional couplers, these numbers are fixed; for bidirectional couplers, they depend on how the coupler is installed in a system. The bidirectional three-port coupler shown in the bottom of *Figure 10-5*, for example, could be either a 1×2 or 2×1 coupler, depending on its use in a network.

The classic T coupler is a 1×2 or 2×1 device. Closely related are 2×2 couplers, which have two input and two output fibers and often function as T couplers with one input or output port unused. In star couplers, n and m are larger, and often equal, because in practice the inputs and outputs are connected to the same terminal devices.

Type of Fiber

Coupling methods differ for single- and multimode fibers.

The type of fiber used is critical in design and selection of couplers. Some coupling techniques only work with single-mode fibers. This is particularly true for integrated-optic couplers, which will be described in more detail in Chapter 12. Others work with multimode fibers. Generally, those that work with multimode fibers also work with single-mode input, although they may not be efficient to use with single-mode fibers.

Power Distribution

Couplers can distribute power evenly or unevenly among output ports.

The distribution of optical power among output ports is another factor in coupler design. In multiport star couplers, the goal generally is to distribute power as equally as possible among output ports. A figure of merit then is port-to-port variation in output power. However, in T or 2×2 couplers, it may be desirable to divide power unequally so, for example, only a small fraction of the signal transmitted along a data bus is coupled to any individual device. In such a case, a 20-to-1 splitting ratio might be chosen. Uneven splitting also may be desirable to avoid exceeding dynamic ranges of receivers.

In some systems, signals should be isolated from each other. For example, in a 2×2 directional coupler, the signal entering through one input fiber ideally should not be transmitted down the other input fiber. In practice, the isolation is not perfect, but the signal attenuation usually is strong, 40 dB or more.

Note that in some couplers, particularly single-mode types, light of different polarizations may be treated differently. They may, for example, divide the nominally unpolarized light transmitted by conventional fibers into two polarized or partially polarized beams. This is not important for most communication applications, but it can have an impact upon sensing systems.

Excess Loss

Excess loss is attenuation of signal above that needed to divide the signal as desired.

Splitting an optical signal among two or more outputs in a passive coupler means that each output sees a reduction in signal from the input level. In a perfect coupler, those would be the only losses experienced by the signal, but in real passive couplers there is an excess loss. This is the difference between the input and the sum of the outputs, usually expressed in decibels.

$$\text{EXCESS LOSS (dB)} = -10 \log \frac{\Sigma \text{OUTPUT}_i}{\text{INPUT}}$$

It can be considered as power wasted in the coupler. (Note that this concept is not meaningful for active couplers.)

Some excess loss is inevitable in passive couplers, but in general it should be as low as possible. Excess losses as small as 0.05 dB are possible in small directional couplers, and couplers with losses below 0.5 dB can be produced routinely. Higher excess losses, of a few decibels, should be expected with multiport star couplers.

Cost

Couplers are not cheap. Simple three- or four-port couplers cost as little as $50 each in large quantities and about $100 each when sold in smaller lots. Multiport star couplers are much more expensive. These prices represent important reductions from early levels and are not forbidding to fiber-optic specialists. However, those used to working with electronic connectors may be surprised to find couplers costing many times the price of connectors, when the differences are much smaller for electronic equipment.

PRACTICAL COUPLERS

The basic concepts described above have been used in many different coupler designs. The field is evolving rapidly, making it impossible to review comprehensively. Many types of couplers have been demonstrated, and we can only sample a few of the major ones below. Although an understanding of coupler design is useful, it is more important for fiber-optic users to know the functional characteristics of couplers than the details of their internal workings.

T and 2×2 Couplers

The function of the T and the closely related 2×2 coupler is to transfer light signals between a pair of fibers. One of the most common ways to make such a coupler is by transferring light directly between two fibers, as shown in *Figure 10-8*. It is not enough just to lay two fibers beside each other. The plastic buffer must be entirely removed first. Then the clad fibers are melted together and pulled to create a tapered region where light can be transferred between the fiber cores. For single-mode fibers, where the core is much smaller than the fiber's outer diameter, some of the cladding may be removed before the fibers are heated and fused together, but this is not necessary for multimode fibers. These couplers are called fused fiber couplers or fused biconical taper couplers.

Evanescent Wave Coupling

The cores themselves need not be fused together. If the distance between them is small enough—say, a few micrometers—a phenomenon called evanescent wave coupling will transfer light between them. This occurs because some light, called evanescent waves, actually travels in the inner part of the cladding. Similar effects are used in integrated-optic couplers of planar waveguide structure, which will be described in Chapter 12.

Couplers are not cheap.

Many T and 2×2 couplers rely on light transfer between fibers.

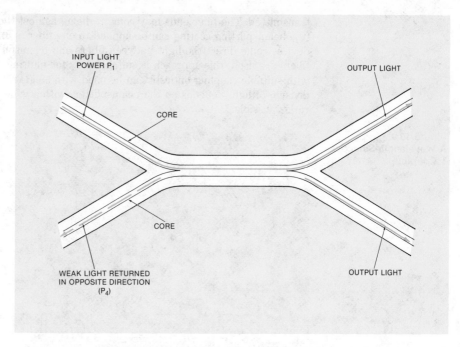

Because this approach starts with two fibers each having two ends, it produces a 2×2 coupler with two inputs and two outputs. To make a pure T coupler with only three ports, one of the input or output fibers is ignored. Because of its design, such a coupler is inherently directional. A typical directivity for a 2×2 coupler is 40 to 45 dB, measured by comparing the input power entering through a single fiber (P_1) to the power reflected back to the other input fiber in the direction opposite the input (P_4 in *Figure 10-8*). Mathematically,

$$\text{DIRECTIVITY (dB)} = -10 \log \frac{P_4}{P_1}$$

Optical Beamsplitters

Wavelength-selective T couplers can be built around a beamsplitter that separates wavelengths.

Another approach, often used in wavelength-selective couplers, relies on optical beamsplitters to divide an input beam into two segments or combine two input beams into a single output. The basic idea is similar to a one-way mirror, which reflects part of the incident light and transmits the rest. In some wavelength-selective couplers, a special coating is applied to a glass plate that makes it transparent at some wavelengths and reflective at others. When the coated plate is turned at a 45° angle to the incoming light, one wavelength is transmitted but others are reflected, as shown in *Figure 10-9*. If the direction light is travelling is reversed, light from both the output fibers shown in the figure would be directed out the input fiber, because λ_1 would still be reflected and λ_2 would still be transmitted. The fiber ends may come right up against the beamsplitter (the beamsplitting coating can be applied to one fiber end), or other optical devices could direct the light between fibers and beamsplitter. If only fibers are used, this approach is suitable only for multimode fibers, but with suitable coupling optics it can be used with single-mode fibers. The division of light depends on the beamsplitter, with a wide range of splitting ratios possible.

**Figure 10-9.
A Wavelength-Selective
T Coupler**

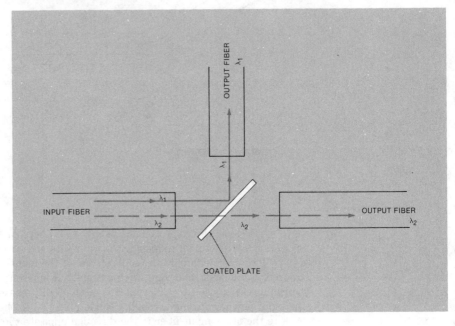

Star Couplers

Star couplers can be made by fusing many fibers to create a mixing region.

A star coupler has different functional requirements than a T or 2×2 coupler. When light is transferred between only two fibers, it can be concentrated in a small region (e.g., the core of a single-mode fiber). However, if light is to be distributed among many fibers, it must be spread out reasonably uniformly over the region where the fibers collect light. Then the cores of the output fibers must collect this light efficiently. This can lead to large excess losses unless the output fibers are bunched tightly together.

Many approaches used in making star couplers resemble those used for T couplers. For example, the cores of many fibers can be fused together to form a mixing region, as shown in *Figure 10-10*. This approach is used to make couplers with over 200 output ports; 64×64 fiber star couplers made using this process are sold commercially. Such fibers are called transmissive stars because light is transmitted through a mixing region from input fibers to output fibers. Efficiency depends on how well the output fibers can collect light from the mixing element.

Figure 10-10.
Fusion of Many Fibers
to Create a Mixing
Region

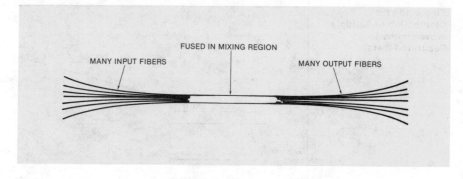

MANY INPUT FIBERS

FUSED IN MIXING REGION

MANY OUTPUT FIBERS

Star couplers can be
transmissive or reflective.
Transmissive stars sepa-
rate input and output fi-
bers and, hence, have
lower excess loss.

Reflective star couplers also are possible. These are made like
transmissive couplers, but the mixing region is cut in half and coated with a
reflective material so light entering through one fiber returns to be
distributed among all the fibers entering the coupler. Unlike a transmissive
star, a reflective star does not discriminate between input and output
fibers. It sends light into all fibers, even those that provided the input.

The choice between reflective and transmissive stars depends on
network architecture. If signals are transmitted and received through the
same fiber, then reflective stars are desirable. However, if input and output
channels are separate, transmissive stars may be the better choice. Because
no light is wasted by being directed to input fibers, transmissive stars have
excess loss at least 3 dB lower and, in practice, are the most popular type.

Most star couplers work with multimode rather than single-mode
fibers. This is largely because the larger-core multimode fibers collect light
much more efficiently. Single-mode star couplers with only a few input and
output fibers can be reasonably efficient, but serious problems remain in
trying to produce multiport star couplers for single-mode fibers.
Fortunately, that is not a serious practical problem because single-mode
fibers are very rarely used in networks that require multiport networks.

Wavelength-Selective Multiport Couplers

Special coatings or grat-
ings can be used to make
couplers that separate
many wavelengths.

Wavelength-division multiplexing may involve transmission of
more than two different-wavelength signals through the same fiber. One
way to separate those wavelengths is by cascading a series of beamsplitting
couplers such as those described earlier and shown in *Figure 10-9*. Another
approach is to use a special coupler designed explicitly for wavelength-
division multiplexing. The most straightforward approach is to use a
diffraction grating to spread out a spectrum of light from the input fiber
and focus specific wavelengths in that spectrum onto fibers in a linear
array, as shown in *Figure 10-11*. If light at five wavelengths, λ_1, λ_2, λ_3, λ_4,
and λ_5, enters through the input fiber, the grating will split the different
colors and direct each to the appropriate output fiber. Conversely, if inputs
and outputs were reversed, the grating would combine the five
wavelengths into a single output at the top fiber.

**Figure 10-11.
Distribution of Multiple
Wavelengths to
Separate Fibers**

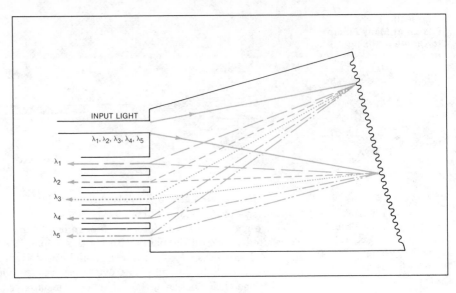

Waveguide Couplers

Passive couplers can be
made using planar
waveguide technology.

Planar waveguide technology, described in more detail in Chapter 12, also can be used for passive couplers. A planar waveguide is a flat, thin-film analog of an optical fiber, which confines light in a thin stripe using phenomena similar to those that guide light in optical fibers. (In a sense, fibers are a special type of optical waveguide.) Thus, it is possible to couple light between planar waveguides much as light can be coupled between fused fibers. For example, a T coupler-like splitter can be made by dividing a planar waveguide in a Y shape so one waveguide splits in half.

Planar waveguide technology is potentially important but remains in the laboratory because its conceptual simplicity is offset by practical problems in making the devices and in transferring light to and from them. It and its uses in couplers and other devices will be described in Chapter 12.

Active Couplers

The most common active
coupler is a special-pur-
pose repeater with an
electrical output to drive
a terminal device.

The conceptually simplest type of active couplers have already been described: receivers mated with transmitters, which detect an optical signal then regenerate it for transmission to the next node in a system. These are essentially special-purpose repeaters, which can contain multiple transmitter elements if they have to drive multiple fibers. Their normal use is in short-distance systems (e.g., local-area networks) so they normally contain pin-photodiode detectors and LED sources. In their most common form, there is one optical output (back to the network) and one electrical output (to a terminal device). Their major uses are in local-area networks, such as the one shown in *Figure 10-1*, which will be described in more detail in Chapter 18.

Other active couplers also have been developed. One example is shown in *Figure 10-12*. This coupler operates with fibers carrying signals in two directions at different wavelengths—780 and 880 nm. In the version at left, an internal long-wavelength pass filter reflects the short-wavelength input to a silicon detector. Long-wavelength light from the LED source in the coupler is transmitted through the central filter to the fiber and through the fiber to the coupler at left. The assembly is made of injection-molded plastic; the active components are installed in the holes made for them.

Figure 10-12.
Active Couplers with
Internal Beamsplitters
(Courtesy ADC Fiber
Optics)

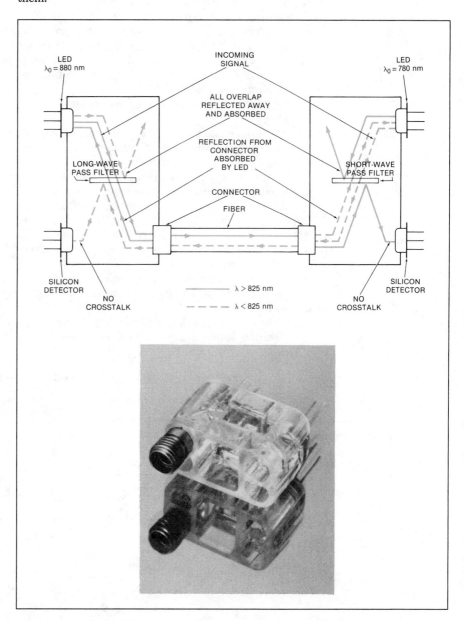

WHAT HAVE WE LEARNED?

1. A coupler connects three or more fiber ends.
2. An optical signal must be divided among output ports, reducing its magnitude.
3. Fiber couplers can be passive or active. Passive couplers are purely optical devices that work only with the signal they receive. Active couplers include a repeater and transmitter to regenerate signals.
4. Couplers are needed for wavelength-division multiplexing, if multiple signals are sent at different wavelengths through the same fiber.
5. Major elements in coupler design include number of ports, directional sensitivity, wavelength selectivity, type of fiber, cost, and power requirements.
6. Total output power from a passive coupler can be no more than the input power.
7. A T coupler has three ports; a star coupler has more than three.
8. Some couplers work only if light is going in one direction; others work if light is going in either direction.
9. An n×m coupler has n input ports and m output ports.
10. Excess loss is attenuation above that needed to divide the signal as desired. It may be 0.1 dB in small directional couplers or a few decibels in multiport star couplers.
11. Many T and 2×2 couplers rely on light transfer between fibers.
12. Wavelength-selective T couplers can be built around a beamsplitter that separates wavelengths.
13. Star couplers can be made by fusing many fibers to create a mixing region.
14. Special coatings or gratings can be used to make couplers that separate many wavelengths.
15. The most common active coupler is a special-purpose repeater with an electrical output to drive a terminal device.

WHAT'S NEXT?

In Chapter 11, we will look at the special tools and accessories used in fiber optics.

Quiz for Chapter 10

1. A fiber-optic signal can be split among two or more devices by:
   a. putting them in parallel across an input fiber.
   b. passing the signal through them in series.
   c. splitting the signal up so some light goes to each.
   d. none of the above.

2. The difference between active and passive couplers is:
   a. only passive couplers contain optics.
   b. active couplers contain light sources and detectors.
   c. active couplers divide signals.
   d. passive couplers can work with no more than two inputs or outputs.

3. What is not required for wavelength division multiplexing?
   a. Wavelength-selective couplers.
   b. Transmission of signals at different wavelengths through the same fiber.
   c. A fiber able to transmit more than one wavelength.
   d. Transmission of signals in different directions.

4. If a T coupler with no excess loss equally divides input power between two output ports, how does the power at one of the output ports compare with the input power?
   a. The same.
   b. 3 dB higher.
   c. 3 dB lower.
   d. 0.5 dB lower.

5. A star coupler divides output equally among 100 ports, with excess loss of 3 dB. The total drop in power between input port and one of the outputs is:
   a. 3 dB.
   b. 10 dB.
   c. 20 dB.
   d. 23 dB.
   e. 100 dB.

6. How many fibers can be connected to a 64×64 star coupler?
   a. 64.
   b. 128.
   c. 1024.
   c. 4096.

7. A 2×2 coupler can be made from a pair of fibers by:
   a. fusing them together so light transfers between their cores.
   b. enclosing them in a common plastic coating.
   c. putting them in the same loose tube in a cable.
   d. attaching them end to end.
   e. all the above.

8. A fused 2×2 fiber coupler is inherently:
   a. directional.
   b. bidirectional.
   c. wavelength-selective.
   d. polarizing.

9. What type of coupler is needed for bidirectional transmission through a single fiber at different wavelengths?
   a. Wavelength-selective star.
   b. Wavelength-selective T.
   c. Transmissive star.
   d. Active directional coupler.

10. What can active couplers do that
is impossible for passive
couplers?
   a. Wavelength-division
   multiplexing at more than
   two wavelengths.
   b. Generate output power
   higher than total optical input
   power.
   c. Interconnect more than three
   terminal devices.
   d. Operate without electrical
   power.

Tools and Accessories

ABOUT THIS CHAPTER

So far we've covered the major components of fiber-optic systems. However, those major components aren't the only equipment used with fiber optics. There is specialized measuring equipment that we'll describe in Chapter 13. And as with other technologies, there also is a wide variety of specialized tools and accessories for fiber-optic systems. These are the sorts of things that are most puzzling to those novices who think that they understand fiber optics.

This chapter will give you a basic introduction to this specialized fiber-optic equipment. Some of the equipment is most likely to be used in development laboratories or in specialized systems. Other equipment is common but doesn't fit under the categories we have covered so far. Even if you don't encounter such equipment, you may hear enough about it to wonder what it is. This chapter will tell you.

FIBER REELS

The way fibers are wound onto reels affects their transmission while on the reel.

Optical fibers fresh from the factory are wound around reels, just as you would expect. However, the reels are not simply another piece of throw-away packaging for everyone but the janitor to ignore. The reel can have a surprisingly large effect on fiber characteristics while the fiber is on the reel. The way the fiber is wound on the reel can lead to microbending losses affecting fiber attenuation. Variations in attenuation can arise from changes in winding tension, the way fiber is laid down on the reel, or the structure of the reel itself. Some fiber manufacturers take these effects seriously enough to write technical notes about the reels they use and expend extensive effort in developing the best possible reel.

FIBER TOOLS

Conventional cable cutters and scissors can cut cable. Fiber coatings can be removed carefully with mechanical strippers or with special solvents.

Specialized tools help prepare optical fibers and cables for use in systems. Conventional cable and wire cutters can remove the outer jackets of fiber-optic cables, and ordinary household scissors or tin snips can cut strength members. However, care must be taken in the delicate jobs of removing fiber coatings and cutting fibers to avoid weakening the fiber and/or causing excess loss at the connector or splice being installed.

Mechanical strippers can remove plastic coatings from fibers, but extreme care must be taken to avoid nicking the fiber. Microscopic surface nicks can be starting points for fiber breaks. Other alternatives for all-glass fibers include using chemical solvents to dissolve the plastic but not the glass, and heating to soften the low-melting-point plastic but not the fiber. Note, however, that solvents and heat cannot be used for plastic-clad or all-plastic fibers.

Fibers are scribed and broken to make smooth ends perpendicular to the fiber axis. For high-performance connectors, the ends must be polished as well.

Cutting the fiber also is a delicate process. Normally fibers are cleaved by scribing with a diamond- or carbide-tipped tool to form a weak point where the fiber will break when bent. Special tools and small jigs are built to break fibers evenly, with ends as close to perpendicular to the fiber axis as possible. Cleanly cleaved fiber ends are necessary for low-loss fusion splices.

An additional step—polishing—is needed for most high-performance optical connectors. Typically, the fiber is inserted and glued into the connector ferrule with a small tip left protruding. This tip is ground and polished to a flat, smooth surface by rubbing it against polishing pads. Grinding and polishing is a multistage process similar to sanding down a rough wood surface. Start grinding with a coarse grit to remove rough edges, then polish with successively smaller grits until you eventually produce a surface flat and smooth enough for use in a connector.

INDEX-MATCHING MATERIALS

In Chapter 8, we discussed the use of index-matching fluids and other index-matching materials to limit reflection losses in some connectors and in mechanical splices. These deserve a bit more explanation because their function is unique to optical systems.

Index-matching fluids reduce surface reflection losses.

The amount of reflection at the interface between two optical materials depends on the difference in their refractive indexes and the angle at which light strikes the interface. As the refractive-index difference increases, reflective (Fresnel) losses occur over a wider range of angles and their magnitude also increases. If we make the simplifying assumption that all the light is going straight through, these losses can be calculated for optical fibers from the formula:

$$\text{LOSS(dB)} = 10 \log_{10} (1 - p)$$

where p is a reflection coefficient given by:

$$p = \left(\frac{n_1 - n_0}{n_1 + n_0} \right)^2$$

where n_1 and n_0 are the index of the fiber and the medium from which light enters it, respectively. This formula is per surface. If two surfaces are involved (as in the glass to air to glass transition for light passing through an air gap between two fibers), total loss is the sum of reflections at two interfaces. If the two interfaces are between the same materials, total loss is double the loss at one (because of the square, the signs don't matter).

The glass used in fibers has a refractive index of about 1.5, much higher than the index of air, which is only slightly greater than 1.0. Thus, there can be significant loss—about 0.3 dB—at glass–air interfaces. That loss can be reduced by filling the space between fiber ends with a transparent fluid or solid with a refractive index close to that of the fiber core. In a glued joint, that role is filled by transparent epoxies; if not, it can be filled by so-called index-matching fluids. Those typically are thick, viscous liquids such as glycerine or silicone grease. They fill the interior of the connector so no air spaces remain between the fiber ends. The index match need not be precise. A refractive index of 1.4 would cut Fresnel loss down to 0.005 dB per surface, or 0.01 dB overall.

MICROLENSES

The earlier descriptions of transmitters, receivers, and connectors glossed over one common internal component of such devices—the microlens. Its role is to couple light between fibers and other devices, as shown in *Figure 11-1*.

**Figure 11-1.
A Microlens Focuses
Light**

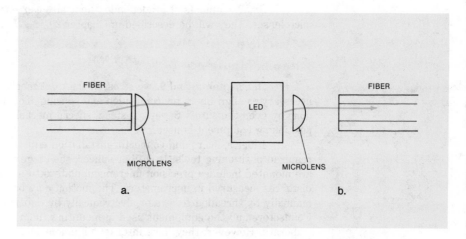

As the name implies, a microlens is a very small lens that focuses light from an optical fiber or a light source onto a fiber core or a detector. They are components you're not likely to see unless you assemble connectors, transmitters, or receivers, but they are vital components. The light emerging from fibers, LEDs, and semiconductor lasers forms a diverging cone-like beam. Microlenses bend those light rays to form a more nearly parallel beam and can focus parallel light rays onto a small spot, such as a fiber core.

Most microlenses are discrete components. However, a microlens can be formed at the tip of a fiber by melting the fiber end. The surface tension of the molten glass rounds the tip into a hemisphere, as shown in *Figure 11-2*, which can act like a lens and focus light.

A microlens is a very small lens that focuses light from an optical fiber or light source.

Figure 11-2.
Rounded Tip of a Fiber

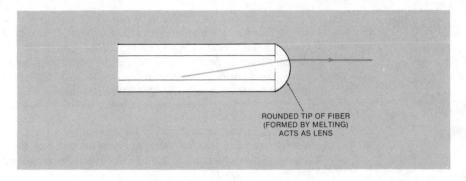

ROUNDED TIP OF FIBER
(FORMED BY MELTING)
ACTS AS LENS

Small segments of graded-index fiber also can serve as microlenses. They will be described in Chapter 20.

MOUNTING AND POSITIONING EQUIPMENT

Micromanipulators

Micromanipulators are used to align fibers.

In Chapters 8 and 9, we stressed the tight mechanical tolerances necessary to align fibers for connection and splicing. This is not a job to be done by even the surest fingers. Instead, special mechanical mounting and positioning equipment is used.

Fiber splicers and connector-installation equipment include the required positioning tools. In a fiber splicer, the stage where the fibers are mounted includes precision micromanipulators to move the fiber ends distances measured in micrometers. The motion may be controlled manually by threaded screws or electronically by digitized controls. Connectors and the equipment used in mounting them appear less elaborate. However, they have internal alignment mechanisms that provide the same tight tolerances.

Mounts

Special micropositioning equipment is made for handling loose fibers in the laboratory.

Special micropositioning equipment made for handling loose fibers in the laboratory is shown in *Figure 11-3*. Mounts which hold the fiber around its circumference have precise adjustments to align it with respect to another fiber, a detector, or a light source. Some simply hold the fiber steady, others come with adjustments in two dimensions, and others allow adjustments in three linear dimensions (X, Y, and Z axes) and two angles. Vacuum chucks hold the fiber in a slot by air pressure, a vacuum pump pulls the air out from underneath the fiber, and atmospheric pressure pushes the fiber into the slit.

Figure 11-3.
Mounts for Holding
Optical Fibers in the
Laboratory *(Courtesy*
Newport Corp.)

Optical Table

If you are working extensively with loose fibers and discrete light
sources, you probably will need an optical table or bench. An optical table
has a precisely flat surface, usually with an array of threaded holes on
which accessories can be mounted. An optical bench is a linear structure on
which you can place mounts to hold optical devices. Optical benches and
tables provide the rigid, stable work areas needed in laboratory work with
components as small as discrete fibers, light sources, and detectors.

MICROSCOPES

Microscopes are needed to
examine or visually align
tiny fiber-optic
components.

Optical fibers are so small that they are hard to see—let alone
inspect. To examine or visually align tiny fiber-optic components, you need
a microscope.

Specialized microscopes come with virtually all fusion splicers.
Most are binocular types, with eyepieces for both eyes, and magnification of
about 50 power. Those microscopes stare down onto an illuminated stage
built to align two fibers end-to-end. They make fibers easily visible even to
the hopelessly nearsighted with thick glasses (I can testify from trying
several at trade shows).

To inspect connector terminations, you need another type of
microscope to look at fiber ends. The usual microscopes for that purpose
are monocular (one-eyed) types with higher magnification—100–400 power.
That magnification is large enough to make a 50-μm fiber core appear
several millimeters in diameter. Most such microscopes are hand-held or
have small stages to hold the connector or fiber end in place.

ATTENUATORS

In Chapter 7, we discussed the fact that fiber-optic receivers require input signals in a certain dynamic range, over which they respond linearly. In Chapter 10, we saw that couplers may not distribute signals evenly—so some output ports may receive more light than others. Later on, we'll learn that some optical measurement instruments can handle only limited input power. Laboratory experiments often need to vary signal level to measure the response of components. All these facts point to the need for attenuators.

Attenuators discard surplus optical power. Although at first that may sound like a silly idea, it isn't. Detectors and receivers work best at certain power levels. They must be designed to accept power much lower than the transmitter output. However, that isn't always the case. In some fiber-optic networks, some terminals will receive much more power than others because there is less fiber and fewer connectors and couplers between them and the transmitter.

Attenuators add to the total attenuation to reduce signal levels to those which the receiver can best handle. Thus, all terminals in a network can use the same type of receiver, even if some receive 20 dB more signal than others. That power differential can be offset by attenuators.

Long lengths of optical fiber can serve as attenuators, but normally other types are used. One type is the gap loss attenuator, which relies on a variation of the separation of a pair of fiber ends. This type can be adjusted by moving the fibers farther apart. The precise adjustment must be calibrated for each device because of the many factors affecting gap loss. However, once the calibration is made, it is good unless the fibers are changed or the device is damaged. Such an attenuator is not sensitive to wavelength.

Another alternative is to use optical filters like camera filters or sunglasses, which transmit only part of the light entering them with the precise level set at manufacture. The optical transmission of filters, like that of other optical devices, can depend on wavelength, although some types have very low wavelength sensitivity.

MODE STRIPPERS OR FILTERS

As we saw in Chapter 4, multimode fibers can transmit light in many distinct waveguide modes. This has some important advantages for the transmission of light from large-area, high-divergence sources. But it creates some important problems for people trying to measure fiber characteristics.

The problems center on the way light is distributed among the fiber modes. Much light entering the fiber goes into high-order modes, which are barely confined in the fiber. As the light travels along the fiber, it leaks out of these modes, and eventually—after travelling a few kilometers—the light assumes an equilibrium distribution among fiber modes. However, if light hasn't travelled that far through the fiber, the distribution of light among high-order modes is uneven and very hard to predict. That complicates measurement of such quantities as fiber, connector, and splice attenuation because they depend strongly on the distribution of light among fiber modes.

To avoid this problem, light can be passed through a mode stripper or mode filter after it enters the fiber. This is a length of fiber bent or otherwise treated so high-order core modes and modes propagating in the cladding will escape.

MULTIPLEXERS

We've briefly discussed multiplexers in earlier chapters, and we'll go into their applications more in the chapters that describe how fiber-optic systems work. But we should stop to explain the basic ideas behind multiplexing and the types of multiplexers used in fiber-optic systems.

Multiplexing is the simultaneous transmission of two or more signals through one information channel. It is ubiquitous throughout much of the international telecommunications network. Many video channels are multiplexed on a cable-television network—you receive them all through a single coaxial cable. In contrast, the telephone wires running to your home carry only one telephone circuit. However, that single-circuit cable runs only to the nearest switching office or concentration point. There the signals carrying voices to and from your phone are merged with other voice signals, forming a multiplexed stream of information carrying many voice circuits.

From the standpoint of fiber-optic systems, we can classify multiplexers into two types: electronic and optical. Electronic multiplexers combine signals electronically, generating a pre-multiplexed signal that drives a fiber-optic transmitter. Optical multiplexers combine two or more signals optically for transmission over the same fiber.

Electronic Multiplexing

Electronic multiplexing equipment was developed for conventional wire-based communications and has been updated for fiber-optic systems. It takes different forms in digital and analog systems.

In digital systems, which are the most widely used, the standard electronic multiplexing method is time-division multiplexing, shown in *Figure 11-4*. In the example shown, four separate bit streams, each at 1.5 Mbit/s, feed into a multiplexer. The multiplexer combines the signals, selecting first one pulse from input 1, then a pulse from input 2, and so on in sequence. Essentially, the multiplexer shuffles the pulses together, although in practice some retiming is necessary because the pulses transmitted at a lower bit rate are too long to stuff into a faster stream of bits. At the other end of the system, a demultiplexer sorts the bits out, putting bit 1 into channel 1, bit 2 into channel 2, and so forth.

Multiplexing is the simultaneous transmission of two or more signals through one fiber.

Electronic multiplexers combine two or more signals to produce an electronic signal that drives a fiber-optic transmitter.

**Figure 11-4.
Time-Division Digital
Multiplexing**

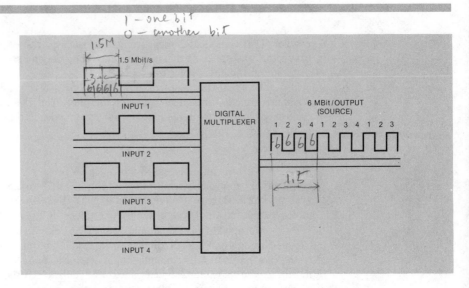

Things are done differently in the analog multiplexer shown in *Figure 11-5*. In this case, each of the four inputs modulates a separate carrier frequency. Suppose each input signal covers a frequency range of 0–1 MHz. The first signal modulates a 10-MHz carrier, generating signals varying in frequency from 10 to 11 MHz. The second modulates a 12-MHz carrier, generating a signal from 12 to 13 MHz. The third and fourth signals modulate carriers at 14 and 16 MHz, respectively, to generate signals at 14–15 and 16–17 MHz. This is called frequency division multiplexing. Note that some dead space is left between channels to avoid interference and that the carrier frequency is much higher than the signal frequency. The signals are sorted out by a demultiplexer at the other end of the system. Electronic bandpass filters divide the multiplexed signals which are mixed with the carrier frequencies to regenerate the original signals.

Neither electronic multiplexing methods requires anything special from the fiber-optic system other than an ability to carry the required signal frequency or data rate. All the signal combining is done electronically; all the fiber system needs to worry about is carrying the one signal that drives its transmitter. It does not matter (from a fiber-optic standpoint) that the input signal is a composite of many different signals.

**Figure 11-5.
Analog Electronic
Frequency-Division
Multiplexing**

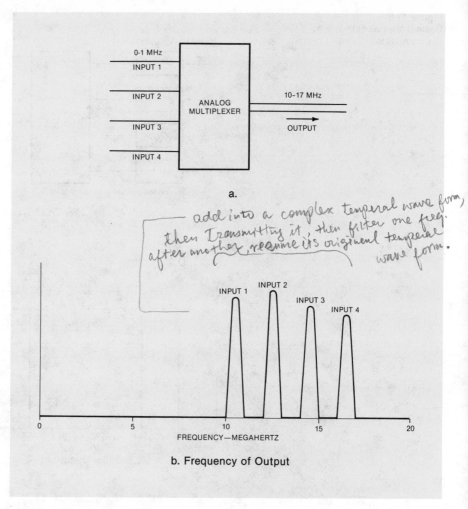

a.

add into a complex temporal wave form, then Transmitting it, then filter one freq. after another, resume its original temporal wave form.

b. Frequency of Output

Optical Multiplexing

Optical wavelength-division multiplexing sends multiple signals down a fiber at different wavelengths.

Electronic multiplexing is valuable because it lets many low-frequency or low-speed signals be combined to use effectively the large transmission capacity of optical fibers. Even more capacity is possible if different optical signals are sent simultaneously at different wavelengths in what is called wavelength-division multiplexing, as shown in *Figure 11-6.* Wavelength-division multiplexing also allows different signals to be sent simultaneously through the same fiber at different wavelengths.

Figure 11-6.
Optical Wavelength-
Division Multiplexing

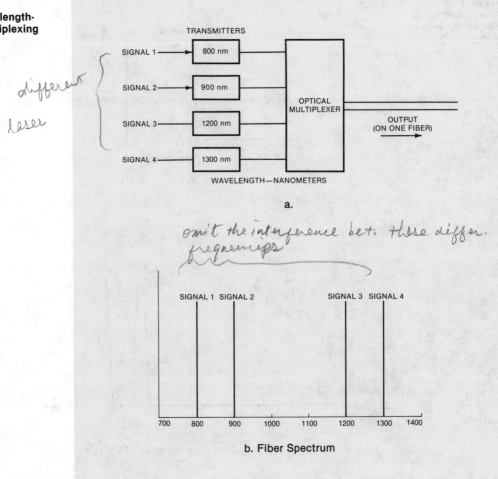

a.

b. Fiber Spectrum

Wavelength-division multiplexing bears some resemblance to analog electronic frequency-division multiplexing. In this case, the carrier frequencies are light waves emitted by semiconductor lasers or LEDs, modulated by separate input signals. Let's assume that we have four separate signals modulating lasers as shown on page 219.

Signal	Light Wavelength, nm
1	800
2	900
3	1200
4	1300

An optical multiplexer (a four-port fiber coupler) combines these signals into one output beam. At the output end, an optical demultiplexer (another coupler) separates the different wavelengths and directs them to separate receivers. Each receiver then generates the electrical signal that was transmitted at one wavelength.

SWITCHES

Optical-switching technology is difficult and in most present systems switching requires conversion of optical signals into an electronic form.

If you're familiar with electrical communications and electronics, you may wonder why it's taken so long to settle down and talk about switches. The reason is that optical switching is a major problem. Semiconductor electronic switches can readily redirect signals, routing them from your telephone through a vast network of switching centers and multiplexed long-distance lines to their destination—on the other side of town or on the other side of the world. There are no fiber-optic analogs of those switches. Researchers working on integrated optics, which will be described in the next chapter, have succeeded in switching optical devices between two states—one highly transmissive and the other absorbing most light entering the device. They've even built some rather primitive optical computers. However, they're a long way from optical devices that can switch as well as electronic devices.

Switches needed for fiber-optic systems fall into two broad categories. Simple switches are needed to re-route signals manually, such as to bypass an inoperative terminal in a local-area network or to redirect calls from one phone to another. High-speed, high-performance switches are needed at telephone-network nodes to interconnect thousands of phone lines.

Slow and simple switches can be made, such as the mechanical type shown in *Figure 11-7*. In that device, a free fiber end is moved mechanically between two slots, so it can collect light from one of two fibers. This sort of opto-mechanical switching has disadvantages that led telephone companies to abandon electro-mechanical switching: it is slow, and repeated flexing can cause fatigue and breakage of the fiber. The tight tolerances in fiber coupling make opto-mechanical switching even more difficult. However, low-performance switches can serve such functions as bypassing failed nodes of a network.

Figure 11-7.
A Mechanical Optical
Switch

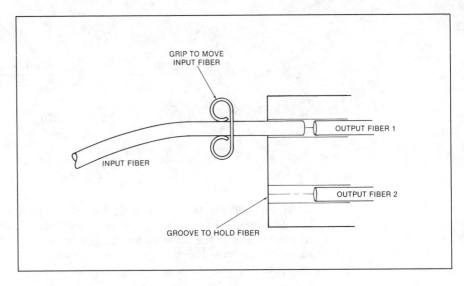

Progress is being made on fast, all-optical switches, but today's capabilities are far short of the needs of telephone switching offices. The short-term solution is to convert optical signals back into electronic form for switching and re-routing, then re-convert them to optical form for transmission. Researchers at laboratories operated by companies such as the American Telephone & Telegraph Corp. and the Nippon Telegraph and Telephone Corp. are working on the high-performance all-optical switches that will be needed for tomorrow's networks.

WHAT HAVE WE LEARNED?

1. The way fibers are wound onto reels affects their transmission while on the reel.
2. Conventional cable cutters and scissors can cut cable. Fiber coatings can be removed carefully with mechanical strippers or with special solvents.
3. Fibers are scribed and broken to make smooth ends and may be polished as well.
4. Index-matching fluids reduce surface reflection losses. They need not precisely match the index of glass, but they should be close.
5. A microlens is a very small lens that focuses light from an optical fiber or light source.
6. Micromanipulators are used to align fibers.
7. Microscopes are needed to examine or visually align tiny fiber-optic components.
8. Attenuators reduce optical power to a desired power range.
9. Mode strippers or filters remove undesired modes from multimode fibers to improve measurement precision.
10. Electronic multiplexers combine two or more signals to produce an electronic signal that drives a fiber-optic transmitter.

11. Optical wavelength-division multiplexing sends multiple signals down a fiber at different wavelengths.
12. Optical switching technology is difficult, and in most present systems switching requires conversion of optical signals into electronic form.

WHAT'S NEXT?

Our next stop as we travel through the world of fiber-optic components is the perpetual Tomorrowland of integrated optics, which we may someday reach.

Quiz for Chapter 11

1. How can properties of a fiber change when it is wound onto a reel?
   a. Adhesion of the fiber to the reel may weaken it.
   b. Microbending can increase fiber attenuation.
   c. There are no significant changes in any fiber property.
   d. Pressure on the fiber inside the reel changes its attenuation.

2. The plastic coating on a fiber can be removed by:
   a. mechanical stripping.
   b. solvents.
   c. peeling with fingers.
   d. a and b.
   e. none of the above.

3. Surface reflection losses when light passes between a glass fiber and air and back into a glass fiber are:
   a. negligible.
   b. under 0.1 dB.
   c. 0.3 dB.
   d. 1 dB.
   e. over 1 dB.

4. Index-matching fluid reduces attenuation when light is transferred between fibers by:
   a. reducing Fresnel reflection losses caused by the large difference in refractive index between glass and air.
   b. preventing light leakage from the connection.
   c. guiding light through the junction.
   d. changing the refractive index of the fiber.

5. A microlens:
   a. is a very small lens that couples light between a source and fiber.
   b. can be formed by melting the end of a fiber.
   c. can be made from a segment of graded-index optical fiber.
   d. is found inside some transmitters.
   e. all of the above.

6. An attenuator can be used to:
   a. limit output power from a laser.
   b. prevent noise from entering into a fiber-optic system.
   c. reduce power level to match a receiver's dynamic range.
   d. prevent excess optical power from damaging fibers.

7. A mode stripper is used to:
   a. discard excess modes in coupling from multimode to single-mode fiber.
   b. discard modes that might cause errors in measuring properties of a multimode fiber.
   c. prevent modal dispersion.
   d. match transmission characteristics of two fibers.

8. Which type of multiplexing is performed optically?
   a. Wavelength-division multiplexing.
   b. Time division multiplexing.
   c. Analog frequency-division multiplexing.
   d. All of the above.

9. Optical multiplexers are:
   a. connectors.
   b. transmitters.
   c. couplers.
   d. switches.

10. Switching in most fiber-optic systems is done:
   a. with special fiber-optic couplers.
   b. by mechanical fiber-optic switches.
   c. by redirecting light in fibers.
   d. electronically after optical signals are converted to electrical form.

Integrated Optics

ABOUT THIS CHAPTER

So far, we've talked about technology that is ready to buy and use. Fiber optics has moved fast and, although some areas may slow down, continued progress seems certain. Much new technology will be improvements on components we've already discussed, including ultra-low-loss fibers and better couplers. However, there's a whole new field waiting around the corner—integrated optics.

Integrated optics is a vague term that means different things to different people. It was a research field even before the first low-loss fibers were made, and sometimes it seems an aging resident of Tomorrowland, one of those fields where breakthroughs always lurk around the corner—or in the next proposal for a research grant. In fact, the focus of integrated optics research has evolved over the years, and some unpromising areas have been abandoned in favor of more fruitful ground. Much of the new generation of integrated-optics research is aimed at improving the performance of fiber-optic systems by developing new components, including optical switches, transmitters, couplers, and receivers.

In this chapter, we'll discuss integrated optics and its potential contributions to fiber-optic communications and related fields of optical information processing. This won't be an exhaustive discussion because, integrated optics remains in the laboratory, but it is important to know about work that could affect the evolution of fiber-optic technology.

INTEGRATED OPTIC CONCEPTS

Integrated optics is a broad field, covering the use of compact devices to perform optical functions.

As a broad term, "integrated optics" has taken on many meanings. One common theme runs through all work in integrated optics: using compact devices to perform optical functions. Beyond that, however, there is much diversity in technology, in goals, and in applications. Not all integrated-optics research is aimed at integrating multiple optical functions in a single device.

Integrated optics includes thin-film optical waveguide devices.

What might be called traditional integrated optics uses thin-film devices to perform optical functions. The thin films incorporate optical waveguides that are rectangular in cross section, as shown in *Figure 12-1*. These waveguides are typically formed on a substrate by diffusing a dopant in from the top. Light can be transferred between waveguides, or the waveguides can be made to perform various functions. In fiber optics, they are most likely to be used to modulate light or couple it between fibers, as is described in more detail below.

**Figure 12-1.
Embedded Thin-Film
Waveguide**

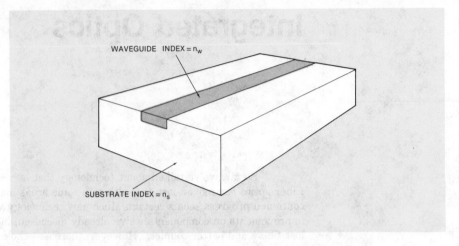

WAVEGUIDE INDEX = n_w

SUBSTRATE INDEX = n_s

Integrated optoelectronics combine optical and electronic functions performed by transmitters and receivers.

A family of devices often called integrated optoelectronics is being developed for use in transmitters and receivers. The goal is to integrate optical and electronic devices in a single device, much as present integrated electronic circuits combine many electronic elements. Devices already demonstrated integrate on a single semiconductor chip the functions of detector and receiver electronics, light source and driver electronics, or a simple repeater. They promise simple, high-performance transmitters and receivers, but some important technological barriers remain to be overcome, as we shall see later in this chapter.

Bistable optical devices can be switched between two stable transmission states.

The newest domain that falls under the heading of integrated optics is optical bistability. Bistable means having two stable states. For optical devices, that means having two different states for transmitting light—one with high transmission and the other with low transmission. This is analogous to a switching transistor, another device that is bistable, with the difference between its two stable states being how much current the device carries (i.e., its effective resistance). Research remains in the early stages, but bistable optics could be the basis for a new generation of all-optical switches.

PLANAR INTEGRATED OPTICS

Waveguide Structures

Light is confined in a planar waveguide in the same way as in a fiber.

A closer look at the planar optical waveguide in *Figure 12-1* shows that in many ways it works like an optical fiber. (In fact, waveguide theory applies to both rectangular and round optical waveguides—planar devices and fibers, respectively.) Light is confined in a planar waveguide by total internal reflection. The waveguide has a refractive index n_w that is higher than that of the surrounding material.

Unlike the core of an optical fiber, a planar waveguide normally is not completely surrounded by identical material of lower refractive index. In *Figure 12-1*, the embedded waveguide is surrounded on three sides by material with a refractive index n_s that is slightly lower than that of the waveguide. On the top is air, with index very close to 1, much lower than n_w.

Several structures can be used in planar waveguides.

There are several other waveguide structures, shown in *Figure 12-2*. Each confines light to the waveguide, but the degree of confinement and the ease of fabrication differ considerably. The choice depends on the application. Because a major use of planar waveguides is coupling light from one channel to another, as we will discuss later, the high confinement of the ridge waveguide—surrounded on three sides by air—often is not desirable. Coupling is enhanced in variations where the sides of the ridge are partly surrounded by material other than air. Examples are the rib and bulge waveguides shown in *Figure 12-2*. The most common waveguide type for couplers is the embedded waveguide, shown in *Figure 12-1*.

Figure 12-2.
Four Other Types of
Planar Waveguides

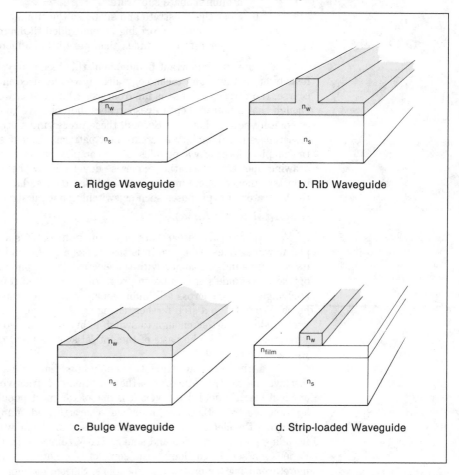

a. Ridge Waveguide   b. Rib Waveguide

c. Bulge Waveguide   d. Strip-loaded Waveguide

Fabrication Processes

Planar waveguides are made by depositing materials on a substrate or diffusing material into a substrate in patterns formed by standard semiconductor fabrication techniques. Processing details can vary. The standard way to make an embedded waveguide follows:

1. Deposit photoresist on a lithium niobate (LiNbO₃) substrate.
2. Use a light source and mask to expose a stripe on the photoresist.
3. Etch away the exposed photoresist to form a stripe, with the unexposed material left behind.
4. Deposit a thin layer of titanium metal on the substrate where the photoresist was etched away.
5. Remove the unexposed photoresist, leaving a titanium stripe on the lithium niobate substrate.
6. Heat the substrate and stripe so the titanium diffuses into the lithium niobate forming an embedded titanium-doped stripe with higher refractive index than pure lithium niobate.

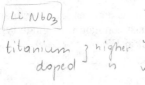

Note two important points about this technology. First, it draws heavily on methods and technology already widely used in electronics. Of course, the correspondence is far from exact, but the processing is similar enough that integrated-optics researchers did not have to develop a completely new technology. Second, these processing methods give researchers the flexibility to create new patterns. Linear stripes are only the simplest form of waveguides. More complex patterns can be created by drawing upon the fabrication expertise developed by the electronics industry. Indeed, some impressively complex devices have been made in the laboratory for purposes such as switching optical signals.

Waveguide Materials

What made researchers pick lithium niobate as a substrate for planar waveguides? One reason is that it has a useful and unusual property: its refractive index changes with the electric field applied to it. Because an optical waveguide's transmission properties depend on refractive index, applying a voltage across a lithium niobate waveguide can change how it guides light. To go a step farther, changing the voltage can change the waveguide properties enough to switch light between two optical waveguides. That is the basis of the optical switches and modulators that we describe below.

Lithium niobate (LiNbO₃) is only one of many materials that displays the electro-optic effect—the variation of refractive index with ambient electric field. However, it is one of the most popular because it has a strong electro-optic effect and is well characterized. Many other materials displaying the electro-optic effect can be used in planar waveguides, including gallium arsenide and similar III–V semiconductors, silicon, and various glasses. In the long term, some of those other materials may be more promising for practical applications. Silicon, gallium arsenide, and other semiconductors are attractive because electronic devices can be made from them.

Couplers

Light in a planar waveguide is not completely confined within the high-index waveguide material. An evanescent wave leaks out into the surrounding lower-index material. If two waveguides are close enough on the same substrate, light can be transferred between them, as shown in *Figure 12-3*. The power in each waveguide varies along the region where they are close enough to couple.

**Figure 12-3.
Transfer of Light Energy
Between Two Coupled
Waveguides**

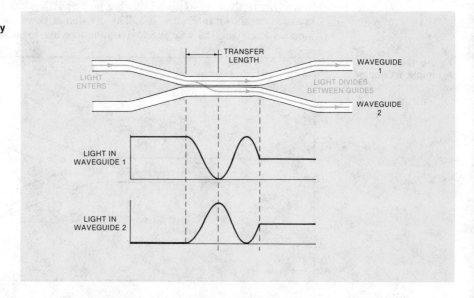

To understand how the coupling process works, consider the pair of coupled waveguides in *Figure 12-3*. Suppose light initially enters the upper one but not the lower one. As light travels along the upper waveguide, more and more of it transfers to the lower one. Once it's travelled a certain distance (sometimes called the transfer length), which depends on the optical characteristics of the waveguide, all the light will have shifted to the lower waveguide. Then the process starts again, this time with the light transferring from the lower waveguide to the upper one. Thus, as the light travels along the two waveguides, the distribution of light energy oscillates back and forth between them. The oscillation stops at the end of the region where the two are coupled, and the final distribution of light is established. (In this oversimplified case, we pretend there aren't any losses, but the real world is not so kind. The amplitude of the light wave decreases as it travels along the waveguide. In fact, losses in planar waveguides are much higher than those in fibers.)

In a practical T coupler made from planar waveguides, light would enter at one end of one waveguide and emerge at the other end of the two parallel waveguides. The ratio of output intensities from the two waveguides would depend on their lengths and optical characteristics. A coupler could be designed to divide the light into fixed proportions (e.g., 75% into one waveguide and 25% into the other) by selecting its length and optical properties.

Evanescent-wave couplers are not the only type of waveguide coupler. The Y coupler, shown in *Figure 12-4*, is a simpler approach to visualize. A single waveguide splits in two, with the output waveguides going off at equal angles to divide light equally between them. This approach can be extended to more than two output waveguides either by making more than two output branches or by putting multiple Y junctions in series, as shown in *Figure 12-5*. The division of power depends on junction angle and the way successive junctions are arranged.

Simple waveguide couplers can be made by making a waveguide branch.

Figure 12-4.
A Simple Y Coupler

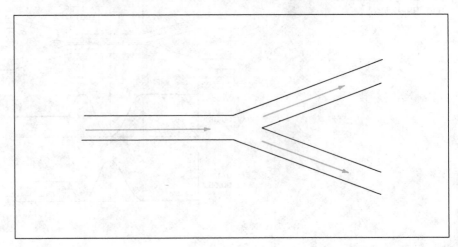

Figure 12-5.
Multiple Y Couplers in Series

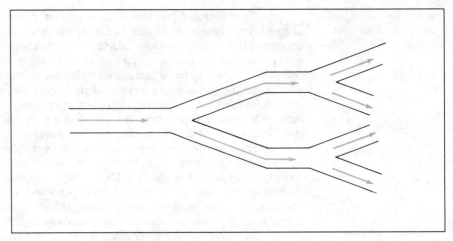

Switches

Static couplers are fine if signals are always split the same way between two outputs. However, if the signal is sometimes routed to one output and sometimes to another, you need switches or active couplers. Switching is one of the strong points of waveguide couplers because it uses a property of waveguide materials we've mentioned only in passing—the sensitivity of their optical properties to electric fields.

Let's start by looking at the evanescent-wave coupler in *Figure 12-3*. We said that the transfer length—the distance light has to travel along the two coupled waveguides before it is transferred completely from one to the other—depends on their optical characteristics. (The details of that dependence are the type of complexity that you don't want to worry about.) The optical characteristics of most materials are fixed, but those of materials such as lithium niobate change when an electrical field is applied to the waveguide.

Applying electrodes across the waveguides so the two experience different electric fields, changes their optical properties—and the transfer length. Suppose that at zero voltage, all of the energy entering waveguide 1 in *Figure 12-6* emerges in waveguide 2 after travelling a length L. Increasing the voltage applied across the waveguides changes their optical characteristics so eventually, at a voltage V, light will first shift from waveguide 1 to waveguide 2, then completely back to waveguide 1 again. Further increase in voltage might reach a point where a third transfer will be made, from waveguide 1 back to waveguide 2. However, that may not be possible in practice because of limits on the device's optical characteristics and operating voltages.

**Figure 12-6.
Transfer of Light Energy
Between Two Coupled
Waveguides**

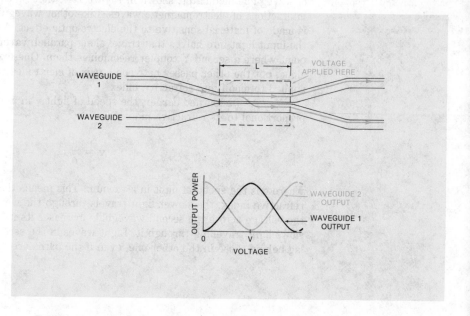

A Y coupler also can be used as a switch if electrodes are placed in the proper places. Changes in the applied electric fields alter the refractive indexes of the two output waveguides, and light tends to travel along the output waveguide with the highest refractive index.

Modulators

Light signals can be modulated by passing them through a waveguide while changing its optical characteristics.

Similar concepts can be used to make waveguide light modulators. From the outside, a light modulator seems to change its transparency, at times transmitting nearly all light entering it, at others transmitting almost none. However, inside a waveguide modulator is a special type of optical switch or coupler. Like a two-port switch, a waveguide modulator divides input light between two waveguides, with the degree of division depending on the electrical modulation voltage. One of those waveguides delivers the (modulated) output light signal. Light entering the other waveguide is thrown away.

One subtlety to note is that modulation can be continuous, while switching is supposed to be either on or off. Thus, modulators must be able to transmit not just 100% or 0% of the input light but also intermediate values and must be able to respond quickly to changes in the modulation signal. Switches need to respond quickly, but they need not have gradations between off and on—much less gradations in transmission that are linearly proportional to their drive modulation. (In practice, optical switches are not ideal switches and may not always switch all the signal to one output port and keep the other output precisely at zero.)

Interactions of electromagnetic waves, as well as optical switch concepts, can be used to modulate light in waveguides.

These principles could adapt the evanescent-wave and Y coupler switches described earlier for use as modulators. There also is another type of waveguide modulator, shown in *Figure 12-7*, that relies on more complex interactions of electromagnetic waves. Like other waveguide modulators, it is made of material sensitive to the electro-optic effect. A Y coupler divides the input light into halves that travel along parallel waveguides to another point where a second Y coupler recombines them. One waveguide is left alone, but the other passes between a pair of electrodes where a voltage is applied to modify its refractive index.

As we learned earlier, the speed of light v in a material is proportional to its refractive index n:

$$v = \frac{c}{n}$$

where c is the speed of light in a vacuum. This means that the larger the refractive index, the slower light travels through the waveguide. Thus, if the voltage across the second waveguide increases its refractive index, it slows light travelling through it. Light waves in the second waveguide will lag behind those in the other one, even if the paths are otherwise equal.

**Figure 12-7.
Modulator that Depends
on Wave Interactions of
Light**

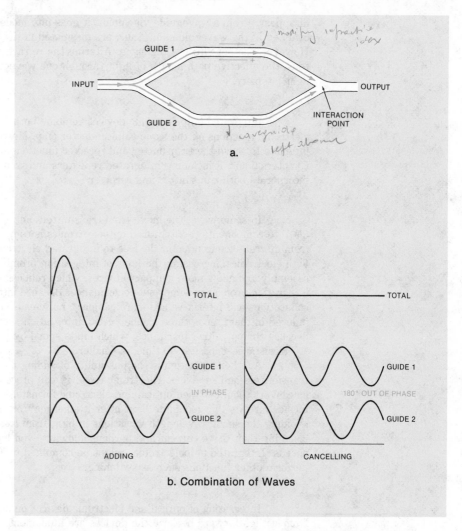

a.

b. Combination of Waves

What happens when the light from the two waveguides is
combined at the output depends on the differences in the path lengths that
they have travelled. If the path lengths are identical, the two beams add
together to give an output equal to the input (neglecting any losses within
the modulator). The same thing happens if the path lengths differ by
exactly an integral number of wavelengths. However, some light is lost if
the path lengths differ by some other value.

The easiest way to visualize this is as the interference of two light
waves, as shown in *Figure 12-7*. If the two waves are 180° out of phase,
they are everywhere equal in magnitude but opposite in sign, so they add
to zero. If they are out of phase by a smaller amount, they add together to
give less light than the original input. The light doesn't magically

disappear. From a waveguide viewpoint, it goes into modes that are not confined in the waveguide and, thus, are dissipated in the substrate. However, the net effect is the same. Altering the refractive index and, thus, the effective path length of light through one waveguide modulates light intensity.

INTEGRATED OPTOELECTRONICS

Integrated optoelectronics combine optical and electronic devices on one semiconductor substrate.

Integrated optoelectronic devices combine both optical and electronic functions on the same semiconductor chip. Mostly, they are intended to serve the transmitter and receiver functions we discussed in Chapters 6 and 7, including integrated repeaters and transceivers that incorporate both transmitter and receiver.

Benefits

In some ways, the move from transmitters and receivers made of discrete components to integrated optoelectronics is comparable to that from discrete semiconductor devices to integrated electronic circuits. As with electronics, increasing the level of integration promises to simplify assembly and allow more compact devices, which reduces costs. Similarly, integration promises to improve performance. Because integrated connections and interfaces are more compact, noise caused by currents induced in them tends to be smaller. Also, integration should raise speed because the parasitic capacitance, which limits speed by its effect on the RC (resistance–capacitance) time, is smaller.

Another attraction is the possibility of adding more functions to transmitters and receivers. Electronic circuitry can process signals as well as convert them between optical and electronic formats. Examples might include circuits that modulate light after it emerges from the source, avoiding the small wavelength variations in light from lasers modulated by changing their drive current and perhaps allowing even higher transmission speeds. Integrated optoelectronics might be combined with waveguides to perform other functions such as switching.

Materials

Optical and electronic devices use different materials, making integration difficult.

Integration of optical and electronic devices on a single chip is far from straightforward because the devices are fundamentally different. For transmitters, they start at the fundamental level of materials. Silicon, the most common electronic semiconductor, is unsuitable for light sources, as is germanium. Most III-V compounds used in semiconductor lasers and LEDs that emit light in the visible and near infrared are not well developed for electronics. Notable progress has been made with gallium arsenide, used in 800- to 950-nm lasers, but GaAs does not emit the longer wavelengths needed for long-distance telecommunications. The best candidate for long-wavelength integrated optoelectronics is indium phosphide (InP), but work on InP electronics remains in the early stages.

Lasers

Most interest in light sources concentrates on lasers, because they are fast enough to take advantage of the high speeds possible with integrated optoelectronics. However, lasers require optical feedback from each end, a need traditionally met by cleaved reflective end facets. That is a problem because semiconductor electronics are planar. Waveguides are hard to make in such chips, and cleaving the chip—the simplest way to form end facets—makes it too small for much integration. Some progress has been made in providing laser feedback in other ways, but integrated transmitters remain small devices in the laboratory and are limited to a few devices plus a laser light source.

Receivers

More progress has been made with integrating receivers than transmitters.

As mentioned in Chapter 7, some receivers already integrate the functions of detection and amplification, such as phototransistors, photodarlingtons, and integrated detector—amplifiers. These are simple devices designed to meet modest performance requirements at a low cost—particularly phototransistors and photodarlingtons. High-performance receivers are much harder to make, but they are easier than transmitters because they do not require waveguide structures or cleaved end facets. Material problems are much less because silicon detectors are usable at wavelengths to about 1000 nm, although other materials are needed at longer wavelengths.

Nonetheless, high-performance integrated receivers remain laboratory devices, with minimization of internal capacitance of the photodetector a critical requirement. The state of the art for receivers and transmitters is fabrication of comparatively simple integrated optoelectronic circuits operating in the gigabit-per-second range.

Future Uses

Near-term uses of integrated optoelectronics are likely to be in fast transmission over short distances at 800–900 nm.

Despite these problems, prospects look good for integrated optoelectronics. Near-term applications are likely to be in high-speed transmission with GaAs devices over short distances at 800–900 nm. Some prototypes have been tested. One transmitter sent a 400-Mbit/s non-return-to-zero signal through 4 km of 50/125 graded-index fiber, with a bit-error rate of 10^{-9}. Such high-speed transmitters could push local-area networks, intraoffice transmission, and cable-television systems to multigigabit-per-second speeds not possible with conventional transmitters using discrete devices. They also could be used in high-performance computers to transmit signals among integrated circuits and logic boards. Longer-wavelength systems, probably based on indium phosphide, could be used for long-distance telecommunications at 1300 or 1550 nm.

BISTABLE OPTICS

The past several years have seen the development of another new family of optical devices able to switch states and perform logic operations. These are called optically bistable devices because they have two stable states—one in which they transmit light well, and one in which they transmit light poorly.

The basic idea of a bistable optical device is shown in *Figure 12-8*, which plots light transmitted by a bistable device versus light entering it. In this example, if light intensity is low, only a small fraction of the light is transmitted. The fraction of light transmitted remains small until the input intensity passes a threshold value, above which it increases rapidly. The name "bistable" comes from the device's two stable states, which transmit different fractions of incident light. (In some bistable devices, transmission is higher at lower powers and drops as power increases.)

Figure 12-8. Relationship of Input Light to Light Transmitted by Bistable Optical Device

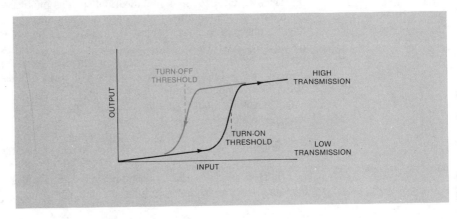

Hysteresis Loop

If the transmitted light intensity simply retraced its path as you reduced input intensity, the device would not be truly bistable (although it still might be useful for optical switching). In a true bistable device, the transmission remains high as the input power is reduced below the turn-on threshold value, until it finally drops when you reach a lower turn-off threshold. A plot of input versus output is not a simple curve; it contains a loop.

If you're familiar with electronics, this loop will look familiar. It's a hysteresis loop, a feature that appears in input/output plots for some circuits and components. The implications in optics are the same as they are in electronics. A device with such a loop can have two stable states at one input level, with depending on whether it approached that state from above or below on the input scale.

Such hysteresis loops allow special operations. Look again at the curve in *Figure 12-8*. Suppose you want to switch the bistable device from low to high transmission. You start by biasing it to an input level that falls within the loop. For an optical device, that means you deliver a certain steady level of light input. To switch to the more transparent state, you add enough extra switching light to pass the turn-on threshold. Turn off the switching light, and the steady optical bias will keep the device in the high-transmission mode. In electronic terms, you have latched into the on state.

To get down to the off level, reduce the optical input further, below the steep dropoff in transmission at the right of the optical hysteresis loop. In electronics, this isn't a big problem; it just takes a negative voltage pulse. In optics, it's more difficult because you don't have negative light, unless you can find a way to generate negative interference effects. That's tricky because it requires light waves that are coherent and precisely adjustable in phase. A more straightforward approach is to modulate the bias as well as the switching input. In this case, the bias light would be reduced below the turn-off threshold to shift to the low-transmission state.

If device threshold is at the right level, another alternative is to modulate two equal-intensity inputs. If both are off, total input is below the turn-off threshold. If only one is on, the input remains below the turn-on threshold. If both are turned on, the device switches on and remains on even if one is switched off. Such devices are useful in logic circuits.

Switching without Hysteresis Loops

Some bistable optical devices lack a true hysteresis loop.

Although the term "bistable optics" might imply the existence of a hysteresis loop, it has come to cover a broader class of devices in which the transmission changes abruptly with incident light intensity, but without an open hysteresis loop, as shown in *Figure 12-9*. These devices are bistable devices with optical parameters chosen to close the hysteresis loop. They can serve as switches because their input–output curves have a sharp turn-on like of a transistor.

**Figure 12-9.
Bistable Device with
Hysteresis Loop Closed**

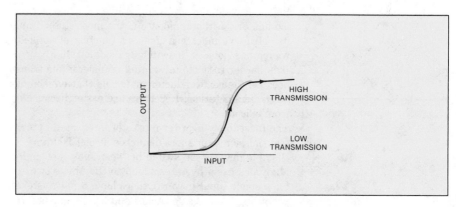

Bistable optical devices can be used like switching transistors.

This means that bistable optical devices can be used much like switching transistors. A small change in optical input, provided by a switching optical signal added to the optical bias level, turns on the device, much as a small signal to the base switches a transistor from non-conductive to highly conductive. Like transistors, bistable optical devices with closed hysteresis loops can be used for switching and logic.

That sounds like great news for the designers of fiber-optic transmission systems. Arrays of bistable devices could be turned into switching networks, working on purely optical signals, so there would be no need to convert signals from optical to electronic form and back for switching. Indeed, developers hope someday to do just that, but we aren't there yet. Bistable optics are still in early development. They are complex and difficult to make and often require high optical powers to operate. Many require light at wavelengths not available from present semiconductor lasers and LEDs. Some have output in a different form than their input—an important drawback because that means that the output of one stage cannot serve as the input to the next. Bistable optics is very much an infant technology.

WHAT HAVE WE LEARNED?

1. Simple integrated optoelectronics exist, including phototransistors, photodarlingtons, and detector-amplifiers, but high-performance integrated optics is a technology of the future.
2. Integrated-optics research includes work on planar waveguide devices, integrated-optoelectronic circuits, and bistable optics.
3. Integrated optoelectronics combine optical and electronic functions performed by transmitters and receivers.
4. The most common type of planar waveguide is the embedded waveguide, formed by diffusing a dopant into a substrate. The technology draws heavily on semiconductor electronics fabrication technology.
5. Light leaks out of a planar waveguide into the surrounding lower-index material so light can be coupled between two waveguides on the same substrate. The light energy switches back and forth as it travels along the waveguide.
6. Planar waveguides are being studied for use as couplers, switches, or modulators. Lithium niobate is the most-studied material, because of the way its refractive index changes as a function of applied electric field. That allows waveguide devices to switch or modulate light.
7. It is harder to make integrated optoelectronic transmitters than receivers because the semiconductor materials that emit light are not well developed as electronic materials. Receivers are easier because detectors can be made from silicon.
8. The most likely near-term uses of integrated optoelectronics are in fast transmission over short distances at 800–900 nm.
9. Bistable devices can switch between modes of high and low transmission. Many possess a hysteresis loop so they have two stable transmission states for a single input power. Others have a flat hysteresis loop. They are promising for use as switches and logic elements, but the technology is young and more work is required.

WHAT'S NEXT?

Now that we have learned about the major components of fiber-optic systems, Chapter 13 will look at how their properties are measured.

Quiz for Chapter 12

1. Light is confined in planar waveguides by:
   a. difference in refractive index between the waveguide and the surrounding material.
   b. cladding layer within the waveguide.
   c. reflective material on the waveguide walls.
   d. dopant layers at the boundaries of the waveguide.

2. Which is not a step in fabrication of an embedded planar waveguide?
   a. Deposition of photoresist.
   b. Formation of pattern on the photoresist.
   c. Removal of exposed photoresist to form a stripe.
   d. Diffusion of exposed photoresist into the substrate.

3. How is light transferred between two parallel planar waveguides that serve as a coupler?
   a. The waveguides must be in contact.
   b. By evanescent waves leaking between the waveguides.
   c. Through a region of material with refractive index between that of the waveguide and substrate.
   d. By reflection at the end of the waveguides.

4. What happens as the length of a parallel-waveguide coupler is increased?
   a. All the light is transferred from one waveguide to the other and remains in the waveguide that collected the energy.
   b. The division of light between the two waveguides becomes perfectly equal—50-50.
   c. Light oscillates back and forth between the waveguides so the amount of light in each waveguide varies with length.
   d. All the light is transferred back to the original waveguide where it remains.

5. Coupled parallel planar waveguides can be made to serve as switches by:
   a. no means known—the division of power is static.
   b. bending the waveguides back and forth with magnets.
   c. changing the amount of power transmitted through them.
   d. altering the electric fields applied to them.

6. What property of electro-optic materials is changed by the application of an electric field?
   a. Light absorption.
   b. Ability to function as waveguides.
   c. Internal reflections.
   d. Refractive index.

7. A waveguide modulator divides light into two outputs in a way that depends on the magnitude of the applied electric field and the degree of modulation. What happens to those two outputs?

a. One is transmitted to the outside world as the modulated beam; the other is thrown away.

b. The two are combined as the modulated beam.

c. One is subtracted from the other.

d. The way the two are combined is controlled by an applied magnetic field.

8. Which of the following is a major problem in making integrated optoelectronic transmitters?

a. It is hard to create waveguides in silicon.

b. Silicon devices do not emit light.

c. Electronic devices absorb the light emitted by lasers and LEDs.

d. Finding a suitable housing to package the semiconductor chips.

9. What is the best-developed material for integrated receivers?

a. Germanium.

b. Gallium arsenide.

c. Indium phosphide.

d. Silicon.

10. The most attractive potential uses of bistable optical devices are:

a. couplers.

b. modulators.

c. switches and logic devices.

d. transmitters and receivers.

Measurements

ABOUT THIS CHAPTER

Measurement capability is as vital for fiber optics as it is for other technologies. You can't blithely assume that manufacturers' data sheets are infallible. Many specify ranges of performance, and some measure characteristics in different ways. Even if everyone uses the same scales and measuring tools, components may be defective, damaged during installation, or fail during use. You must know properties of components when you're building a system, and you must be able to check those properties when installing or repairing it. You also must be able to check if the system works properly—and if it isn't, find out why so it can be repaired.

This chapter discusses the basics of optical and fiber-optic measurements, the quantities that are measured, and the capabilities of important measurement tools. We also will examine some pitfalls of fiber-optic measurements and discuss misconceptions that you might carry over from other fields. We emphasize optical measurements because they are what make fiber optics different from other fields.

ABOUT MEASUREMENT

Measurement terms and techniques must be defined carefully so everyone knows what's being measured.

You can't start making measurements by pulling out the fiber-optic version of a yardstick and measuring away. Proper measurements require careful definition and control so everyone knows exactly what's being measured and what it means. If you don't, you end up with units like the cubit. Though it recurs throughout the Bible and other literature of ancient times, nobody is quite sure how long a cubit was. In fact, scholars believe the cubit had many different definitions, just as the foot would if each of us defined the unit as the length of our own foot. To avoid such problems, specialists develop standard techniques and definitions so measurements made in one place yield the same results as those made elsewhere.

Specialists at organizations such as the National Bureau of Standards devote their lives to developing, perfecting, and standardizing measurement techniques. The Department of Defense and industrial groups such as the Electronic Industries Association and the American National Standards Institute have programs to establish standard measurement techniques. To outsiders, much of this work may seem like the splitting of technological hairs. Indeed, the level of detail involved is far more than you need to know at this stage of your involvement in fiber optics, and we won't delve into it that deeply in this chapter.

Nonetheless, such exhaustive standardization does have practical importance. Without precise and generally accepted measurement standards, you can't be sure what performance specifications and other quantities mean. For example, although the theoretical numerical aperture of multimode fibers does not depend on length, the actual numerical aperture does. For short lengths, the NA (which is calculated from angular measurements) may be significantly more than the theoretical value. That can have a major impact on system design. We will look at those details after first looking at the basic concepts behind light measurement.

BASICS OF OPTICAL MEASUREMENT

Fiber-optic measurements involve light and other quantities, such as the variation of light with time.

Most important fiber-optic measurements involve light, in the same way that important electronic measurements involve electric fields and currents. There are some exceptions, such as the length and diameter of optical fibers and cables, the sizes of other components, and the electrical characteristics of transmitter and receiver components. Because this is a book about fiber optics, we will only mention such measurements in passing. However, we will talk about measuring light in conjunction with other things because you cannot quantify the properties of optical fibers by considering only light. For example, to measure the dispersion of light pulses travelling through an optical fiber, you must observe how light intensity varies as a function of time, a task that requires time as well as light measurement.

When you're working with light, you need to know what can be measured. The most obvious quantity is optical power, which like electrical voltage is a fundamental measuring stick. However, that is rarely enough; it usually must be measured as a function of other things, such as time, position, and wavelength. Wavelength itself is important because optical properties of materials, light sources, and detectors all depend on wavelength. Other quantities that sometimes are important are the phase and polarization of the light wave. You need to learn a little more about these concepts before going into more detail on measurement types and procedures.

Optical Power

Defined

Specialized terminology makes fine distinctions about quantities related to optical power.

We have an intuitive feeling for the idea of optical power (measured in watts) as the intensity of light. However, a closer look shows that optical power and light intensity are rather complex quantities and that you need to be careful what you talk about. *Table 13-1* lists the most important quantities, which are described in more detail below.

**Table 13-1.
Measurable Quantities
Related to Optical
Power**

Quantity and Symbol	Meaning	Units
Energy (Q)	Amount of light energy	joules
Optical power (P or ϕ)	Flow of light energy past a point at a particular time (dQ/dt)	watts
Intensity (I)	Power per unit solid angle	watts per steradian
Irradiance (E)	Power incident per unit area	watts per square centimeter
Radiance (L)	Power per unit angle per unit projected area	watts per steradian-square meter
Average power	Power averaged over time	watts
Peak power	Peak power in a pulse	watts

Power measures the rate
of flow of light energy.

What we call power (ϕ) measures the rate at which electromagnetic waves transfer light energy. It is a function of time because this rate can vary with time. Mathematically, it is expressed as:

$$\phi = \frac{dQ}{dt}$$

or

$$POWER = \frac{d(ENERGY)}{d(TIME)}$$

Sometimes called radiant flux, this optical power is measured in watts, which is equivalent to joules (a measure of energy) per second. It is the same type of power that you measure electrically or thermally in watts. (Note, however, that the power ratings of light bulbs measure their input electrical power requirement—not the amount of light output, which is much lower.)

Like electrical power, optical power is a measurable manifestation of more fundamental quantities. In the case of electrical power, it is voltage (V) and current (I):

$$P = VI$$

or power = volts × amperes. This relationship can also take other forms, using the relationship V = IR (voltage = current × resistance):

$$\text{POWER} = I^2R = VI = \frac{V^2}{R}$$

As we saw earlier, light and other electromagnetic radiation can be described as a wave comprised of oscillating electric and magnetic fields. The wave has a characteristic amplitude A and oscillates with a particular period or frequency ν, as shown in *Figure 13-1*. The frequency, in turn, determines the wavelength λ, which equals the speed of light c divided by the frequency ν:

$$\lambda = \frac{c}{\nu}$$

**Figure 13-1.
Properties of an
Electromagnetic Wave**

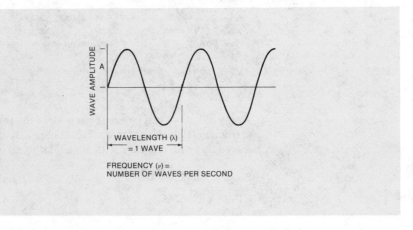

Compared to Electrical Power

Optical power is proportional to the square of the light wave amplitude and to the wave's frequency.

For light and other forms of electromagnetic radiation, the power is proportional to the square of the electromagnetic wave amplitude A at a given wavelength. Power also is proportional to the oscillation frequency of the electromagnetic wave ν. Thus in physical terms, power in a light wave is proportional to:

- The square of the field amplitude (A^2)
- The oscillation frequency of the electromagnetic wave ν
- The inverse of the wavelength $1/\lambda$.

The latter two relationships should look familiar. Earlier, when we discussed the basic nature of light, we mentioned that the energy in a photon increased with frequency and decreased with wavelength. The optical power measures the flow of photons (energy) per unit time and, thus, depends on photon energy. (Photon energy is given by Planck's law, $E = h\nu$, where h is called Planck's constant.)

Look above and you can see some similarities between optical and electrical power. Optical power is proportional to the square of electromagnetic wave amplitude, which measures the electric field in the wave. Electrical power is proportional to the square of the voltage across a resistance R. That similarity should not surprise you, because optical and electrical power are just different versions of the same thing.

There are important differences from a measurement standpoint. Electric field (voltage) and the flow of electrons (current) can be measured directly. However, there is no easy way to measure amplitude of an electromagnetic wave directly. Instead, optical detectors are best suited to measure power directly.

Electrical field and current can be measured directly, but in optics it is power that can be measured directly.

Measurement Quirks

Before we go deeper into measuring various forms of optical power, we warn you about a couple of potentially confusing measurement quirks. In electrical measurements, the decibel power ratio can be defined in terms of voltage and current. These are in the form:

The definition of the decibel power ratio is different when power and voltage are measured.

$$\text{POWER RATIO (dB)} = 20 \log \frac{V_1}{V_2}$$

or

$$\text{POWER RATIO (dB)} = 20 \log \frac{I_1}{I_2}$$

That definition differs from the optical definition of power ratio in decibels for powers P_1 and P_2:

$$\text{POWER RATIO (dB)} = 10 \log \frac{P_1}{P_2}$$

Why the different factor preceding the log of the ratio? Because electrical power is proportional to the square of voltage or current. In optics, we measure power directly. All three formulas are correct, but be careful which one you use. The formula that gives decibels in terms of measured power works both for light and electricity; the ones that give decibels in terms of voltage or current work only for electricity.

A second potentially confusing point is measurement of optical power in some peculiar-seeming units. Normally, power is measured in watts or one of the metric subdivisions of the watt—milliwatts, microwatts, or nanowatts. Sometimes, however, it is convenient to measure power in decibels to simplify calculations of power level using attenuation measured in decibels. That is a problem because the decibel is a dimensionless ratio, so it cannot measure power directly. However, power can be measured in decibels relative to a predefined power level. In fiber optics, the usual choices are decibels relative to one 1 mW (dBm) or to 1 μW (dBμ). Negative numbers mean powers below the reference level; positive numbers mean higher powers. Thus, +10 dBm means 10 mW, while −10 dBm means 0.1 mW.

Optical power can be measured in decibels relative to one milliwatt (dBm) or one microwatt (dBμ.)

Such measurements will come in very handy when we talk about system design in Chapter 14. Suppose, for instance, that you start with a 1-mW LED, lose 3 dB coupling its output into a fiber, lose another 10 dB in the fiber, and lose 2 dB in each of three connectors. You can calculate that simply by converting 1 mW to 0 dBm and subtracting the losses:

INITIAL POWER	0. dBm
Fiber coupling loss	−3. dB
Fiber loss	−10. dB
Connector loss	−6. dB
FINAL POWER	−19. dBm

This ease of calculation and comparison is the real virtue of the decibel-based units.

Types of Power Measurement

Optical power can be measured in terms of spatial or angular distribution, but for fiber optics the most relevant value is total power, which is measured by light detectors.

As *Table 13-1* indicates, optical power can be measured not just by itself but also in terms of its distribution in angle or space. In many cases (e.g., measuring the brightness of illumination), it is important to know not just total power but also power per unit area. The main concern of fiber-optic measurements is with total power (in the fiber or emerging from it) or power as a function of time, but you should be aware of other light-measurement units to make sure you know what you're measuring.

Light Detectors

Light detectors measure total power incident on their active (light-sensitive) area, a value often given on data sheets. This quantity is the product of active area times the power per unit area (assuming light intensity and detector response are uniform). It also is the quantity sought in most fiber-optic applications. Fortunately, the light-carrying cores of most fibers are smaller than the active areas of most detectors. As long as the fiber is close enough to the detector and the detector's active area is large enough, virtually all the light will reach the active region and generate an electrical output signal.

Irradiance and Intensity

Irradiance E is power per unit area. Intensity I is power per unit solid angle.

Things are more complicated in measuring optical power distributed over a large area, which may not all be collected by the detector. Then another parameter becomes important, irradiance E, the power density per unit area (e.g., watts per square centimeter). You cannot assume that irradiance is evenly distributed over a given area unless the light source meets certain conditions (e.g., that it is a distant point source such as the sun and that the entire area is at the same angle relative to the source). Total power P from a light source is the irradiance E collected over area A. This can be expressed as an integral:

$$P = \int E \, dA$$

taken over the entire illuminated surface. If the irradiance E is uniform over the entire area, this becomes:

$$P = EA$$

where A is the area.

The term "intensity" (I) has a special meaning in light measurement—the power per unit solid angle (steradian) with the light source defined as the center of the solid angle. This is a measure of how rapidly light is spreading out from its source. Unfortunately, the term is often misused (used in place of "irradiance") to indicate power per unit area so it is wise to be certain what is meant.

Peak and Average Power

When output of a light source or optical fiber varies with time, measured power can differ with time. Power is an instantaneous measurement. The highest power level reached in an optical pulse is called the peak power. The average level of optical power received over a comparatively long period (say over a minute), is average power. For digitally modulated fiber optic light sources operating at a 50% duty cycle (i.e., on half the time), average power is half the peak power. However, the peak power may be thousands of times the average power for a laser that emits only a few very short (but quite powerful) pulses per second. Most meters made specifically for fiber optics measure average power.

Power is an instantaneous measurement that varies with time.

Energy

At the beginning of this section, we saw that power measures the flow of energy (which in light measurement is symbolized by the letter Q, not E). Most fiber-optic measurements are of power rather than energy, but you should recognize the relationship of the two quantities. Energy can be measured in joules or (equivalently) watt-seconds. Total energy Q delivered is the integral of power P over time:

Energy of a pulse is the product of average power in the pulse times the pulse duration.

$$Q = \int P(t)\, dt$$

where P(t) is in general a function that varies with time. If the average power over an interval t is P, this can be simplified to:

$$Q = P \times t$$

Thus, average energy in a pulse that lasts t seconds is P watts; the pulse energy is $Q = Pt$ joules.

As this formula indicates, each light pulse contains an amount of energy Q. If you go deeply into communication theory, you will find that the ultimate limits on detection of pulses and communication capacity are stated as minimum pulse energy required to deliver a bit of information.

Energy per pulse is measured in some experiments with ultra-high-performance laboratory systems. However, in practical fiber-optic systems, the emphasis is on measuring power.

Radiometry and Photometry

Radiometry measures light in the whole electromagnetic spectrum. Photometry measures only light visible to the human eye.

Fiber-optic measurements are only a small part of light measurement. We have only skimmed the surface of that broader field because not much of it is critical if you're only working with fiber optics. However, you should appreciate the critical difference between the related fields of radiometry and photometry.

Radiometry and photometry are often used interchangeably, but they are not synonyms. Strictly speaking, photometry is the science of measuring light visible to the human eye. If the light isn't visible by the human eye, it doesn't count in photometric measurements. Radiometry, in contrast, is measurement of light in the whole electromagnetic spectrum, whether or not people can see it. Radiometry measures power in watts. The corresponding unit in photometry (power visible to the human eye) is the lumen. Radiometric measurements give equal weight to light at visible and invisible wavelengths. Photometric measurements weigh the contributions of different wavelengths according to eye sensitivity. Thus, 550-nm light, where the eye is most sensitive, counts much more on a photometric scale than light of equal power at a wavelength where the eye is less sensitive (e.g., 450 nm in the blue). Photometry ignores invisible ultraviolet and infrared light.

Fiber-optic measurements are made on a radiometric scale. The only time to use a photometric scale is in measuring light visible to the human eye. Some people in the industry are inexcusably sloppy in their terminology and call instruments photometers even though they measure radiometric units. Perhaps they'll learn better if you insist you want only radiometers. (A radiometer–photometer is a common instrument calibrated in both radiometric and photometric units.)

Most fiber-optic power meters are calibrated specifically for certain wavelengths used in fiber systems. Power meters measure average power.

Finally, two important notes about how the real world works. One is that strictly speaking most fiber-optic power meters are not radiometers because they are calibrated specifically for certain wavelengths used in fiber systems (e.g., 1300 nm). The other is that fiber-optic power meters, radiometers, and photometers all measure average power. Their response times are slow, and analog and digital meter displays cannot track instantaneous power fluctuations.

Wavelength

Wavelength is important because it affects the interaction between light and matter, including detector response. It can be measured directly.

The major importance of wavelength is not its subtle effect on power as described above but its more visible effects on the interaction between light and matter. Because detector response depends on wavelength, you need to know wavelength to calibrate power measurements (except with a radiometer that has uniform spectral response over a wide range). Wavelength also is important in its own right because of its effect on other interactions (e.g., light transmission by fibers).

As described in Chapter 4, fiber loss depends on wavelength, as does that of all other materials, including air and water. The difference between opaque and transparent materials is quantitative rather than qualitative. Transparent materials absorb little visible light. Opaque materials absorb visible light strongly. At non-visible wavelengths, materials transparent to visible light (e.g., air), become opaque, and some opaque materials become transparent.

Thus, wavelength must be known to characterize light waves. The most straightforward ways to determine wavelength are by measuring the angle at which light is bent or scattered by prisms or diffraction gratings because that angle depends on wavelength, as shown in *Figure 13-2*. Prisms refract light of different wavelengths at different angles because the refractive index of glass is a function of wavelength. Diffraction gratings scatter light at different angles depending on its wavelength and the spacing of lines on the grating. The most precise measurements are made by comparing the light being studied with light of a known wavelength.

**Figure 13-2.
Prisms and Diffraction
Gratings Spread out
Light by Wavelength**

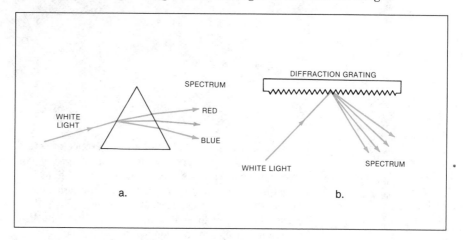

Phase

Phase measures a light wave's progress in its oscillation cycle.

Like other types of waves, electromagnetic waves have a property called phase. Envision a light wave as a sine wave, which goes through a cycle of 360° or 2π radians before repeating itself, as shown in *Figure 13-3*. The phase of a light wave is a measure of its progress in its oscillation cycle.

Phase of a light wave can be measured only by comparing it to other light waves of the same wavelength. This measurement relies on the interference of two waves, a phenomenon shown in *Figure 13-3*. When two light waves come together at a point, their net amplitude is the sum of their amplitudes, which can be either positive or negative. If the peak of one wave arrives at the same time as the valley of an equal-amplitude wave, the two cancel and the net amplitude is zero. However, if the peaks arrive at the same time, the two waves add.

**Figure 13-3.
Phase and Interference
of Light Waves**

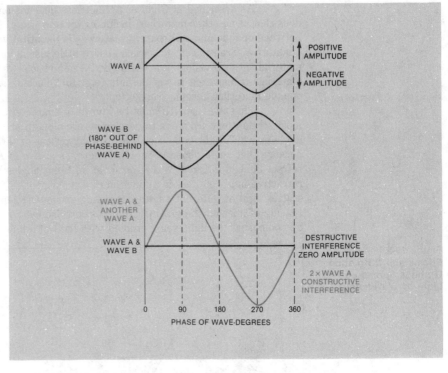

In the real world, subtle interference effects normally are overwhelmed by the superposition of many light waves with different wavelengths and phases. They can show up if light is coherent (i.e., in phase and of the same wavelength) like the light emitted by a laser. Phase effects are of little practical importance in today's fiber-optic systems because laser light loses its coherence in fibers, but they may be more important in coherent communication systems.

Polarization

Polarization is the alignment of the electric fields that make up light waves.

Like phase, polarization is a property of light waves that only rarely affects fiber optics. Earlier, we saw that a light wave is made of electric and magnetic fields oscillating perpendicular to each other and to the direction the wave is travelling. Polarization is the alignment of these fields, specifically of the electric field. (Because the electric and magnetic fields are always perpendicular, if one is aligned, both are.) If light is made up of waves with their electric fields all in the same plane, it is linearly polarized. If the field direction changes regularly along the light wave, it is elliptically or circularly polarized. And if the light is made up of waves with electric fields not aligned with each other, it is unpolarized.

You measure polarization by passing light through a polarizer, which transmits light only if its electric field is aligned in a particular direction. The fraction of light transmitted by the polarizer indicates the degree of polarization in the direction of the polarizer.

As we saw earlier, very few fibers preserve the polarization of light entering them. Polarization measurements can be valuable in those exceptional cases of polarization-maintaining and single-polarization fibers. They also can help assess performance of certain optical components (e.g., couplers). Polarization measurements could prove very important in advanced sensing and coherent communication systems that use polarization-preserving fiber, but now these uses are rare.

Timing

As we saw in Chapter 7, the electrical output from a detector reproduces the input optical signal. Monitoring and recording this electrical output makes it possible to measure the variation in an optical signal in time.

In practice, a detector does not respond instantaneously to a change in optical input. Each detector has its own characteristic response time, because it takes time for its output to rise and fall even in response to an instantaneous change in optical signal. Thus, the rise and fall times of detector output pulses depend both on rise and fall times of the optical pulse itself and on detector response time. Slow detectors thus cannot measure fast optical pulses.

Note that pulse repetition rates in even the slowest fiber-optic systems are extremely fast on a human scale—thousands per second—so variations in signal level are not seen in real time. They are recorded on an oscilloscope or other display in a way that lets people see events too fast for them to perceive otherwise.

Position

Optical power varies with position as well as time, and suitable optical detectors can sense this differential. The cells in the retina of your eye can tell light regions from dark spots—otherwise, you couldn't read this page. Retinal cells have small areas, giving your eye a very fine resolution. Electro-optical light detectors can be made with large or small active areas. The detectors used with fiber optics generally have a single active cell a millimeter or more across. Other detectors, used to record images, may have many discrete active cells, like the retina of your eye. Some special types have electrical output that varies with the position of a spot of light on the sensitive area.

As mentioned previously, a single-element detector measures the total power incident on its entire active area. It can measure the distribution of optical power by scanning an illuminated surface. Alternatively, a multi-element array of detectors can measure power distribution over its surface. Use of either technique allows measurement of light distribution, for example, as it emerges from an optical fiber. However, you should realize that assumptions made about the pattern of light distribution can impact the results. The coarser the resolution relative to the size of the measured structure, the larger the effect of the assumptions. And as you would expect, measurements with a detector with 500-μm active area can tell nothing about light distribution in a 50-μm fiber core.

QUANTITIES TO BE MEASURED IN FIBER OPTICS

Some vital characteristics of fiber-optic systems can be measured only indirectly by calculations based on more fundamental measurements.

Some fundamental optical quantities we have talked about so far in this chapter are themselves important in fiber-optic systems. Optical power level is one example. However, other factors important in fiber-optic systems (e.g., transmission bandwidth and numerical aperture) are measured only indirectly by measuring more fundamental quantities such as power distribution in time and space and inferring other characteristics from them.

Making such indirect measurements requires some implicit and explicit assumptions. Before we go into the details of those measurements, it's time for a homily about making unwarranted assumptions.

Measurement Assumptions

Unwarranted assumptions can cause you to discard essential data.

Assumptions can be dangerous in making measurements. You may think that they're absolutely essential to make a measurement task manageable. Unfortunately, their real achievement may be in oversimplifying the job by throwing away data that you should consider.

Modal distribution changes along the length of a multimode fiber.

One example is assumptions about light distribution in a fiber. It is reasonable to assume light is distributed in the same way along the length of a single-mode fiber, but not in a multimode fiber. Light may take a kilometer or more to distribute itself stably among the modes a multimode fiber can transmit, and that mode distribution can be rearranged by a splice or connector. This mode distribution affects many things, including loss within the fiber, numerical aperture and light-acceptance angle, transfer of light between fibers, and the distribution of light emerging from the fiber.

Loss of a connector or splice can depend on the direction light is travelling.

Another logical (but false) assumption is that loss of a connector or splice is the same for light going in either direction. Suppose that light distribution in the cores is uniform, and the only difference between fibers is that one has a 49 μm core and the other has one 51 μm across—within normal manufacturing tolerances. When light goes from the smaller fiber into the larger one, there are no geometrical losses caused by the difference in core size. However, if light was going in the opposite direction, the core-diameter difference would cause an added 0.3-dB loss.

Pitfalls become subtle in more sophisticated measurements. We cannot go through them all here, but the important point to remember is think measurement techniques through carefully and exercise care in making assumptions.

Optical Power

Measurements of optical power require knowing the wavelength and duty cycle.

Optical power is the quantity most often measured in fiber-optic systems. The power may be output from a light source, power emerging from a length of optical fiber, or power in some part of a system. The wavelength must be known to measure output power properly with a detector calibrated for that wavelength. Duty cycle—the fraction of the time the light source is on—should also be known to interpret properly measurements of average power. The usual assumption is 50% (half on, half off) for digital modulation, but under certain circumstances that may be far off (e.g., if a series of ones are being transmitted in NRZ code so the transmitter is continually sending at its high level).

A beam in the air can be measured directly by stopping it with a detector, or it can be sampled by taking a known fraction with a beam splitter. However, you can't directly measure the optical power within an optical fiber unless you cut into the fiber. You also can't transmit power after you've measured it—light that is measured is absorbed by a detector and is no longer available. Thus, to sample the power level in a transmitted signal, you need a beam splitter that will divert a small part of the light to a detector and transmit the rest.

Note the importance of making sure that the power level in a system is not too high as well as not too low. Recall from Chapter 7 that too much power can overload a receiver and cause distortion in analog systems (comparable to the effects of putting too much power into an audio speaker). In digital systems, similar effects occur but are manifested as an increase in bit-error rate.

Attenuation

Attenuation is −10 log(P_{out}/P_{in}).

Attenuation is the most important property of passive optical components because it determines what part of an optical signal is lost within the component and how much passes through. It is always a function of wavelength, although the wavelength sensitivity varies widely. In fibers, the variation with wavelength is large; in some other components it is negligibly small.

Attenuation measurements require comparison of input and output levels P_{in} and P_{out}, respectively. It is measured in decibels as:

$$\text{ATTENUATION (dB)} = -10 \log \frac{P_{out}}{P_{in}}$$

The negative sign is added to give attenuation a positive value because the output power is always lower than the input power for passive devices.

Precise measurements of fiber attenuation rely on cutting back fibers to compare power emerging from short and long lengths.

Precise fiber attenuation measurements are based on the cut-back technique shown in *Figure 13-4*. The power transmitted through the desired length of fiber (10 km in the figure) is measured first. Then the fiber is cut to a short length (about a meter), and the power emerging from that length is measured, using the same light source and meter. Taking the ratio of those power measurements eliminates input coupling losses (which occur in both measurements) while leaving the intrinsic transmission loss (which is only present in the long-fiber measurement). In the figure, the measured output power rose from −20 dBm to −10 dBm, indicating that fiber loss was 10 dB, or 1 dB/km.

**Figure 13-4.
Cut-Back Method of
Attenuation
Measurement**

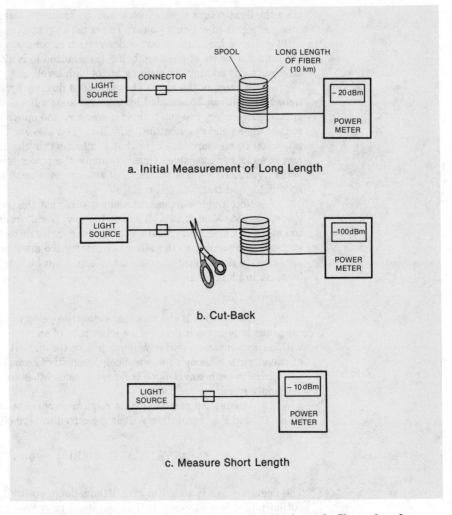

a. Initial Measurement of Long Length

b. Cut-Back

c. Measure Short Length

That method is much more accurate for single-mode fibers than for multimode fibers because of the mode-distribution changes we learned about earlier. Accurate measurement of true long-distance attenuation of multimode fibers requires insertion of a mode filter in the fiber segment that will remain after cutting back. This removes the high-order modes that gradually leak out of the fiber, leaving the lower-order modes that can travel long distances. Be aware, however, that this will not accurately measure the loss that light will experience in short multimode fibers because that loss depends on propagation of high-order modes that are eliminated from measurements by adding a mode filter.

Similar measurements can be made on cables with mounted connectors by replacing the short cut-back fiber segment with a short jumper cable (including a mode filter if desired). That approach simplifies measurements by avoiding the need to cut fibers at a modest sacrifice in accuracy.

One special problem with single-mode fibers is that light can propagate short distances in the cladding, throwing off measurement results by systematically underestimating input coupling losses. To measure true single-mode transmission and coupling, fiber lengths should be at least 20 or 30 m.

Continuity of Fiber

Checking fiber continuity is important in testing system function. The simplest test is shining a light through the fiber and monitoring the output.

A major concern in installing and maintaining fiber-optic cables is system continuity. If something has gone wrong with the system, you need to check whether the cable can transmit signals. If it can, you know you have another problem. If it can't, you need to find out where the break or discontinuity is. In some cases, the break may be obvious—a cable snapped by a falling tree limb or a hole dug by a careless contractor. However, such damage is not always obvious, and the cable route may not be readily accessible.

The simplest test of fiber continuity does not require elaborate equipment. A technician on one end shines a flashlight into the fiber, and one on the other end looks to see if any light emerges. That quick-and-dirty test can be checked by measuring cable attenuation. Sites of discontinuities can be located with optical time domain reflectometers, as described later in this chapter.

Dispersion

Pulse dispersion is determined by sending a short pulse down a fiber and measuring how long it is at the end of the fiber or system.

Pulse dispersion is an important characteristic of fibers that can limit bandwidth and data rate. As we saw in Chapter 4, it is the sum of modal dispersion (present only in multimode fibers), material dispersion (intrinsic to the material from which the fiber is made), and waveguide dispersion (dependent on refractive-index profile). All three quantities depend on wavelength, although the dependence is very weak and indirect for modal dispersion. Only material and waveguide dispersion are present in single-mode fibers, where they add to make chromatic dispersion. That quantity depends not merely on operating wavelength but also on the range of wavelengths being transmitted.

Sophisticated laboratory instruments can isolate the components of fiber dispersion. However, in practice, the main concern is total dispersion. That can be measured directly in the time domain by sending a short optical pulse down the fiber and measuring how much it has spread at the output, as shown in *Figure 13-5*. Note that it is important to know the central wavelength and the range of wavelengths being transmitted because of the way dispersion changes with wavelength. For multimode fibers, it is important to know modal distribution launched into the fiber.

**Figure 13-5.
Measuring Fiber
Dispersion**

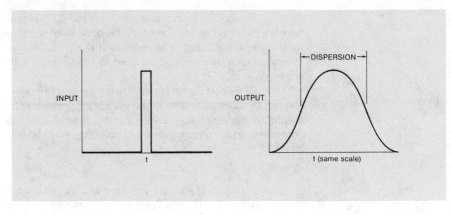

Bandwidth and Data Rate

Bandwidth in analog sys-
tems and data rate in digital systems are
essentially the inverse of
pulse dispersion in a fi-
ber. They can be mea-
sured indirectly as
dispersion or directly as
bandwidth or data rate.

Bandwidth in analog systems and data rate in digital systems are
essentially the inverse of pulse dispersion in a fiber. Although these
quantities can be measured indirectly as dispersion, they also can be
measured directly. Frequency measurements can be made by comparing
the strengths of signals at various frequencies. The result is a plot of signal
strength versus frequency, such as the one shown in *Figure 13-6*. This plot,
with level response over a broad range of frequencies, then a rapid decline
past a certain level, is characteristic of fiber-optic systems. Typically, the
upper limit on bandwidth is specified as the point at which signal strength
has dropped by 3 dB.

**Figure 13-6.
Plot of Received Signal
Strength as a Function
of Frequency**

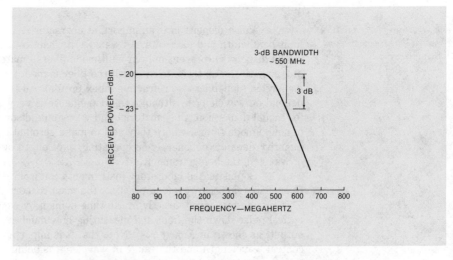

Maximum digital data rate can be extrapolated from analog
bandwidth or pulse dispersion characteristics, or it can be estimated in
other ways. One method is to study the shapes of output pulses in the eye

pattern described below. Another is to define the maximum bit rate as the highest possible with bit-error rate no more than a certain acceptable level, which is also described below.

Bit-Error Rate

Quality of digital transmission is assessed by measuring the fraction of bits received incorrectly, the bit-error rate.

The bit-error rate is a straightforward concept used in assessing the performance of many digital systems, including fiber-optic types. It involves generation of a randomized bit pattern and comparison of the signal emerging from the system being tested with the original pattern. Counting both total bits and the number of errors detected gives the bit-error rate—the fraction of bits received incorrectly.

As might be expected, the bit-error rate increases as received power drops, as well as when the system approaches other performance limits, such as maximum transmission rate. It provides a quantitative assessment of the performance of a digital communication system (or of other digital electronics). The upper bit-rate limit of a system can be defined as the speed at which bit-error rate exceeds a certain level, typically 10^{-9} (one bit in a billion) for a telephone voice transmission system, and 10^{-11} or 10^{-12} for data transmission.

Eye Pattern Analysis

An eye pattern is the superposition of the waveforms corresponding to a series of received bits. It indicates quality of digital transmission.

One of the most popular ways to assess performance of a digital fiber-optic link in the laboratory is by studying the eye pattern, shown schematically in *Figure 13-7*. To produce an eye pattern, an optical transmitter is hooked to a signal source (e.g., a generator of pseudorandom data strings), and the output of the receiver is displayed on an oscilloscope screen. The output of the signal source triggers the scope.

Figure 13-7.
An Eye Pattern
Schematic (*Courtesy Amp Inc.*)

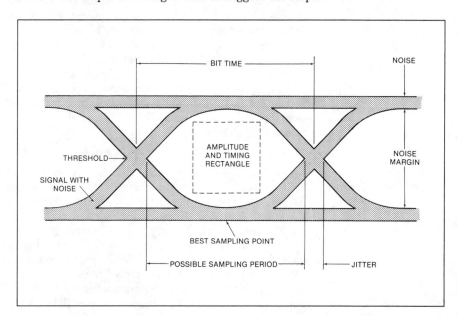

The display is the superposition of the waveforms corresponding to a series of received bits. The better the transmission quality and the more uniform the received signal, the more open the eye will appear. Signs of the eye closing indicate the likelihood of transmission errors as successive bits interfere with each other.

Careful interpretation of the eye pattern can yield important data on fiber link performance. Data signals can be sampled at any point within the eye, but the best point is in the middle. Some important points for interpreting eye patterns are:

- Height of the central eye opening measures noise margin in receiver output
- Width of the signal band at the corner of the middle of the eye measures the jitter (or variation in pulse timing) in the system
- Thickness of the signal line at top and bottom of the eye, which is proportional to noise and distortion in the receiver output
- Transitions between top and bottom of the eye pattern, which show the rise and fall times of the signal that can be measured on the eye pattern.

Mode Field and Core Diameter

Mode field diameter is the region occupied by light in a single-mode fiber.

As we saw in Chapter 4, fiber core diameter can vary because of manufacturing tolerances. In addition, mode field diameter—the diameter of the region occupied by light propagating in a single-mode fiber—is somewhat larger than the core diameter. These quantities can be measured.

Practical interest in the mode field and core diameters depend on the distribution of light, and measurements thus are based on light distribution. One approach is to scan across the end of the fiber with another fiber of known small core diameter, observing variations in light intensity transmitted by the scanning fiber. Other approaches rely on observing the spatial distribution of light near to or far from the fiber, the near-field and far-field intensity patterns. Those distributions of optical power can be used to calculate the core diameter.

A related quantity important for both single- and multimode fibers is the refractive-index profile, the change in refractive index with distance from the center of the fiber. This is measured in the same way as mode field and core diameter.

Numerical Aperture and Acceptance Angle

Numerical aperture is not measured directly; it is deduced from measurement of the acceptance angle.

As we learned in Chapter 4, the numerical aperture measures how light is collected by an optical fiber and how it spreads out after leaving the fiber. It measures angle, but not directly in degrees or radians. Although NA is widely used to characterize fiber, it is not NA that is measured but the fiber acceptance angle, from which NA can be deduced.

Numerical aperture and acceptance angle are most important for multimode fibers. As mentioned earlier, measured numerical aperture depends on how far light has travelled through the fiber because high-order modes gradually leak out as light passes through a fiber. The measured numerical aperture will be larger for shorter fibers, which carry a larger complement of high-order modes than it will be for long fiber segments. Measurements are made by observing the spread of light emerging from the fiber.

Cut-Off Wavelength

One important characteristic of single-mode fibers is cut-off wavelength, the wavelength at which the fiber begins to carry a second waveguide mode. Fibers have a theoretical cut-off wavelength calculated from their core diameter and refractive-index profile, but the measured effective cut-off wavelength is slightly different. As with core and mode-field diameter, cut-off wavelength is a laboratory rather than a field measurement.

Normally, the cut-off wavelength is measured by arranging the fiber in a test bed that bends the fiber a standard amount. Fiber attenuation as a function of wavelength is measured twice. First, the fiber is bent in a manner that causes the second-order mode to leak out almost completely. Second, the fiber is arranged so it transmits both first- and second-order modes. These two measurements are compared, giving a curve such as the one in *Figure 13-8*, which shows excess loss as a function of wavelength. In this case, λ_c is the effective cut-off wavelength, which is defined as the wavelength above which second-order mode power is at least a certain amount below the power in the fundamental mode. The measurement finds this value by locating the point where excess loss caused by stripping out the second-order mode is no more than 0.1 dB.

Figure 13-8. Measurement of Effective Cut-Off Wavelength *(Courtesy Douglas Franzen, National Bureau of Standards)*

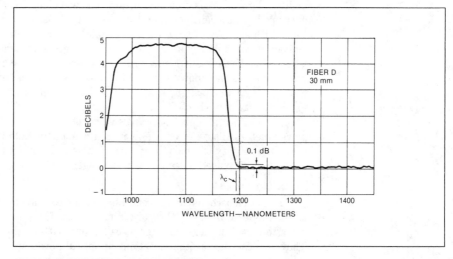

MEASUREMENT INSTRUMENTS

We started by describing fiber-optic measurements in general terms and by specifying the quantities that we needed to measure. Specialized instruments can perform these measurements easily. Many are analogous to instruments used for other optical or telecommunication systems. We'll look at the most important of these instruments below.

Optical Power Meters

The simplest optical measurement instruments are optical power meters, calibrated detectors with suitable electronics. Most are compact and portable, with digital readouts on both decibel and watt scales. Priced at a few hundred dollars and up, they are invaluable tools that can be adapted for many measurements.

Be careful in selecting power meters to note the type of detector used and the wavelengths for which it is calibrated. Many power meters incorporate internal photodetectors, which are usable at most common wavelengths, but some have silicon detectors intended only for use at wavelengths less than 1.1 μm. Some allow use of external detectors.

The most common wavelengths at which meters are calibrated are 665, 820, 850, 1300, and 1550 nm (the 665 nm is for plastic fibers). Some are calibrated for the 633-nm wavelength of helium–neon lasers and the 780-nm wavelength of some diode lasers. Some meters are calibrated at closely spaced points (e.g., every 10 nm) across a broad range of the spectrum. However, not all meters usable at multiple wavelengths are calibrated at all of those wavelengths.

Attenuation Meters

An optical power meter can be combined with a calibrated light source to serve as an optical attenuation meter or an optical loss test set. The power meter measures the drop in optical power from the level emitted by the source to the level received at the detector. By using multiple light sources emitting at different wavelengths, attenuation can be measured at different points in the spectrum. Most commercial instruments come with a choice of fixed-wavelength light sources, but versions that can emit light continuously adjustable in wavelength sometimes are used in the laboratory. These instruments measure attenuation by comparing measured power level to the known output of the light source and are calibrated for use with a particular light source. Optical continuity checkers are simplified versions that only check for the presence or absence of light.

Calibrated light sources also are sold separately for use with power meters. These often are more convenient for field measurements and come in a variety of types. Variables include laser or LED sources, continuous or pulsed output, and wavelength.

Oscilloscopes

Oscilloscopes are as useful in measuring waveforms transmitted through optical fibers as they are in other waveform measurements. They are often used in measuring the so-called eye pattern shown in *Figure 13-7*. They also can display the variations of optical power with time.

Bit-Error-Rate Meters

Specialized instruments are made for the bit-error-rate measurements we discussed earlier. Some models are portable; others are designed for laboratory use. They are generally similar to instruments made for use in non-optical systems, although they have been adapted for use specifically with fiber optics.

Optical Time-Domain Reflectometer

The optical time-domain reflectometer (OTDR) is one of the most powerful (and unfortunately, among the most costly) fiber-optic measurement instruments. It is a sort of optical radar that sends a short light pulse down a fiber and observes the small fraction of that light scattered back to it. An actual OTDR plot is shown in *Figure 13-9*. Such plots give signal strength as a function of distance down the fiber. The slope indicates the fiber loss. Discontinuities in the curve indicate positions of losses (e.g., connectors) and the size of the step indicates the amount of loss.

The optical time-domain reflectometer sends pulses down a fiber to spot discontinuities and indicate their magnitude. It also can measure attenuation from one end of a fiber, but the accuracy of OTDR measurements is controversial.

Figure 13-9.
Data Recorded by an OTDR *(Courtesy Photon Kinetics)*

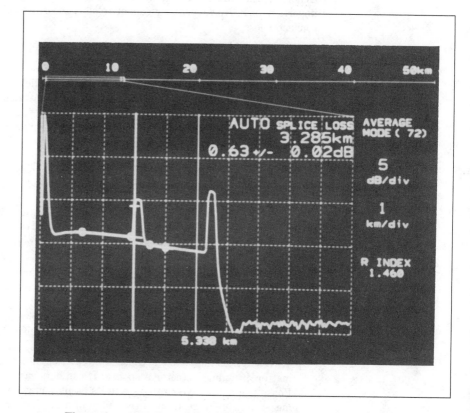

The major attraction of the OTDR is its convenience and ability to spot cable faults remotely. It requires access to only one end of the fiber in cable segments tens of kilometers long. Timing how long it takes light to travel from the instrument to a point in the fiber and back can locate flaws and junctions in the fiber. That's invaluable if you have to play detective and find where a fiber is broken in a long cable. Just plug your OTDR into one end of the cable and send a pulse of light down it. If you see a smoothly declining curve, the cable is okay. But if there is a large, sharp drop in the curve, the fiber is damaged. The instrument can locate a break to within a matter of meters, although it cannot spot breaks within a few meters of the instrument itself. Because their main use is in field measurements, OTDRs normally are packaged for field use.

The first OTDRs required enough knowledge of their operation to interpret analog waveform plots with more detail than the cleaned-up example shown in *Figure 13-9*. Newer versions can automatically locate splice and connector junctions and measure their attenuation. Measurements are made by averaging returns of many optical pulses, significantly improving signal-to-noise ratio and measurement quality.

Some controversy remains over accuracy of OTDR measurements. Generally, they are good enough to locate faults in the field, a very important task. However, they cannot reliably measure splice and connector loss.

Specialized Instruments

Most other measurements we've described above are typically performed in a laboratory setting rather than in the field. Unless you become heavily involved in fiber-optic research and development, you are unlikely to try to measure directly such quantities as spectral dispersion or characteristics of a fiber preform. Some special instruments are made for those tasks, but they are beyond the scope of this discussion.

WHAT HAVE WE LEARNED?

1. Precision is essential in defining what is to be measured so everyone is making the same measurements.
2. The basic quantity being measured in fiber optics is light. Measurements give information such as variations in optical power in time and space. Optical power is itself the rate of change in light energy with time.
3. Like electrical power, optical power is the square of a field amplitude, but the lightwave field (unlike voltage) is not easy to measure directly.
4. Optical power can be measured in decibels relative to 1 mW (dBm) or 1 μW (dBμ).
5. Pulse energy is the average power in the pulse times the length of the pulse.
6. Most fiber-optic power meters are calibrated specifically for certain wavelengths used in fiber systems. Power meters measure average power.
7. Wavelength is important because it affects the interaction between light and matter, including detector response. It can be measured directly.
8. Phase is a light wave's progress in its oscillation cycle of 360°. It can be measured only in relation to the phase of other light waves of the same wavelength.
9. Some vital characteristics of fiber-optic systems can be measured only indirectly by calculations based on more fundamental measurements.
10. Attenuation measurements require comparison of input and output power levels. The most accurate way to measure loss of lengths of fiber is to compare output power before and after cutting the fiber back to a short length.
11. Pulse dispersion is determined by sending a short pulse down a fiber and measuring how long it is at the end.

12. Bandwidth in analog systems and data rate in digital systems are essentially the inverse of pulse dispersion in a fiber. They can be measured indirectly as dispersion or directly as bandwidth or data rate.

13. An eye pattern is the superposition of the waveforms corresponding to a series of received bits. It indicates quality of digital transmission.

14. Numerical aperture is not measured directly; it is deduced from measurement of the acceptance angle.

15. Cut-off wavelength is measured by observing the wavelength above which power transmitted in the second waveguide mode is no more than a minimum level. This is done by measuring wavelength dependence of excess loss caused by stripping out the second-order mode.

16. Specialized instruments have been developed for the most common fiber-optic measurements.

17. Optical power meters are calibrated detectors with suitable electronics and readouts calibrated for particular wavelengths used in fiber-optic systems.

18. An attenuation meter is the combination of a light source and optical power meter calibrated for use together.

19. The optical time-domain reflectometer sends a short light pulse down a fiber and measures the light reflected back to it. In this way, it can spot discontinuities from one end of the fiber and indicate their magnitudes. It also can indicate fiber attenuation, but accuracy of OTDR measurements is controversial.

WHAT'S NEXT?

Now that you understand the nuts and bolts of fiber optics and the basics of measurement, Chapter 14 will give you a look at how systems are designed.

Quiz for Chapter 13

1. Optical power is:
   a. the light intensity per square centimeter.
   b. the flow of energy.
   c. a unique form of energy.
   d. a constant quantity for each light source.

2. Optical power is proportional to:
   a. the square of the electromagnetic field amplitude.
   b. the frequency of the electromagnetic wave.
   c. the inverse of the wavelength $1/\lambda$.
   d. all the above.
   e. none of the above.

3. Which measures power per unit area?
   a. Irradiance.
   b. Intensity.
   c. Average power.
   d. Radiant flux.

4. Which is not a directly measurable optical quantity?
   a. Wavelength.
   b. Field amplitude.
   c. Power variation in time.
   d. Phase of light waves.
   e. Polarization of light.

5. Attenuation is:
   a. the difference in input and output powers measured in milliwatts.
   b. the ratio of output power divided by input power (in decibels).
   c. the increase in power provided by a repeater.
   d. a comparison of power levels at different wavelengths.

6. The cut-back technique of fiber measurement is accurate because:
   a. fiber length can be measured precisely after a segment is cut off.
   b. it does not require connectors.
   c. the input-coupling effects are the same for the full fiber and the short segment remaining after the cut-back.
   d. it does not require splices.

7. Mode field diameter and refractive index profile can be measured by:
   a. observing the far-field intensity pattern produced by a fiber.
   b. timing pulse dispersion through a long length of the fiber being studied.
   c. calculations based on fiber length and attenuation measurements.
   d. measuring the variation in light intensity across the end of the fiber.

8. Numerical aperture is measured:
   a. directly.
   b. indirectly from the refractive index of core and cladding.
   c. indirectly by measuring acceptance angle.
   d. indirectly by measuring attenuation and modal dispersion.
   e. only in theoretical calculations.

9. Cut-off wavelength of a single-mode fiber is measured by:
   a. finding at what wavelength stripping out the second mode increases excess loss by no more than 0.1 dB.
   b. observing modal pattern emerging from the fiber as transmitted wavelength is changed.
   c. coupling light into only the fiber's second mode and detecting at what wavelength the fiber starts transmitting light.
   d. measuring power distribution in the fiber.

10. An optical time-domain reflectometer is most valuable for:
    a. precise characterization of connector loss.
    b. extending repeater spacing.
    c. measurement of fiber attenuation in the laboratory.
    d. location of cable faults in the field.

14

System Design Considerations

ABOUT THIS CHAPTER

Now that we've learned about the nuts and bolts of fiber optics, it's time to see how to put those components together to make systems. We will do this on two levels. In this chapter, we'll learn about general system concepts. In Chapters 15 through 20, we'll look at specific systems and applications for fiber optics.

On top of any list of fiber-optic system design considerations is the loss budget, making sure that enough signal reaches the receiver for the system to operate properly. Other considerations close behind in importance are designing systems that are cost-effective and that offer the required transmission capacity—whether in the form of analog bandwidth or digital bit rate. This involves looking at trade-offs for which there are rarely hard-and-fast rules, such as the choices between fibers of various core diameters and light sources of various power levels.

This chapter won't prepare you for heavy-duty system design. However, it will prepare you to evaluate system designs and develop system concepts that should work. It concentrates on simple designs to give you a clear idea of how systems work.

VARIABLES

Design of fiber-optic systems requires balancing many cost and performance goals.

Design of a fiber-optic system is a balancing act. You start with a set of performance requirements, such as sending 100 Mbit/s through a 5-km cable. You add some subsidiary goals, sometimes explicitly, sometimes implicitly. For example, you may demand cost as low as possible, less than another alternative or no more than a given amount. Your system might need an error rate of no more than 10^{-9}, for example. It should meet certain reliability standards, such as operating without interruption for at least five years.

You must look at each goal carefully to decide how much it is worth. Suppose, for instance, you decide that your system absolutely must operate 100% of the time. You're willing to pay premium prices for transmitters, receivers, and super-duper heavily armored absolutely gopher-proof cable. But how far should you go? If that is an absolute must because of national security and you have unlimited quantities of money, you might buy up the entire right of way, install the cable in ducts embedded in a meter of concrete, and post guards armed with tanks and bazookas to blow away anyone who comes near the cable with a backhoe. If its purpose is just to keep two corporate computers linked together, you might stop with the gopher-proof cable or with laying a redundant cable along a second route different enough from the first that nobody with a backhoe would knock out both.

That somewhat facetious example indicates how many variables can enter into system design. In this chapter, we will concentrate on the major goals of achieving specified transmission distance and data rate at reasonable cost. Many design variables enter into the equation. Among them are:

- Light source output power (into fiber)
- Coupling losses
- Spectral linewidth of the light source
- Response time of the light source and transmitter
- Signal coding
- Splice and connector loss
- Type of fiber (single- or multimode)
- Fiber attenuation and dispersion
- Fiber core diameter
- Fiber NA
- Operating wavelength
- Receiver sensitivity
- Bit-error rate or signal-to-noise ratio
- Receiver bandwidth
- System configuration
- Number of splices, couplers, and connectors
- Type of couplers (active versus passive)
- Costs.

Many of these variables are interrelated. For example, fiber attenuation and dispersion depend on operating wavelength as well as the fiber type. Coupling losses depend on factors such as fiber NA and core diameter. Some interrelationships limit the choices available (e.g., the need to achieve low fiber loss may require operation at 1300 or 1550 nm).

Some variables may not give you as many degrees of freedom as you might wish. For example, you may need to interconnect a dozen computer terminals. You have enough flexibility to pick any of the possible layouts in *Figure 14-1*, but you still have to connect all 12 terminals, and that requires enough optical power or active couplers to drive them all.

POWER BUDGETING

Usually the first task in designing a fiber-optic system is power budgeting. That is, you need to make sure that subtracting all of the system's optical losses from the power delivered by the transmitter leaves enough power to drive the receiver at the desired bit-error rate or signal-to-noise ratio. That design should leave some extra margin above the receiver's minimum requirements to allow for system degradation and fluctuations (e.g., degradation of a laser light source or addition of a splice to repair a broken cable).

**Figure 14-1.
Three Ways to
Interconnect a Dozen
Terminals**

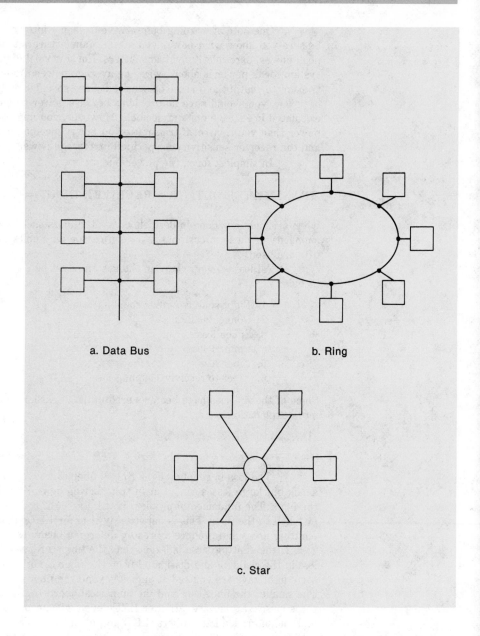

a. Data Bus

b. Ring

c. Star

One note of warning before we get deeply into power budgeting: be sure you know what power you mean. Manufacturers tend to specify peak power, particularly for transmitters. For a normal digital fiber-optic system, peak power is about twice the average power. As long as transmitter output and receiver sensitivity are specified as the same type of power, you should have no problems because power budgets are calculated in relative units of decibels. However, you might have 3 dB less power than you thought if transmitter output is specified in peak power and the receiver sensitivity is specified in average power.

In simplest form, the power budget is:

> The difference between transmitter output and receiver input equals the sum of system losses and margin.

$$\text{TRANSMITTER OUTPUT} - \text{RECEIVER INPUT} = \Sigma \text{ LOSSES} + \text{MARGIN}$$

when arithmetic is done in decibels or related units such as dBm. The simplicity of these calculations is a main reason why units such as dBm and dBμ are used.

All losses everywhere in the system must be considered. These include:

1. Light-source-to-fiber coupling loss
2. Connector loss
3. Splice loss
4. Coupler loss
5. Fiber loss
6. Fiber-to-receiver coupling loss

Some of these losses have been covered in detail earlier, but others deserve more explanation.

Light Source Coupling

Matching LEDs to Fibers

> Significant losses can occur in coupling light from sources into fibers.

The loss in coupling light from a fiber to a receiver typically is small, but large losses can occur in transferring light from the source into the fiber. The fundamental problem is that the light sources often are not matched to the fiber. This is particularly true for LEDs that have large emitting areas and produce a broadly diverging beam, as shown in *Figure 14-2*. If the emitting area is larger than the fiber core, some emitted light is lost in the cladding and dissipated from the fiber. Another problem is that some light rays are emitted at angles beyond the fiber's acceptance angle. The smaller the fiber core and the numerical aperture, the more severe these losses. Losses are large for LEDs even when coupled to 62.5/125 multimode graded-index fibers. LED output in the 1-mW range can be reduced to about 50 μW—a 13-dB loss—by losses in getting the light into the fiber. Losses are even higher in coupling to single-mode fibers.

Figure 14-2.
Difficulty in Coupling
LED Output into a Fiber

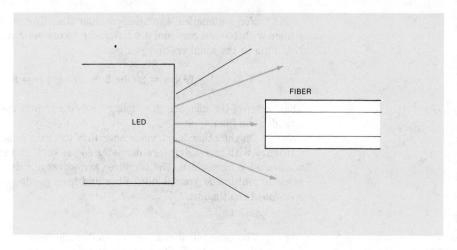

Laser Sources

A good LED can couple 50 μW into a 62.5/125 multimode fiber, but a good laser can transfer a couple of milliwatts into a single-mode fiber.

Semiconductor lasers deliver more power into optical fibers. Their advantages are smaller emitting area, smaller beam spread, and higher output power. Where a good LED might couple 50 μW into a 62.5/125 multimode fiber, a good semiconductor laser could transfer a couple of milliwatts into a single-mode fiber.

If this makes lasers seem like better light sources, it's largely because they are. Semiconductor lasers can be modulated faster and deliver more power into a fiber than an LED. However, cost counts in the real world, and LEDs are cheaper than lasers. They also last longer, and most LEDs don't require the cooling and stabilization equipment needed by laser transmitters—additional factors that help make them cheaper than lasers.

Fiber Choice

Increasing a fiber's core diameter and/or numerical aperture increases the amount of light it collects.

Another factor in this coupling equation is the choice of fiber. Looking at *Figure 14-2* shows that increasing the fiber's core diameter and/or numerical aperture should increase the amount of light it collects. The difference in coupling loss between two fibers—fiber 1 and fiber 2—is roughly:

$$\text{LOSS DIFFERENCE (dB)} = 20 \log \frac{D_1}{D_2} + 20 \log \frac{NA_1}{NA_2}$$

where the Ds are core diameters and the NAs are numerical apertures of the two fibers. This relationship is valid as long as the emitting area is larger than the fiber core, and no optics are used to change effective size of the emitting area or the effective NA of the emitter.

For a numerical example, consider the differences in coupling into a fiber with 100-μm core and 0.3 NA and into one with a 50-μm core and 0.2 NA. Plug in the numbers and you get:

$$\text{LOSS} = 20 \log 2 + 20 \log 1.5 = 9.6 \text{ dB}$$

That factor-of-ten increase in coupling loss clearly indicates the need to consider the fiber type.

On the other hand, remember that the gain in light-collection efficiency with increasing core diameter comes only at a sacrifice in transmission bandwidth. The sacrifices are worst in moving from single-mode to multimode graded-index fiber and from graded-index multimode to step-index multimode.

Single-Mode Fibers

Loss in coupling light from an LED into a single-mode fiber is about 19 dB higher than coupling into a 50/125 fiber.

The example above was for multimode fibers. Carry it a step further to a single-mode fiber with a nominal core diameter of 10 μm and NA of 0.11, and you will immediately see a big problem. The loss in coupling into such a single-mode fiber is:

$$\text{LOSS} = 20 \log 5 + 20 \log 1.8 = 19.2 \text{ dB}$$

above the loss in coupling LED output into 50/125-μm fiber.

If your first impulse is to say, "Forget it," you are in good company. For many years, developers did just that. However, there are ways to ease that coupling problem. Instead of just butting the fiber end against the LED or aiming LED output in the general direction of the fiber end, developers can focus the light onto the fiber end, with tiny optics. Losses remain significant, but they are far below the exceedingly high levels predicted by our simple cookbook formula. Such developments open up the possibility of single-mode fiber systems with LED transmitters.

Laser output couples more efficiently into fibers because of the smaller emitting area and beam spread.

What about coupling diode laser output into fibers? As we implied above, it is a much more efficient process than LED coupling. Lasers emit light from a smaller spot (typically smaller across than the core of a single-mode fiber), and laser light does not spread out as fast as that from an LED. As shown in *Figure 14-3*, this makes coupling much simpler. In addition, lasers emit higher powers, so they can deliver a milliwatt or more into a fiber. Thus, a laser is the obvious choice if lots of power is needed because of high system losses (e.g., from transmission over long distances or because of the need to distribute signals to many terminals).

**Figure 14-3.
Laser Output Is Easier
to Couple into a Fiber**

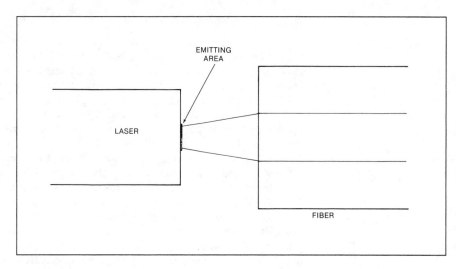

Fiber Loss

Fiber loss roughly equals attenuation multiplied by transmission distance. However, transient loss of 1–1.5 dB occurs when large-area LEDs excite high-order modes that leak out in the first few hundred meters of multimode fibers.

    The simplest approximation to loss in an optical fiber is to multiply the attenuation (in decibels per kilometer) by the transmission distance. However, for multimode fibers and LED sources, this is only an approximation. An LED with emitting area and numerical aperture larger than the fiber core excites high-order modes that leak out as they travel along the fiber. This transient loss runs about 1 or 1.5 dB and is concentrated in the first few hundred meters of fiber following the excitation source.

    This means that for the first few hundred meters fiber loss is higher than calculated by multiplying the specified loss by distance travelled. Over a kilometer or so (depending on how the specified loss was derived), the actual loss becomes roughly equal to the calculated loss. If signals are sent longer distances through multimode fibers uninterrupted by connectors, the actual loss may be lower than the calculated value.

Fiber-to-Receiver Coupling

Losses normally are small in transferring light from fibers to receivers.

    One of the rare places where the fiber-optic engineer wins is in coupling light from a fiber to a detector or receiver. The light-sensitive areas of most detectors are larger than most fiber cores, and their acceptance angles are larger than those of multimode fibers. Of course, if you're determined to screw things up, you can find a detector with a light-collecting area smaller than the core of large-core multimode fibers. However, with minimal care, you shouldn't run into this problem.

Receiver Sensitivity

In much of the discussion that follows, receiver sensitivity is taken as a given. That is, we assume a receiver must have a minimum power input to work properly. Things aren't quite that simple because there are trade-offs between received power, speed, and bit-error rate or signal-to-noise ratio. These are indicated for a digital receiver in *Figure 7-3*. As data rate increases, a receiver needs more input power to operate with a specified bit-error rate. If the data rate is held fixed but the input power is decreased, the error rate can increase steeply.

These trade-offs are not always useful. Error rate can increase steeply as input power decreases. At the margin of receiver sensitivity, a 1-dB drop in power can increase error rate by a factor of 1000 or more! You can gain more by lowering data rate, particularly near the receiver's maximum speed. However, many system designs do not allow much flexibility in transmission speed. As we will describe later in this chapter, some gains in sensitivity are possible by switching bit encoding schemes, but the simplest course may be using a more sensitive receiver.

Other Losses

Splices, connectors, and couplers can contribute significant losses in a fiber-optic system. Fortunately, those losses generally are easy to measure and calculate. Connectors, couplers, and splices have characteristic losses that you can multiply by the number in a system to estimate total loss. If connector loss is 0.7 dB, the total loss of four connectors is about 2.8 dB. However, there are two potential complications.

One is the variability of loss, particularly for connectors and splices. A given connector may be specified as having maximum loss of 1.0 dB and typical loss of 0.6 dB. The maximum is the specified upper limit for that type of connector; no higher losses should show up in your system. The typical value is an average, meaning that average connector loss should be 0.6 dB but that individual connectors may be higher.

If a system contains four such connectors, its total connector loss can be calculated in two ways. The worst-case approach is to take the highest possible loss (1 dB) and multiply it by the number of connectors to get 4 dB. On the other hand, on the average the loss at each connector is 0.6 dB so the most likely (average) value of the losses of four connectors in series is 2.4 dB. The prudent approach for so few connectors is to take the worst-case value for attenuation, but in some cases with many connections (particularly with splices) it is more realistic to take the average loss.

Connectors can cause further complications for multimode fibers because of the transient losses mentioned when we discussed fiber loss. If they are in the first part of the fiber cable, near a light source, they may accelerate transient losses by effectively stripping away higher-order modes. However, once light has travelled far enough through the fiber to reach an equilibrium distribution of modes (a kilometer or so), a connector can redistribute some light to higher-order modes, which tend to leak out of the fiber—a milder form of transient losses than experienced with light sources.

These hard-to-quantify modal losses associated with connectors (and to a lesser degree, with splices) for multimode fibers generally fall under the heading of concatenation effects that arising from joining two separate fiber lengths. They add to the uncertainties of calculating system attenuation.

Margin

System margin is a safety factor to allow for aging of components and system modifications and repairs. Typical values are 3–10 dB.

One quantity that always figures in the loss budget is system margin, a safety factor for system designers. This allows for the inevitable uncertainties in counting losses. You cannot rigidly specify component performance; real devices operate over ranges. The effects of concatenation, putting lengths of fiber end-to-end, are very hard to calculate and predict and might vary with temperature and physical stresses applied to the fiber. Devices age, generally emitting less power and becoming less sensitive. Cables may be damaged or broken so the fibers in them must be spliced, adding to overall attenuation.

Adding these uncertainties and degradations can lead to trouble unless there is a margin of error to account for them. Depending on the application, the performance requirements, the cost, and the ease of repair, this may be 3–10 dB.

To see how loss budgeting works, let's step through the three simple examples shown in *Figures 14-4, 14-5,* and *14-6. Figure 14-4* (example A) shows a short system transmitting 20 Mbit/s between two points in a building. *Figure 14-5* (example B) shows a telephone system carrying 400 Mbit/s between two switching offices 40 km apart. *Figure 14-6* (example C) shows an intrabuilding network delivering 10-Mbit/s signals to ten terminals. These examples are arbitrary and were picked to be illustrative, not representative, of actual design practice. Note that, in considering only loss budgeting, you don't directly consider the system data rate. Actual system design must consider transmission speed, as will be described later in this chapter.

Example A

Transmission over short distances in a building can be dominated by connector losses if there are several connectors in the system.

In *Figure 14-4*, we only need to transmit signals through 200 m of fiber, but we want to use a network of fiber cable already installed in the building. That means that we must route the signal through patch panels with connectors. In the example, we have six connectors, three on each floor: one linking the terminal device to the cable network for that floor and two at the patch panel. The light source is an LED that couples −13 dBm into an 85/125 fiber with attenuation of 5 dB/km at the LED's 850-nm wavelength. The loss budget is as follows:

LED power to fiber	−13.0 dBm
−Connectors (6 @ 1.0 dB)	6.0 dB
−Fiber loss (200 m @ 5.0 dB/km)	1.0 dB
−System margin	10.0 dB
Required receiver sensitivity	−30.0 dBm

The calculation shows that the dominant loss is from the connectors. The fiber loss may underestimate transient loss, but the large system margin leaves room to accommodate a slightly higher level of fiber loss.

**Figure 14-4.
Example A**

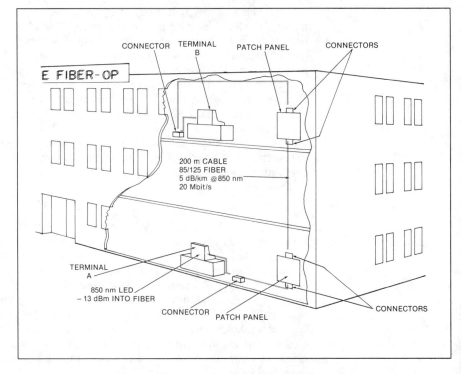

The calculated receiver sensitivity is a reasonable level, and system margin could be improved by picking a more sensitive receiver. This calculation started with a given loss, system margin, and input power, but it could be performed in different ways to extract other values. For example, you could calculate the input power required for a certain level of internal losses, receiver sensitivity, and system margin.

Example B

Losses in a 40-km telephone system are dominated by attenuation of the fiber, even though it is only one-tenth of the level in the short-distance system.

Loss sources in the telephone system in *Figure 14-5* are quite different, as shown by the following sample calculation:

Laser power into fiber	0.0 dBm
Fiber loss (1300 nm) (40 km × 0.5 db/km)	20.0 dB
Splice loss (20 × 0.1 dB)	2.0 dB
Connectors (2 @ 1.0 dB)	2.0 dB
Receiver sensitivity	−32.0 dBm
System margin	8.0 dB

The dominant loss is in the 40 km of fiber. Fiber attenuation is lower than in the first example because operation is at 1300 nm. Input power is higher because the long transmission distance justifies use of a more costly laser. Similarly, a more sensitive receiver can be cost-justified. The 20 splices join together segments of a buried or overhead cable. One connector (actually, a mated pair) is on each end, allowing connection of transmitter and receiver through patch panels. Adding a more sensitive receiver or a more powerful receiver could raise system margin.

Figure 14-5.
Example B

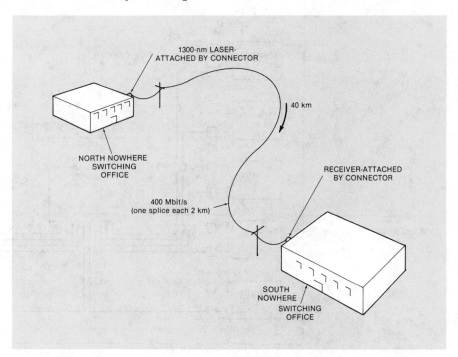

UNDERSTANDING FIBER OPTICS

Coupling losses are the
largest in a multiterminal
network.

Example C

In *Figure 14-6*, there is a different complication—the need to divide the light signal among ten separate receivers. We'll meet this need with a 10×10 directional star coupler, which divides the input signal from any of ten input fibers among ten output fibers. (For simplicity, not all the transmission lines to the star coupler are shown.) We will assume it has excess loss of 3 dB and divides the rest of the input signal equally among ten output ports, for total input-to-output attenuation of 13 dB. As shown in the calculation below, this one device dominates the loss budget:

LED transmitter (850 nm)	−13.0 dBm
Fiber loss (100 m @ 5.0 dB/km)	0.5 dB
Connector loss (2 @ 1.0 dB)	2.0 dB
Coupler loss (includes its own connectors)	13.0 dB
Receiver sensitivity	−30.0 dBm
System margin	1.5 dB

Figure 14-6.
Example C

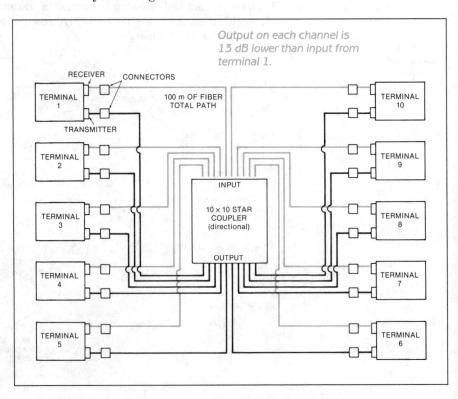

The calculated system margin is unacceptably low because the total system loss—15.5 dB—nearly equals the 17-dB differential between transmitter output and receiver sensitivity. This configuration should have a more sensitive receiver or a transmitter able to transfer more power into the fiber. Note that, in this case and in example A, reducing fiber attenuation to zero would make only a marginal improvement in the loss budget. A much larger improvement could come from using a larger-core fiber that could collect more light from the LED source. The largest losses are in other components—particularly in the star coupler in this example. Indeed, such problems, combined with the high cost of star couplers, have led developers to concentrate on other network architectures, as will be described in Chapter 18.

SIGNAL CODING SCHEMES

Signal coding schemes affect transmission capacity.

The other major consideration in fiber-optic system design is transmission capacity, bit rate, speed, or bandwidth. Those terms all have roughly the same meaning—how much information a fiber-optic system can carry per unit time. Before we take a close look at how to calculate a system bandwidth budget, we should look at signal coding schemes and how they affect system capacity.

Analog Intensity Modulation

Analog transmission relies on intensity modulation of the lightwave carrier.

Analog systems rely on simple intensity modulation of the lightwave carrier. That is, the transmitter output power is proportional to signal amplitude. This is the same as amplitude modulation in radio, and it works well for fiber transmission because there are few noise sources to cause the equivalent of AM radio static. In fact, there isn't much alternative. The high frequencies of light waves make phase and frequency modulation impractical for commercial use. Laboratory coherent fiber-optic systems are basically variants on amplitude-modulated transmission, which use a different decoding scheme. (To be more precise, they are variants of the heterodyne transmission scheme used to increase sensitivity and reduce noise in AM radios.)

Because practical analog systems rely on the same modulation scheme, bandwidth considerations are similar. The practical measure of bandwidth is the highest modulation frequency in the analog signal.

Digital Modulation

Series of digital bits can be encoded in several different ways.

Digital modulation is not simply off-on-off-on. There are several important digital modulation techniques, as shown in *Figure 14-7*. Each has its own distinct characteristics:

- NRZ (no-return-to-zero) Coding—Signal level is low for a 0 bit and high for a 1 bit and does not return to zero between successive 1 bits.
- RZ (return-to-zero) Coding—Signal level during the first half of a bit interval is low for a 0 bit and high for a 1 bit. Then, in the second half of the bit interval, it returns to zero for either a 0 or 1.
- Manchester Coding—Signal level always changes in the middle of a bit interval. For a 0 bit, the signal starts out low and changes to high. For a 1 bit, the signal starts out high and changes to low. This means that the signal level changes at the end of a bit interval only when two successive bits are identical (e.g., between two zeros).

- ● Miller Coding—For a 1 bit, the signal changes in the middle of a bit interval but not at the beginning or end. For a 0 bit, the signal level remains constant through a bit interval, changing at the end if it is followed by another zero but not if it is followed by a one.
- ● Biphase-m or Bifrequency Coding—For a 0 bit, the signal level changes at the start of an interval. For a 1 bit, the signal level changes at the start and at the middle of a bit interval.

Other coding schemes have been developed, but they are used only rarely in fiber-optic systems.

**Figure 14-7.
Digital Data Codes
Used in Fiber-Optic
Transmission**

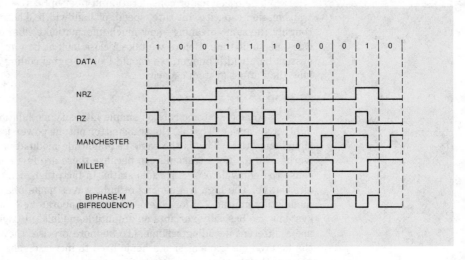

NRZ coding is the most common in fiber systems. Some other coding schemes require higher transmission capacity but are more resistant to errors.

NRZ coding probably is the most common in fiber systems, but each scheme has its advantages and disadvantages. Some, including RZ, Manchester, and bifrequency coding, can make two transitions during a bit interval. Thus, to carry 20 Mbit/s, they have to operate at 40 Mbit/s and need higher system capacity. Offsetting that requirement for higher capacity is improved performance. For instance, the frequently switched Manchester code generates its own clock, while NRZ-coded signals could suffer loss of timing or signal-level drift during a long string of 0 or 1 bits. In general, the more sophisticated codes are more robust and more resistant to transmission errors.

Because of their different natures, these codes are used in different applications. NRZ coding is often used in telephone circuits, where bit-error rates of 10^{-9} are acceptable and stretching transmission distance and speed are important goals. On the other hand, Manchester coding might be used for intrabuilding computer data transmission because the short distances leave plenty of extra bandwidth and bit-error rates generally must be in the 10^{-12} range.

BANDWIDTH BUDGET

Bandwidth budget calculations are not as straightforward as loss budgets, but some components, such as connectors and splices, can be ignored.

Calculating bandwidth budget is both simpler and more complex than calculating loss budget. The simplicity comes from being able to ignore components such as connectors and splices, which have no significant impact on system bandwidth. The complexity comes from the nature of the relationships that limit transmission speed, even after making some simplifying assumptions.

In our earlier description of system loss budget, there was no need to define systems as explicitly digital or analog. Both types of systems experience the same attenuation. Bandwidth and transmission speed considerations are somewhat different; it can make a difference whether the system is analog or digital, just as it matters what type of digital coding is used. For simplicity, we take digital NRZ coding in the examples that follow. While details may differ slightly for other types of systems, the principles are the same.

For a signal to be received correctly, the overall time response of a system must be less than the bit time.

One other initial assumption is needed to simplify calculations of bandwidth budget. We assume that we can calculate everything we need to know from time response to signal inputs without looking directly at frequency response. That is reasonable because there is a characteristic time per bit T at any transmission rate R:

$$T = \frac{1}{R}$$

where R is the speed in bits per second. Thus, the bit time is 1 ns at 1000 Mbit/s. (Roughly the same relationship exists for analog signals; there is a characteristic time that is the inverse of the frequency. Thus, a 1-ns time response corresponds to a 1-GHz analog signal bandwidth.)

Overall Time Response

For a signal to be received correctly, the overall time response of a system must be less than the bit time. Time response in this case means the longer of rise or fall time of the signal emerging from the system. If the time response is too long, successive pulses start overlapping and the system starts performing poorly. (The same principle applies to analog transmission.) Thus, a system that transmits 1,000,000 bit/s must have a time response faster than 1 μs (one-millionth of a second).

The overall time response of a system is the square root of the sum of the squares of response times of transmitter, receiver, and fiber.

The choice of time response simplifies calculations. The overall time response of a system can be calculated from the time responses of individual components using the formula:

$$t^2 = \Sigma t_i^2$$

where t is the overall time response and t_i is the time response for each component.

Connectors, splices, and couplers do not affect time response significantly. There are three components that do: the transmitter, the fiber, and the receiver. That means that the characteristic response time t of a fiber-optic system is:

$$t = \sqrt{t_{trans}^2 + t_{fiber}^2 + t_{rec}^2}$$

that is, the square root of the sum of the squares of the response times of the transmitter, receiver, and fiber.

Fiber Response Time

Fiber response time is
the square root of the
sum of the squares of mo-
dal and chromatic
dispersion.

Rise and fall times of transmitters and detectors are given on data sheets and are ready to plug into the formula. Fiber response time must be calculated for the length of fiber from the specified value. It is the sum of two components—modal dispersion and chromatic dispersion. The latter must itself be calculated using the sum-of-squares relationship. Modal dispersion requires knowing only the specified value and the length of fiber, but calculation of chromatic dispersion also requires knowing spectral width of the transmitter light source. In practice, chromatic dispersion can be significant even for multimode fibers, and you should not ignore it.

The best way to see what is involved in response time calculations is to look again at a couple of examples considered earlier in this chapter. In example A, we considered transmission of 20 Mbit/s through 200 m of 85/125 fiber using an 850-nm LED. A typical commercial 85/125 fiber has modal bandwidth of 200 MHz-km, which is equivalent to a modal dispersion of 5 ns/km. For a 200-m length, that corresponds to a 1-ns time response, according to the formula:

$$t_{modal} = DL$$

or

$$\text{TIME RESPONSE} = \text{MODAL DISPERSION} \times \text{LENGTH}$$

where t_{modal} is time response caused by modal dispersion, D is modal dispersion, and L is length.

To that must be added the effect of chromatic dispersion, $t_{chromatic}$, calculated from the formula:

$$t_{chromatic} = D_c \times \Delta\lambda \times L$$

or

$$\text{TIME RESPONSE} = \text{CHROMATIC DISPERSION} \times \Delta\lambda \times \text{LENGTH}$$

where D_c is chromatic dispersion, $\Delta\lambda$ is the range of wavelengths in the source output, and L is the length of the fiber. For a typical commercial 85/125 fiber, the chromatic dispersion is 110 ps/nm-km at 850 nm, which combined with a 50-nm linewidth for a typical 850-nm LED gives a time response of 1.1 ns. Thus, chromatic dispersion is larger than modal dispersion.

The two terms add together by the sum-of-squares formula:

$$t_{fiber} = \sqrt{t_{modal}^2 + t_{chromatic}^2}$$

giving a fiber time response of 1.5 ns.

Receiver Response Time

Combining this with the response times of a typical commercial LED transmitter (8 ns) and receiver (20 ns) gives an overall pulse response of about 22 ns—well within the range required for transmission at 20 Mbit/s (a 50-ns bit time). This example also indicates the typical case for short-distance transmission, where the prime limitation on transmission speed is receiver response time.

Matters are quite different for long-distance transmission. In example B, a 400-Mbit/s signal is sent through 40 km of single-mode fiber, using a 1300-nm semiconductor laser source. Because the fiber is single-mode, there is no modal dispersion. Although 1300 nm is the nominal wavelength for zero chromatic dispersion, do not blithely assume that the actual dispersion is precisely zero. Manufacturers specify a maximum value, typically about 3.5 ps/nm-km, reflecting tolerances inevitable in manufacturing. Combined with a typical laser spectral width of 3 nm, that gives total dispersion of 420 ps (0.4 ns) over 40 km. Typical rise and fall times for lasers and pin photodiodes at that wavelength are 0.5 ns. The resulting total response time is:

$$T = (0.5^2 + 0.5^2 + 0.4^2)^{0.5} = 0.8 \text{ ns}$$

which is more than adequate for transmission at 400 Mbit/s.

Importance of System Design

These examples may make fiber dispersion seem only a minor consideration. However, that impression is only an artifact of the examples chosen. Suppose that the first example involved transmission over 4 km instead of 200 m. The total fiber dispersion then would be 20 times greater (30 ns) larger than the receiver time response. Adding that together with receiver and transmitter response by the sum-of-squares method gives a total response time of 37 ns. The system would still operate (if it had enough power to go that distance), but it would be closer to its margins.

Dispersion becomes an even more serious problem when using certain fibers at certain wavelengths. Step-index multimode fibers with cores 100 μm or more in diameter have much higher modal dispersion than graded-index fibers. (However, both have comparable levels of chromatic dispersion.) Although some types have bandwidths in the 100 MHz-km range, other 100/140 fibers have bandwidths of 20 MHz-km, and some larger-core fibers have even smaller bandwidths.

Dispersion also is a serious problem in step-index single-mode fibers at the 1550-nm wavelength of minimum attenuation, as we saw in Chapter 4. Now that you have seen how total dispersion is calculated, you can understand the importance of developing light sources with very narrow linewidth and fibers with very low chromatic dispersion.

COST/PERFORMANCE TRADE-OFFS

Cost is a primary consideration in real-world system design. It is up to the user to make cost-performance trade-offs for specific applications.

So far, we have only mentioned in passing one of the most important considerations in real-world system design—cost. Minimizing cost is an implicit goal in all system design; a few guidelines for doing so follow. However, no book can give hard-and-fast rules for the tough job of making trade-offs between cost and performance. Ultimately, it is your judgment as a system user or designer whether pushing bit-error rate from 10^{-8} to 10^{-9} is worth an extra $1000. What I will do here is give you some examples and ideas to apply in working situations.

In earlier chapters, we have examined some trade-offs that can affect cost and performance. One is the choice of making connections with splices or couplers. Another is the choice between powerful, fast, and expensive semiconductor laser sources and cheaper and longer-lived LEDs.

There are some other trade-offs that we have skimmed over in describing the basic technologies. Examples include the lower transmission losses at 1300 nm versus the higher cost of transmitters and receivers for that wavelength and the relatively low marginal cost of adding extra fibers to heavy-duty cables. We will touch on many of these points in the simple examples that follow, then we'll collect some other points not covered in these examples as a set of rough-and-ready (but not necessarily complete) guidelines.

Examples

Importance of Light Coupling

For short-distance transmission, choice of a more-costly fiber with better transmitter-to-fiber coupling efficiency can allow use of a less-expensive transmitter.

Let's go back again to example A, this time making some more variations on the basic theme of transmitting 20 Mbit/s through 200 m of fiber. Consider the choice among three fiber types, with the following properties:

	Attenuation, dB/km	Coupling Loss, dB	Cost/m (of cable)
Type 1	5	−13	$0.50
Type 2	6	−10	$0.55
Type 3	7	−3	$0.60

If you look only at cost per meter and attenuation, Type 1 looks like the best choice. However, the much higher coupling loss (from a standard LED transmitter) means that transmission over 200 m will cause overall loss of 14 dB versus 11.2 for Type 2 and just 4.4 for Type 3. If the high coupling loss is tolerable in your design, Type 1 is the best choice—if you must use a particular transmitter. Suppose, however, that the lower overall fiber attenuation lets you use a $50 transmitter rather than a $100 model. Then you can save $50 on the transmitter by spending an added $20 (10 cents more per meter over 200 m) on cable.

Wavelength-Division Multiplexing

To take another variation on this example, let's consider the advantages of transmitting signals two ways (bidirectionally) through the same fiber. Suppose that wavelength-division-multiplexing couplers replace one connector on each end of the system and have the same signal attenuation as the connectors (because they split signals by wavelength rather than dividing them in half). Also assume that pairs of transmitters at two wavelengths cost as much as a pair at the same wavelength. That simplifies the cost trade-off to coupler versus cable. If you cut the number of fibers in a cable from two to one, you do not cut the cable cost in half—let's assume that you reduce cost from $0.60/m to $0.45/m. That means that over a 200-m distance, you can save $30 on cable by wavelength-division multiplexing so coupler price has to be under $15 each to justify paying for two couplers. If it isn't, you're better with two-fiber cable.

Of course, performance also can come into the equation. Suppose in this last example that you have to thread the cable through a space too tight for a two-fiber cable. Then you don't have much choice about using couplers.

Installation Costs

Installation costs also can enter the picture. Suppose in the last example that installation of two-fiber cable is possible but that it will take a technician a day (at a cost of $100) to do so. That means you can justify paying up to $130 for the pair of couplers.

All of these examples are simplified, but they show the basic idea of how costs can be calculated. Even more than system design, cost calculations and cost–performance analysis depend on the situation. Some guidelines follow, and the following chapters include some examples of how fiber-optic systems work. However, success ultimately depends on your ability to apply your basic knowledge of fiber optics, algebra, and costs to solve real-world problems.

Guidelines

Don't forget to apply
common sense in system
design. Labor is
never free.

We'll start our list of guidelines with a few common-sense rules:

- Your time is valuable. If you spend an entire day trying to save $5 on hardware, the result will be a net loss.
- Installation, assembly, operation, and support are not free. For a surprising number of fiber-optic systems, installation and maintenance costs are more than what you pay for the hardware. You may save money in the long term by paying extra for hardware that is easy to install and service. (People who have students or other unpaid peons available to do the work are rare exceptions to this rule—and even they should remember that they may have to clean up the mess if the students don't do it right.)
- It can cost less to pay an expert to do it than to learn how yourself. Unless you need to practice installing connectors, it's much easier to buy connectorized cables for your first fiber-optic system.
- You can save money by using standard mass-produced components, rather than designing special-purpose components optimized for a particular application.

You also should learn some basic cost trade-offs that people often face in designing fiber-optic systems.

- The performance of low-loss fiber, high-sensitivity detectors and powerful transmitters must be balanced against price advantages of lower-performance devices.
- Low-loss, high-bandwidth fibers generally accept less light than higher-loss, lower-bandwidth fibers. Over short distances, you can save money and overall attenuation by using a higher-loss, more-costly cable that collects light more efficiently from LEDs.
- The marginal costs of adding extra fibers to a cable are modest and much cheaper than installing a second parallel cable. However, if reliability is important, the extra cost of a second cable on a different route may be a worthwhile insurance premium.
- LEDs are much cheaper and longer-lived than lasers, but they produce much less power and are harder to couple to small-core fibers. Their broad range of wavelengths and their limited modulation speed can limit system bandwidth.
- Fiber attenuation contributes less to losses of short systems than losses in coupling light into and between fibers.
- Active couplers allow construction of larger networks than passive couplers because they can generate optical power, while passive couplers only consume and divide light signals.
- Topology of multiterminal networks can have a large impact on system requirements and cost because of their differences in coupler requirements.
- Light sources and detectors for 1300 nm are more expensive than those for the 800- to 900-nm window, although fiber and cable for the longer wavelength may be less expensive.

- Fiber and cable become a larger fraction of total cost—and have more impact on performance—the longer the system.
- Balance the advantages of eliminating extra components with the higher costs of the components needed to eliminate them. For example, as we showed above, it's hard to justify two-way transmission through a single fiber over short distances unless wavelength-division-multiplexing couplers are cheap, large installation savings are possible, or system requirements permit only a single fiber.
- In general, avoid repeaters because they require maintenance as well as capital expense.
- Think of future upgrade possibilities. Fiber-optic technology is still developing, and your own transmission requirements are likely to increase. (To paraphrase one of Parkinson's Laws, "Communication requirements expand to fill the available capacity.") If a small extra expenditure now can open the way for much larger capacity in the future, it's probably worthwhile. As we will see in Chapter 15, telephone companies routinely install cables with spare fibers to allow room for future expansion because that small investment in extra fibers can save the high costs of buying and installing additional cables later.
- Leave room for repair and expansion. Although fiber-optic cables rarely fail by themselves, as we learned in Chapter 5 they can fail with human or other intervention. A careless foot can pull a fiber out of a connector or an ice storm can take out overhead cables. It costs much less to allow for the possibility beforehand than to make up for it afterward. The same applies to leaving room for minor expansion (e.g., the addition of a few terminals to a network) without having to replace the system completely with a larger one.
- Remember coupling losses between transmitter and fiber and fiber and receiver. Some specification sheets list only transmitter output—not power actually coupled into an optical fiber. Coupling losses from fiber to receiver are smaller and, if properly designed, should be no more than those of a good connector pair.

As you grow more familiar with fiber optics, you will develop some of your own guidelines, based on your own experience. Indeed, many of the ideas above really are only the application of common sense.

WHAT HAVE WE LEARNED?

1. Design of fiber-optic systems requires balancing sometimes-conflicting performance goals as well as costs.
2. The system loss budget is calculated by adding all system losses and subtracting them from the transmitter output power. The result equals the minimum power required by the receiver plus the system margin.

3. Significant losses can occur in coupling light from sources into fibers. LEDs in particular often emit light over a broader area and in a broader angle than many fiber cores can accept. Semiconductor lasers transfer light into fibers more efficiently. A good LED can couple 50 μW into a 62.5/125 multimode fiber, but a good laser can transfer a couple of milliwatts into a single-mode fiber.

4. Coupling efficiency depends on the type of fiber as well as the source. The larger the core diameter and numerical aperture, the more light the fiber will collect from a large-area LED source. Lasers, with small emitting area and small beam spread, can transfer light efficiently to smaller-core fibers.

5. Fiber loss roughly equals attenuation multiplied by transmission distance. However, transient loss of 1–1.5 dB occurs when large-area LEDs excite high-order modes that leak out in the first few hundred meters of multimode fibers.

6. There are trade-offs among received power, speed, and bit-error rate or signal-to-noise ratio for receivers, but these often are not useful.

7. Total loss from connectors, couplers, and splices is their characteristic loss multiplied by the number in the system. This can be either the worst case, calculated by multiplying maximum loss by number, or the most likely, calculated by multiplying average loss by number of devices.

8. System margin is a safety factor to allow for aging of components and system modifications and repairs. Typical values are 3–10 dB.

9. Analog transmission relies on intensity or amplitude modulation of the light wave carrier from an LED or semiconductor laser.

10. Series of binary bits can be encoded in several different ways, including non-return-to-zero coding, return-to-zero coding, Manchester coding, Miller coding, or biphase-m coding.

11. Bandwidth budget calculations are not as straightforward as loss budgets, but some components, such as connectors and splices, can be ignored. These calculations are based on the system time response. For a signal to be received correctly, the overall time response of a system must be less than the bit time (i.e., less than the inverse of the bit rate or frequency).

12. Response time of a system is the square root of the sum of the squares of component response times. Calculations must include transmitter and receiver response times and both modal and chromatic dispersion of fibers.

13. Cost is a primary consideration in real-world system design. It is up to the user to make cost-performance trade-offs for specific applications.

14. Installation and operating costs must not be neglected. In some cases, installation costs may be much more than that of the hardware. Small added expenses in hardware may mean major savings in installation.

15. Common sense is vital in making cost–performance trade-offs. Always try to anticipate future events, ranging from expansion to ways to repair a system in case of failure.

WHAT'S NEXT?

In Chapter 15, we will look at the use of fiber optics in long-distance telecommunications systems.

Quiz for Chapter 14

1. A large-area LED couples 10 μW (10 dBμ) into an optical fiber with core diameter of 100 μm and numerical aperture of 0.30. What power should it couple into a fiber with 50-μm core and NA of 0.2?
   a. 10 dBμ.
   b. 9.5 dBμ.
   c. 3 dBμ.
   d. 1.0 dBμ.
   e. 0.4 dBμ.

2. A connector is specified as having loss of 0.6 dB $\pm$0.2 dB. What is the maximum connector loss in a system containing five such connectors?
   a. 0.6 dB.
   b. 3.0 dB.
   c. 4.0 dB.
   d. 5.0 dB.
   e. None of the above.

3. A 10-Mbit/s signal must be sent through a 100-m length of fiber with eight connectors to a receiver with sensitivity of -30 dBm. The fiber loss is 4 dB/km, and the average connector loss is 1.0 dB. If system margin is 5 dB, what is the minimum power that the light source must couple into the fiber?
   a. -13.0 dBm.
   b. -13.4 dBm.
   c. -16.0 dBm.
   d. -16.6 dBm.
   e. -20.0 dBm.

4. A telephone system is designed to transmit 90 Mbit/s through 40 km of cable with attenuation of 0.5 dB/km. The system contains two connectors with 1.5-dB loss, a laser source that couples 0 dBm into the fiber, and a receiver with sensitivity of -34 dBm. How many splices with average loss of 0.15 dB can the system contain if the system margin must be at least 8 dB?
   a. None.
   b. 10.
   c. 20.
   d. 30.
   e. 40.
   f. None of the above.

5. A fiber-optic network uses a transmissive star coupler with excess loss of 1 dB to distribute signals to 30 terminals. The signals must travel through 200 m of fiber with 5-dB/km loss and through four connectors with 0.6-dB loss. If receiver sensitivity is -30 dBm and system margin 5 dB, how much power must the light source couple into the fiber?
   a. $+9.4$ dBm.
   b. 0.0 dBm.
   c. -5.8 dBm.
   d. -24.2 dBm.
   e. None of the above.

6. The time response of a system transmitting a 400-Mbit/s signal must be less than:
   a. 1 ns.
   b. 2 ns.
   c. 2.5 ns.
   d. 4 ns.
   e. none of the above.

7. Response time of a fiber-optic system is:
   a. the square root of the sum of the squares of the response times of components.
   b. the sum of the response times of components.
   c. the average of the response times of the components.
   d. not directly dependent on component response times.

8. What is the time response of a system using an 850-nm LED with 10-ns rise time to transmit though 2 km of 100/140 fiber with 50-ns/km modal dispersion to a receiver with 25-ns rise time? Neglect chromatic dispersion.
   a. 100 ns.
   b. 104 ns.
   c. 135 ns.
   d. 200 ns.
   e. None of the above.

9. Look again at the above example to calculate total chromatic dispersion along the system. Assume a value of chromatic dispersion of 110 ps/km-nm and an LED bandwidth of 40 nm. What is the chromatic dispersion?
   a. Under 1 ns.
   b. 4 ns.
   c. 5 ns.
   d. 8.8 ns.
   e. None of the above.

10. Your system design requires a transmitter that delivers −13 dBm into a fiber. You have a choice between two fibers—one that costs $0.50/m and requires a $100 transmitter and one that costs $0.60/m and can operate with a $30 transmitter. All other things being equal, using the more costly fiber will save money for distances shorter than:
    a. 70 m.
    b. 200 m.
    c. 500 m.
    d. 700 m.
    e. 1000 m.
    f. none of the above.

Long-Distance Telecommunication Applications

ABOUT THIS CHAPTER

Now that we have learned about fiber-optic hardware, it is time to look at its major system applications. This chapter concentrates on the most successful and widely accepted use of fiber optics—long-distance telecommunications transmission, largely (but not entirely) as part of the global telephone network. Fiber optics is so successful that some observers believe it threatens the role of communications satellites. Satellites will not become totally obsolete, but fibers appear better for many communication jobs.

In this chapter we will discuss the world of long-distance telecommunications and the role of fiber optics in that world. This includes both land-based systems and submarine cables, some of which will span the oceans. The growing uses of fibers in local telecommunications—the most important future trend—are the subject of Chapter 16. The division between the two fields is somewhat arbitrary—"long" is a relative term. For our purposes, we will define "long-distance systems" as those systems that operate at high speeds between major population centers—for instance, from Boston to New York not between New York and its suburbs. These transmission systems form the backbones of national and international telecommunication networks.

BASIC TELECOMMUNICATIONS CONCEPTS

Telecommunications

Telecommunications is a broad term, encompassing telephony, telegraphy, and other forms of electronic (and optical) communications over long distances.

Before looking at how optical fibers are used in long-haul telecommunications, I want to clarify what we're talking about. The term "telecommunications" is deliberately broad. It dates back to the era when communication specialists were trying to lump telephones and telegraphs together under one heading. As the telegraph industry withered away, telephony became most of telecommunications. However, the new word had caught on as a way to describe electronic communications over long distances and was useful because new forms were emerging. Telex became an accepted way to send messages around the globe. Facsimile systems began transmitting images of documents. Computer data communications grew rapidly. Much of that traffic is routed partly or totally through the telephone network, but it is all telecommunications, even as much of it becomes optical rather than electronic.

Different Networks

Many separate telecommunication networks exist; some (but not all) interconnect with the telephone system.

Meanwhile, the need to send data, video signals, and other types of information over long distances generated new telecommunication systems. Many are specialized for particular purposes. Packet-switching networks carry chunks of data around the world in ways not useful for telephone conversations, but adequate and less costly for data transmission. Utility companies have strung their own telecommunication lines along their rights of way to let system controllers monitor the status of equipment at remote sites. Military agencies have specialized telecommunication systems, designed for use in normal times and in national emergencies. Many of these systems interconnect with the telephone network as shown in *Figure 15-1*, but some don't.

Figure 15-1. Interconnection of Telecommunication Networks

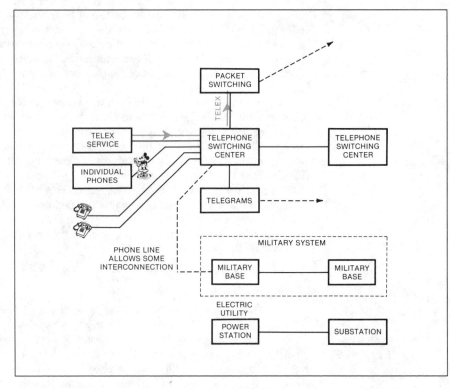

Taken together, these telecommunication systems form a global network, crossing international boundaries. Access is far from perfect. Some underdeveloped countries have so little equipment it may be hard to make a telephone call to the capitol and impossible to reach outlying areas. Totalitarian countries may restrict or monitor calls to their citizens. International calling can be expensive and can leave you facing language barriers. Nonetheless, from my desk, I can telephone England or Australia or use my computer to send a Telex to Japan, by dialing codes not much more complex than I need to phone the neighbors.

Transmission System

Many signals are merged together for transmission over long distances and separated at the other end. The telephone system has a hierarchy of transmission systems, with lower-speed systems feeding into faster ones.

The nuts and bolts of international telecommunications are the transmission system. The basic approach relies heavily on the economies of scale. When transmitting many signals over similar routes, it costs less to multiplex them and send them over one transmission line than to send them over separate lines. The telephone industry has developed a hierarchy of transmission rates, with lower-speed systems feeding into those operating at higher speeds. Because the telephone network goes almost everywhere (at least in developed countries), other systems generally interface with it at some level. The network includes underwater cables, satellite links, and microwave transmission, as well as cables on land and under ground.

The higher-speed parts of the transmission hierarchy carry signals the farthest; the slower the speed, the shorter the distance.

The higher speeds of the transmission hierarchy carry signals over the longest distances, and slower speeds generally operate over shorter distances. Technologically, the clearest examples of long-distance systems are those that extend across vast open spaces (i.e., oceans or deserts).

The functional differences also are visible from an organizational viewpoint in the American telephone industry. Long-distance telephone lines are operated by AT&T, MCI, U. S. Sprint, and other long-distance carriers. Short-distance or regional lines are owned and operated by regional telephone companies. (Those boundaries that now exist between corporations are not nearly as clear as they sound, and drawing lines was a difficult task when AT&T broke up. Some telephone facilities still have lines taped down the middle, with one side belonging to AT&T and the other to the regional Bell operating company.)

The telephone network is the most likely base for the evolution of an Integrated Services Digital Network.

We will concentrate on the telephone network because it is the most pervasive telecommunication system. It also is the likely base for the evolution of an Integrated Services Digital Network (ISDN), which would carry many telecommunication services in digital form. Planners see an ISDN as a logical goal because it could bring any service to any location (within constraints of network transmission capacity). We'll talk more about ISDN in Chapter 16.

It was not too many years ago that the telephone system was built around analog technology. Analog voice signals were combined in analog multiplexers to give higher-frequency signals. Much analog equipment remains in use, but it's being phased out in favor of a new generation of digital telephone technology. We will talk only about that new generation, because fiber optics is a major part of it.

The Digital Telephone Transmission Hierarchy

The hierarchy of digital telephone transmission rates is carefully organized, but the actual telephone network is not nearly as neat.

The hierarchical arrangement of the digital telephone network is shown in *Figure 15-2*. It shows an idealized way in which signals at each step of the hierarchy are combined for the highest possible degree of multiplexing. In practice, the highest degree of multiplexing is not always needed. For example, some 400-Mbit/s systems may operate with only one or two 45-Mbit/s T3 or (equivalent) DS3 inputs. *Figure 15-2* shows maximum available capacity.

**Figure 15-2.
Multiplexing in the
North American Digital
Transmission Hierarchy**

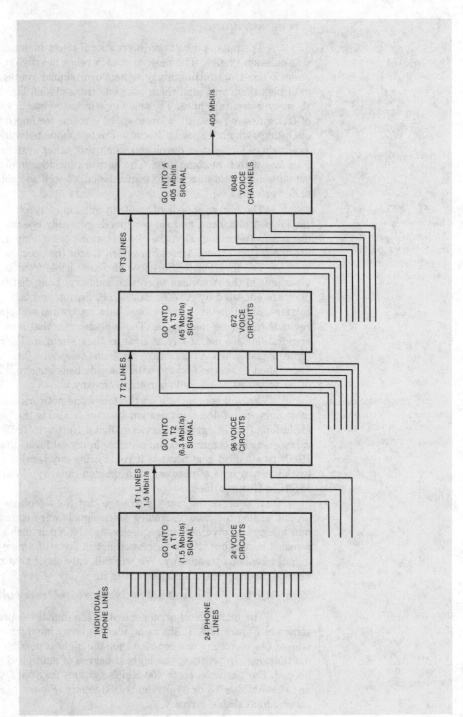

Ideal Case

Ideally, this hierarchy is a symmetrical tree-like structure, where low-capacity systems feed into higher-capacity ones running between major population centers and transmission distance generally increases with speed. As usual, the real world is far from ideal because population and telecommunications users are not evenly distributed. Many interconnections don't fall into the ideal pattern of successively higher transmission rates. Particularly in rural areas, long-distance transmission is not always at the highest level of the hierarchy. And the separation of local and long-distance transmission creates even more complications.

Worldwide Standards

Different digital telephone hierarchies are used around the world.

Multiple standards exist for the digital telephone hierarchy. International digital telephony standards are set by the International Consultative Commission on Telephone and Telegraph, an arm of the International Telecommunications Union known as CCITT from the French-language version of its name. These standards, listed in *Table 15-1*, are used in Europe. North American telephone companies follow de facto standards, also shown in *Table 15-1*, which AT&T established in the days when it was the telephone company. (Japan has its own standards.)

**Table 15-1.
Transmission Rates in
North America and
Europe**

Rate Name, American	Data Rate	Voice Circuits
single line	56,000 bit/s	1
T1 or DS1	1.5 Mbit/s	24
T2 or DS2	6.3 Mbit/s	96
T3 or DS3	45 Mbit/s	672
T3C or DS3C	90 Mbit/s	1344
T4 or DS4	274 Mbit/s	4032
400 Mbit/s	405 or 417 Mbit/s*	6048
565 Mbit/s	565 Mbit/s*	8064 (56 kbit/s equiv)
810 Mbit/s	810 Mbit/s*	12,098
1700 Mbit/s	1700 Mbit/s*	24,192
European (CCITT standard)		
single circuit	64,000 bits/s	1
Level 1	2.048 Mbit/s	30
Level 2	8.448 Mbit/s	120
Level 3	34.304 Mbit/s	480
Level 4	139.264 Mbit/s	1920
Level 5	565.148 Mbit/s	7680

*Actual line rate depends on design and overhead bits and is not standardized.

A close look at *Table 15-1* shows that specifications for the North
American telephone hierarchy become hazy at high speeds. When AT&T
established its T carrier hierarchy well over a decade ago, 274 Mbit/s
seemed as fast as possible for practical transmission. Fiber optics quickly
passed that mark, but the arrival of competition on the long-distance scene
meant that no one entity could set transmission rates in North America.

High-Speed Systems

Long-distance telephone companies use speeds around 400 Mbit/s
for much of their nationwide networks but all bit rates are not identical.
The differences near 400 Mbit/s lie not in the number of conversations
carried (or voice circuits of capacity) but in the amount of overhead bits
added to aid in system operation.

Higher-speed systems are coming. Some American systems
operate near 565 Mbit/s but, as can be seen in *Table 15-1*, they do not carry
the same number of voice circuits as European versions. Equipment for
810-Mbit/s transmission also has been introduced recently. Nor is that the
last step. Both AT&T and the Nippon Telegraph and Telephone Corp. have
announced plans to install 1.7 Gbit/s systems. We'll talk more about that
technology later.

All this confusion reflects the state of flux in the telephone
industry. Long accustomed to careful, orderly, centralized planning for long-
term needs, the industry had planned to shift from aging analog equipment
to digital systems. Deregulation and the rapid development of fiber optics
changed the adjectives preceding the word "planning." The industry has
shifted from analog to digital equipment faster than was planned, and the
opening up of new markets—caused by competition for long-distance
telephone business—stimulated new technology.

Transmission Alternatives

By design, this book focuses almost exclusively on fiber optics.
However great you think fiber optics are (and having followed the field for
the past decade, I think they're pretty darn good), they are not the only
transmission medium suitable for telecommunications. The global network
also includes:

- Twisted wire pairs for individual telephone circuits, some analog
  multiplexing, and digital transmission to 1.5 Mbit/s (or even higher
  over short distances)
- Coaxial cable for analog multiplexing and some high-speed digital
  systems
- Microwave ground links for high-speed analog and digital
  transmission
- Microwave satellite links, for high-speed analog and digital
  transmission.

These systems all remain in use, but some are not being installed in new places. Installation of multiplexed analog systems has virtually stopped; the industry is firmly committed to digital technology. Likewise, no one is laying new coaxial cable for long-distance telecommunications—fiber wins hands down. Others, however, are being installed in quantity. Twisted wire pairs can carry your home phone circuit as well as up to 1.5 Mbit/s over modest distances. Microwave and satellite systems also have their places and will remain in use.

Types of Long-Distance Transmission

The telephone network is assembled from many point-to-point transmission links. Cables are good for point-to-point transmission, but satellites are better for reaching spread-out points with little traffic.

To understand what goes into picking telecommunications technologies, you need to learn about the functions and architectures of telecommunication networks. The telephone network is made of many elements linking pairs of points, called point-to-point links. These run from homes to switching offices and between pairs of telephone-company facilities. A cable is an ideal choice for such a system because it can efficiently carry signals at high speeds between two points.

However, cables cannot solve all telecommunication problems. Satellites are a better choice for connecting hard-to-reach points that are not on major telecommunication channels, as shown in *Figure 15-3*. Places such as Madagascar or the Falkland Islands don't generate enough communication traffic to justify laying an expensive cable to them. A satellite dish is a much cheaper way to stay in touch with the rest of the world.

**Figure 15-3.
Different Media Serve
Different
Telecommunication
Needs**

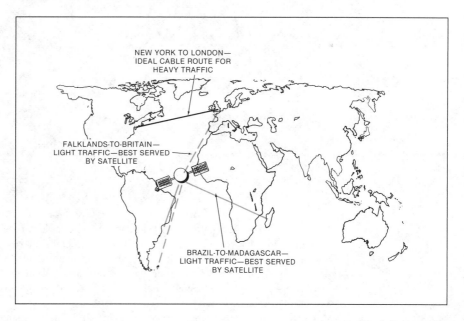

NEW YORK TO LONDON—
IDEAL CABLE ROUTE FOR
HEAVY TRAFFIC

FALKLANDS-TO-BRITAIN—
LIGHT TRAFFIC—BEST SERVED
BY SATELLITE

BRAZIL-TO-MADAGASCAR—
LIGHT TRAFFIC—BEST SERVED
BY SATELLITE

Satellites and other radio-frequency transmission systems have another strength—broadcasting one signal to many points. An antenna can do the job without any physical connection to the people receiving the signal. The antenna can be on a hilltop, a tall building, or even a satellite. The lack of physical connection is important because it allows mobile communications to people in cars, planes, or boats.

A careful look at this pattern shows why fiber optics cannot make all satellites obsolete. They can economically send signals where cables cannot reach. Fibers can't serve car phones or light-traffic routes (e.g., from the Falklands to Britain). But they are attractive for heavily trafficked routes such as New York to London.

LONG-HAUL FIBER SYSTEMS ON LAND

Long-distance telecommunication networks on land interconnect major population centers. One example is shown in *Figure 15-4*, the high-speed fiber-optic backbone of AT&T's long-distance network in the United States. Other long-distance carriers have or are building similar backbone networks.

**Figure 15-4.
AT&T Long-Distance
Backbone System**
(Courtesy AT&T)

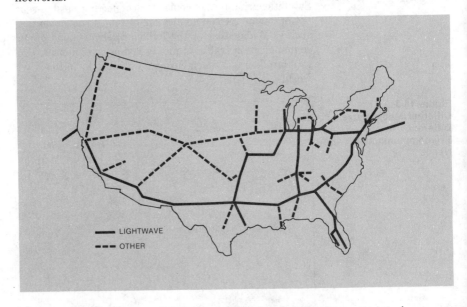

LIGHTWAVE
OTHER

The fiber-optic network (solid lines) does not reach every state in the union. Each cable includes multiple fiber pairs operating at about 400 Mbit/s or more. The backbone system also includes some other transmission media (dashed lines), including microwave links. Other systems, including lower-speed fibers, branch out from the fiber network to carry signals to other regions.

Other networks serve the same function on a smaller scale. *Figure 15-5* shows the first regional fiber-optic backbone system. It carries telephone and cable-television signals to the 50 largest towns in the

Canadian province of Saskatchewan. The population served is modest—about one million people live in the 220,000-mi^2 (570,000-km^2) province, and about one-third of them are in the two largest cities, Regina and Saskatoon —so system capacity is much smaller than that of AT&T's U.S. system.

Figure 15-5.
Saskatchewan's
Regional Fiber-Optic
Backbone System
(Courtesy Saskatchewan
Telecommunications)

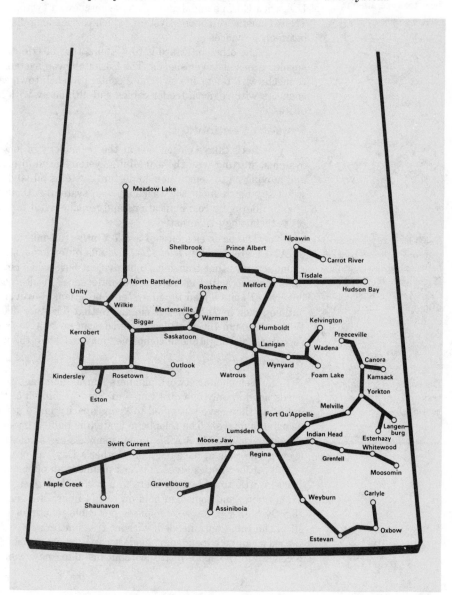

Regional networks must interface with systems transmitting to distant areas and with local services that operate on smaller scales.

However large the region, any regional network must have two sets of interfaces, which don't appear on the maps in *Figures 15-4* and *15-5*. One connects the network to transmission lines to distant points. AT&T's network, for example, connects with cables and satellites that connect the U.S. with Europe and Japan. The Saskatchewan system connects with the Trans Canada Telephone System, which carries long-distance traffic between provinces.

The other interface is to systems that distribute signals over smaller areas at lower speeds. The Saskatchewan system, for instance, connects with telephone switching centers in each town. AT&T's system connects with regional feeder cables and ultimately with local switching offices.

Present Technology

Let's take a closer look at the technology of long-distance fiber systems, starting with the 400-Mbit/s systems used in backbone networks and moving on to some other important systems on land and at sea. We won't say much about the Saskatchewan system. Although it was a major step in fiber-optic communications and remains vital to the province, much of its technology is outdated.

A 400-Mbit/s system can transmit signals 40–50 km between repeaters using single-mode fiber, an In-GaAsP laser coupling −2 dBm into the fiber, and an InGaAsP PIN photodiode with −32 dBm sensitivity.

The basic elements of a 417-Mbit/s transmission system developed by AT&T are listed in *Table 15-2*. Details differ for systems from other manufacturers that transmit an equivalent 6048 voice circuits. Small differences in actual bit rate reflect differences in internal coding and "overhead" bits added to aid system operation. However, the overall outline given in *Table 15-2* is similar to that for other 400-Mbit/s systems. With a 1310-nm InGaAsP laser, signals can be sent 40–50 km without repeaters, although phone companies may not push the upper limit.

System Capacity

Demands for telephone transmission have risen continually.

Telephone-network planners learned long ago that transmission needs head in only one direction—up. Since the birth of the telephone industry, they have struggled to keep capacity up to demand. Worse yet, demand is uneven. The telephone system is built to meet demand during the business day. At 3 A.M., few insomniacs are awake to use the network; if everyone calls at once (e.g., on Mother's Day), all circuits may be busy.

Fiber optics gives network planners an opportunity to plan ahead. It costs little to add extra fibers to a cable. Most designs have room for extra fibers, and some can hold up to 144 fibers. Installation costs are only slightly higher (a few extra splices per cable junction). The extra fibers can sit in the cable until needed, without transmitters or receivers. Then capacity can be expanded at modest cost just by attaching transmitters and receivers to the spare fibers, without installing new cable.

Table 15-2. Characteristics of AT&T's 417 Mbit/s System

Fiber	Single-mode step-index, 8-μm core, 125-μm cladding
Wavelength	1310 $\pm$20 nm from InGaAsP laser
Spectral width	Max 2 nm
Detector	InGaAsP pin photodiode
Line code	NRZ (scrambled)
Transmitter output into fiber (worst case)	-2 dBm
Receiver sensitivity (3 X 10^{-11} bit error rate)	-32 dBm
System loss (includes margin)	30 dB
Connector loss	1.5 dB
Fiber loss	23 dB (equal to about 50 km or 31 mi at 0.46 dB/km)
Margin	5.5 dB

Many cables include spare fibers for future expansion.

Though that sounds eminently sensible, it wasn't that obvious to the industry gurus who tried to predict the market for fiber optics. For a while, they estimated transmitter and receiver sales by counting the amount of fiber and cable sold and dividing that total by the average length of a fiber-optic system. Those estimates didn't agree with sales of transmitters and receivers, creating a mystery until one market analyst finally learned what telephone companies were doing.

Another way to upgrade capacity is adding faster transmitters and receivers to existing cables. Telephone companies in the United States and Japan are planning to quadruple data rates on present 400-Mbit/s single-mode fiber systems to 1.7 Gbit/s. AT&T already has demonstrated such an upgrade in Pennsylvania; it does not require changing the fiber or repeater spacing.

Both these techniques can work with the step-index single-mode fibers (with zero dispersion near 1300 nm) already installed. Phone companies prefer this approach because it avoids the expense of installing new cable. Transmitters and receivers used in such upgrades will either operate at 1300 nm or will use 1550-nm light sources with near-zero spectral width so they will not suffer from the high chromatic dispersion inherent in present single-mode fibers at that wavelength. Wavelength-division multiplexing would allow simultaneous transmission at both wavelengths.

Upgrading Transmission Capacity

It is instructive to look at the 1.7-Gbit/s system developed by AT&T Bell Laboratories to see how transmission speed can be upgraded. The transmitter uses a buried-heterostructure laser of InGaAsP emitting in multiple longitudinal modes. Spectral widths of individual lasers are 1 or 2 nm, small enough to limit chromatic dispersion as long as they are within 5 nm of the fiber's 1310-nm zero-dispersion wavelength. The transmitter also includes circuits to control bias and temperature and an integrated driver circuit made from fast gallium-arsenide electronics.

Both types of receivers tested at Bell Labs used avalanche photodiodes, but the detectors were of different composition—one germanium, the other InGaAs. Both receivers had GaAs preamplifiers and monolithic silicon amplifiers. The decision circuit is a monolithic integrated circuit using direct-coupled gallium-arsenide metal-semiconductor field-effect transistors (MESFETs).

In Pennsylvania, field tests with a step-index single-mode fiber system showed that under normal conditions—with powers of -23 to -28 dBm at the receiver—system bit-error rate averaged under 10^{-12} for 8 to 15 hours. Those measurements were taken with eight regenerators in series, with each pair separated by 23.1 km of fiber. Tests of individual repeaters showed that bit-error rates of 10^{-11} were possible with either receiver at an average power below -32 dBm. AT&T is pleased enough with the results to plan to install such high-speed equipment in its nationwide network within the next couple of years. Specifications of the commercial 1.7-Gbit/s system are listed in *Table 15-3*.

SUBMARINE CABLES

Cables are not confined to land. Some telecommunication cables must cross the two-thirds of the globe covered with water. Some underwater cables are short, crossing a river or the few kilometers of seawater separating an island from the mainland. Others must run hundreds or thousands of kilometers across seas and oceans.

Fiber optics neatly fits the requirements for submarine cables, especially those running long distances. The high costs of making, laying, and operating a cable demand high transmission capacity. Because repeaters are potential trouble points that are extremely hard to reach under the ocean, the cable should have as few repeaters as possible. The cable should transmit digital signals cleanly to be compatible with modern equipment. Those specifications veritably call out "fiber optics."

History of Submarine Cables

Telegraph Between Britain and France

The need for electrical cables first arose with the electric telegraph in the early 19th century. It wasn't long before engineers laid waterproof cables underwater to carry telegraph signals. In 1850, the first submarine cable was laid in the open ocean, between Britain and France, but it carried only a few messages before a fisherman caught it and hauled it to the surface. He thought it was a rare type of seaweed!

**Table 15-3.
Characteristics of
AT&T's 1.7 Gbit/s
System**

Fiber	Single-mode step-index, 8-μm core, 125-μm cladding
Wavelength	1310 ±**5** nm from InGaAsP laser
Spectral width	Max 2 nm
Detector	**Avalanche photodiode**
Line code	NRZ (scrambled)
Transmitter output into fiber (worst case)	−2 dBm
Receiver sensitivity (3 × 10^{-11} bit-error rate)	−32 dBm
System loss	30 dB
Connector loss	1.5 dB
Fiber loss	23 dB (equal to about 50 km or 31 mi at 0.46 dB/km)
Margin	5.5 dB

Differences from the 417-Mbit/s system appear in bold face.

That experience taught submarine cable engineers an important lesson—waterproof isn't enough. The biggest dangers to cables in shallow waters are fishing trawlers and anchors. The next telegraph cable laid between Britain and France in 1851 was armored with ten 7-mm wires of galvanized iron. That protected it from fishermen well enough that it worked for many years thereafter.

Transatlantic Telegraph

The next bold step was to lay a cable across the entire Atlantic Ocean, which is far deeper and wider than the English Channel. Two efforts failed in 1857, but the following year a third cable worked for about three weeks. Then an operator fried the cable by putting 2000 V across it. The Civil War and other problems slowed progress, and it was not until 1866 that the first transatlantic telegraph cable was working regularly.

Telephone Cables

Submarine telephone cables were not made until the mid-20th century because they needed submerged electronic repeaters.

The telephone came along not long afterward, but submarine telephone cables were much harder to build. One key problem was the need for a repeater to boost telephone signals so the cable could carry them long distances. Mechanical relay amplifiers could handle the dots and dashes of the telegraph, but electronic amplifiers are necessary to handle voices. The required electronic technology wasn't available until well into the 20th century. Instead, the first transatlantic phone calls were made by wireless —a short-wave radio known for its crackle, static, and fade-out.

Britain laid the first submarine telephone cable with an underwater repeater in 1943 between Holyhead and Port Erin. The first such cable in North America—and the longest-operating—was laid between Key West, Florida, and Havana, Cuba, in 1950. The first transatlantic telephone cable, the TAT-1 (TransATlantic-1) cable between Britain and Canada, began operation in 1956.

Those early telephone cables were made of coaxial cable, which offers the highest bandwidth of any metal cable. However, coaxial cable has limitations. Its attenuation increases with the square root of transmission frequency and decreases as the inside diameter of its outer conductor increases:

$$\text{ATTENUATION} = C \times \frac{\nu^{1/2}}{D}$$

where C is a constant depending on cable characteristics, ν is the transmitted frequency and D is the inner diameter of the cable's outer conductor.

The transmission frequency should be high to give high cable capacity, but that requires either a smaller repeater spacing (to compensate for higher losses) or a larger cable diameter (to reduce transmission losses). Either one is a problem. Repeaters are costly, can fail, and require electrical power. Thicker cables are hard to lay in the ocean.

Transmission capacity and repeater spacings were limited in early submarine coaxial cables.

Satellites

The TAT-1 cable, which carried 36 telephone circuits, was 1.6 cm in diameter and had repeaters 70.5 km apart. The last transatlantic coaxial cable, TAT-7, which was put into service in 1983, is 5.3 cm in diameter and requires a repeater each 9.5 km. By using special speech interpolation techniques, engineers can crowd 4200 analog phone circuits onto the cable, but that was the practical limit of coax technology. Meanwhile, satellites had become important for transoceanic communications. The first transatlantic communications satellite, Intelsat I ("Early Bird"), was launched April 6, 1965. It could carry 240 phone calls, nearly twice as many as any extant transatlantic cable. By 1970, satellites offered more transatlantic voice circuits than cables, and cables were in trouble.

Because satellites offered many advantages over co-axial cable systems, cable developers sought a different technology.

Cable Versus Satellite

Cable developers were not about to concede defeat by satellites, although coax had reached its limit. Satellite channels do have some disadvantages. The round trip to and from geosynchronous orbit 37,000 km (22,000 mi) above the earth takes a quarter of a second. That delay is barely perceptible, but it can get annoying if a telephone call makes two bounces. Microwave transmission to and from satellites can be intercepted by electronic eavesdropping or ferret spy satellites. Phone calls sent over analog satellite channels can suffer echoes that sometimes sound louder than the person at the other end. Other problems can also limit transmission quality.

Politics play a role, too. Transoceanic cables are owned by international consortia of telecommunication companies and government communication authorities (e.g., AT&T, British Telecommunications, and the West German Post Office, which is responsible for that nation's telecommunications). So far, all international satellites have been owned by the International Telecommunications Satellite Organization, Intelsat. Although the same telecommunication organizations own shares of Intelsat, their control over satellites is less direct than over cables.

Costs of satellite and cable transmission are hard to quantify, but are roughly comparable.

Economics obviously are important, but they are very hard to sort out. Cables and satellites have different lifetimes, require different terminal equipment, have different operating costs, cost different amounts, and have different capacities. Advocates of satellites and cables each have economists who can prove that their approach is better. International telecommunications organizations eventually struck a predictable compromise: where both satellites and cables were available, they would divide transmission between them.

Keeping this balance obviously required a new generation of cable technology. That's where fiber optics came in. With coaxial cable technology coming to a dead end in the mid-1970s, cable developers turned to optical fibers.

Undersea Cable Requirements

There are three types of submarine cables: short unrepeatered, moderate-distance repeatered, and long-distance repeatered transoceanic.

Although we have talked mostly about transoceanic submarine cables, there are two other important types: unrepeatered cables and moderate-distance cables. All three have somewhat distinct requirements.

Unrepeatered cables are typically under 40 or 50 km long and part of a national telecommunications network. Examples include cables running between islands in the Japanese archipelago or between a country's mainland and an offshore island. From the network standpoint, these serve the same function as land cables but happen to run under water.

Moderate-distance cables run from 50 up to several hundred kilometers, typically between nations separated by water. Classic examples are cables running underwater between Britain and the European mainland. These are major links in the international telecommunication network, but their design requirements and operating characteristics differ from those of transoceanic cables.

Transoceanic cables run thousands of kilometers between continents and contain many repeaters. These are the most ambitious and expensive; they cost hundreds of millions of dollars.

Similarities

All three cables have some common design requirements. Any cable must be able to withstand seawater. The repeater housings must be watertight. The repeaters themselves should operate for long periods on the sea floor without maintenance. The cable should have high enough capacity to meet expected growth in demand for a few years—now averaging 20–30% a year on the busy transatlantic route.

Differences

Cables laid in deep water
must withstand high
static pressure; those laid
in shallow water must be
armored.

There also are some important differences. Cables laid in deep water must withstand high static pressures and the process of cable laying. Those laid in shallow water must be armored to protect against fishing trawlers and anchors. The longer the cable, the more care must be taken in repeater design, to keep pulse jitter or transmission errors from accumulating as signals pass through many repeaters. (There are 125 repeaters in TAT-8, the first transatlantic fiber-optic cable, which is due to begin service in 1988.) Reliability is also a larger concern with longer cables because of their high cost and the difficulty of repair. TAT-8 is designed to last for 25 years, with no more than three failures requiring recovery of the cable from the ocean bottom. In a system with as many repeaters as TAT-8, that imposes stringent requirements on component reliability. As we will see, the TAT-8 design includes considerable redundancy to allow alternate signal routings if any components fail.

Although cables are made rugged, they must be laid delicately— fed out slowly from a special ship so they can settle gradually to the ocean bottom. Thick, stiff cable is hard to load and vulnerable to damage. Cable can be laid directly on the deep-sea bed, but in shallow areas it may be buried under the bottom to protect it from shipping and fishing traffic.

To see how these design requirements are met, we will look closely at the design for TAT-8 and a couple of shorter cable systems.

Design of a Transatlantic Cable

TAT-8 will transmit 278-
Mbit/s signals at 1300 nm
through single-mode fiber.

TAT-8 is the first ocean-spanning fiber-optic cable and the one on which the most design information is available. A similar design is being used for TRANSPAC-3, which will run from the west coast of the United States to Guam and Japan via Hawaii. We will look specifically at AT&T's SL (submarine lightwave) system. This will be used for the 5600-km part of TAT-8 from Tuckerton, New Jersey to a branch point off the European coast, shown in *Figure 15-6*, where the cable will split between France and Britain. The cable contains six single-mode fibers, arranged in three pairs. Two pairs will be operating continuously; the third pair is being held in reserve.

Transmission Rates

Speech compression
makes TAT-8 capacity
about 40,000 voice cir-
cuits, although raw digital
capacity equals about
8000 circuits.

The light sources are InGaAsP lasers emitting at 1300 nm, where fiber attenuation in the high-quality single-mode fibers is about 0.4 dB/km. The lasers are modulated at 295.6 Mbits/s, which includes 278 Mbit/s of signal plus added overhead bits to monitor transmission. This corresponds to two channels at the CCITT 140-Mbit/s data rate. Together the two parallel operating fiber pairs can carry 557 Mbit/s. Normally, that would equal about 8000 voice channels, but the builders of submarine cables use sophisticated speech-compression techniques so one digital phone circuit can carry five conversations.

Speech compression may not be used in all long-haul submarine cables, despite the capacity improvement. Some digital signals require the entire 56- or 64-kbit/s speed of a single digital voice channel and cannot be compressed in the same way as voices. One example is facsimile transmission.

Figure 15-6.
Plan for TAT-8 *(Courtesy AT&T)*

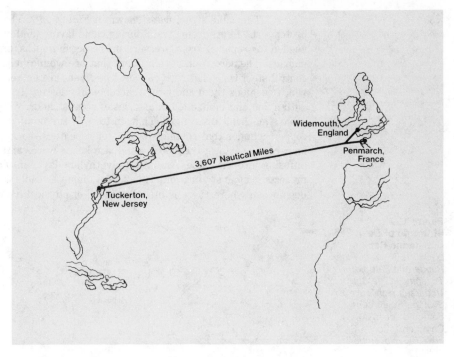

Cable Performance

Performance requirements for the fiber and cable are high. *Table 15-4* shows how single-mode fibers performed in an early test where a sample cable was laid in the ocean.

Table 15-4.
Fiber Characteristics* for TAT-8

Average loss before cabling (includes splices)	0.40 dB/km
Average increase in loss from cabling	0.008 dB/km
Average loss increase from environmental effects	0.005 dB/km
Average loss variation @ 1290–1330 nm	5%
Fiber loss assumed in design	0.45 dB/km
Minimum dispersion wavelength	1311.5 ±2nm
Dispersion @ 1290–1330 nm	<2.0 ps/km-nm
Average loss change due to hydrogen effects	0.002 dB/km in 25 yr
Average loss change due to radiation effects	≤0.007 dB/km in 25 yr

*Measured in laboratory tests by AT&T.

In the TAT-8 cable, six fibers are wound around a central steel wire, and that core is enclosed in a welded copper cylinder jacketed in plastic. Armor is added to segments laid in shallow water.

The cable structure is shown in *Figure 15-7*. The structure protects the fibers from strain during cable laying (and during recovery if needed for repairs), from pressure in the ocean depths, and from external damage. The core contains six fibers that are wound helically around a central steel wire clad with copper. The fibers are embedded in a plastic, which cushions them and limits microbending losses. This core is covered with nylon and embedded in a series of steel strands, which provides the cable's mechanical strength. That, in turn, is surrounded by a welded copper cylinder that conducts power for repeaters—constant current of 1.6 amperes (A) at up to ±7500 V—and also keeps water and hydrogen from diffusing into the cable. An outer polyethylene layer insulates the cable and protects it against abrasion. This 2.1-cm diameter cable can be used in the deep ocean where there is little likelihood of physical damage.

**Figure 15-7.
Structure of Deep-Sea
Submarine Cables**
*(©1984 IEEE, from Peter
Runge and Patrick R.
Trischitta, "The SL
Undersea Lightwave
System,"* IEEE Journal on
Selected Areas in
Communications, *SAC-2,
784–793 [November
1984])*

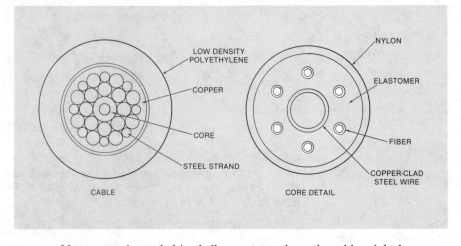

More armor is needed in shallow waters where the cable might be snared by fishing trawlers or otherwise attacked while in place. The standard design allows for one or, as shown in *Figure 15-8*, two layers of armor made up of steel wires 4.19 mm in diameter. Two layers of armor more than double cable diameter to 5.1 cm.

The worst strain applied to a deep-sea cable would be during recovery to make repairs.

The cable structure cannot completely protect the fiber from strain —measured as the percentage by which the fiber would lengthen in response to the pull on it. AT&T engineers have calculated that the worst strain—0.87%—would occur if the cable was hauled up from the ocean depths to repair a fault. Initial cable laying would strain the fibers by 0.44%. Residual strain after the cable is first laid would be only 0.04%, but following repairs the calculated strain after relaying is 0.21%.

Figure 15-8.
Armour Around Shallow-
Water Segment of
TAT-8 (© *1984 IEEE, from*
Ali Adl, Ta-Mu Chien, and
Tek-Che Chu, "Design
and Testing of the SL
Cable," IEEE Journal on
Selected Areas of
Communications, *SAC-2,*
864–872 [November
1984])

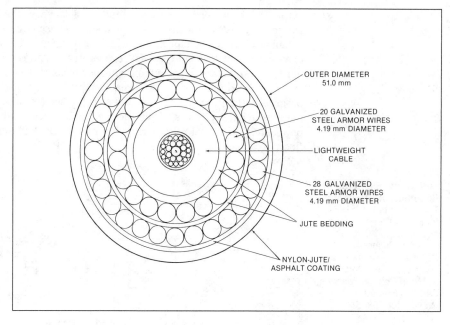

The loss budget for the cable is shown in *Table 15-5.*

Table 15-5.
Loss Budget for TAT-8
Segments (AT&T
Design)

Transmitter output into fiber	−2 dBm
Fiber attenuation	23 dB
Penalties (dispersion, timing errors, etc.)	1 dB
Allowance for component aging	5 dB
Receiver sensitivity (for BER of 10^{-9})	−31 dBm

The budget is conservative and assumes that the receiver requires input 3 dB higher than the −34.2 dBm design goal. That conservatism reflects the desire for a 25-year system lifetime. The 23 dB for fiber loss allows repeater spacing of either 51 km (if fiber attenuation is 0.45 dB/km) or 57.5 km (if attenuation is 0.4 dB/km).

One regenerator channel of a TAT-8 repeater is shown in the block diagram of *Figure 15-9.* Light is detected by an InGaAs pin detector, which is coupled to a monolithic silicon transimpedance amplifier. The pin detector was chosen over an APD because of its lower operating voltage, simpler operating requirements, and expected longer lifetime. Because receiver

TAT-8 regenerators allow for four redundant lasers, three of which are standbys in case of failure.

output can have peak-to-peak amplitude of 3–300 mV, the amplifier includes automatic gain control circuits to keep signals reaching the decision and retiming circuits at uniform, predictable levels. Output from those circuits drives the laser transmitter. The regenerator design can house four 1300-nm InGaAsP lasers to provide redundancy. This is done because the lasers are considered the most failure-prone component, although impressive progress in laser reliability may make such a conservative approach unnecessary. Electronics within the regenerator select which laser to power, and an optical relay carries the light from the laser to the output fiber.

Figure 15-9.
Functional Elements of Regenerator Channel in TAT-8 Repeater (©1984 IEEE, from Peter Runge and Patrick R. Trischitta, "The SL Undersea Lightwave System," IEEE Journal on Selected Areas in Communications, SAC-2, 784–793 [November 1984])

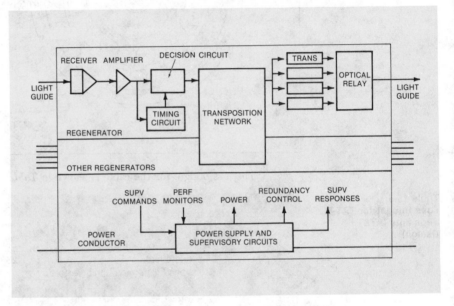

Each repeater includes three regenerators operating in each direction, but only two pairs are used; the third regenerator pair is kept in reserve along with the third fiber pair in case of failures. The repeaters also include circuits that respond to supervisory signals embedded in the data stream. These circuits monitor component performance and allow remote switching among components.

Repeater housings also are crucial because electronics and salt water don't mix. Much of the housing technology comes from coaxial cable repeaters. Care must be taken to dissipate heat generated by the electronic devices efficiently, because component lifetimes—particularly of semiconductor lasers—decrease sharply with temperature. This heat dissipation requirement can limit the number of working fibers in a submarine cable. Fortunately, water at the ocean bottom normally is cold, about 5°C and is a good heat sink.

Fiber-optic repeaters are housed in cases adapted from designs for coaxial cables.

The UK-Belgium 5 Cable

The 112-km fiber-optic cable laid in 1986 between Broadstairs, England and Ostend, Belgium has many similarities with TAT-8 and other deep-sea systems, but it also has some important differences. Like TAT-8, UK-Belgium 5 was designed (by Standard Telephones and Cables Ltd. of Britain) to suffer no more than three failures in a 25-year lifetime. It uses the de facto standard of 1300-nm transmission through single-mode optical fiber. It contains three fiber pairs, each transmitting 280 Mbit/s (twice the CCITT rate of 140 Mbit/s). No channel multiplication equipment is used (at least initially) so capacity is 11,520 voice channels.

A loss budget for UK-Belgium 5 is shown in *Table 15-6*, based on the original assumption that the cable would be 122 km long. (The final route was 10 km shorter.) Note that the shorter length of the system put less pressure on designers to minimize repeater spacing. They chose to include three repeaters about 30 km apart, rather than push to stretch repeater spacing further.

Table 15-6.
Loss Budget for UK-Belgium 5 Cable

Transmitter output	−3.7 dBm
Section loss (for three repeaters)	17.5 dB
Repair allowance	3.5 dB
Receiver level	−24.7 dBm
Receiver sensitivity (10^{-9} BER)	−34 dBm
System margin	9.3 dB

The shorter length eased several design requirements. Although the repeaters require essentially the same current level as TAT-8 repeaters (1.7 A), the voltages required are much lower because there are fewer of them. The smaller number of repeaters also lessens demands on timing accuracy and repeater lifetime. Timing errors accumulate over distance in each repeater. For instance, if timing error per repeater is ±0.5 ns, the maximum error over three repeaters would be ±1.5 ns but it would accumulate to ±62.5 ns over the 125 repeaters in TAT-8. (Realistically, the maximum accumulated jitter would be unlikely to be that large, but it would nonetheless be much larger than in the shorter cable.) Similar considerations apply to repeater lifetime. Suppose one repeater failure is likely every 150 years. In a three-repeater cable, that means one failure is likely in 50 years. However, if the cable has 125 repeaters, one failure is likely each 1.2 years! That's why TAT-8 designers allowed for multiple lasers in each regenerator, while designers of UK-Belgium 5 didn't bother.

Differences also exist in demands on the cable structure itself. All UK-Belgium 5 lies in busy shipping lanes and never plunges to the several-kilometer depths of the deep ocean. Thus, the entire length must be armor-protected, but the severe static pressures at the bottom of the ocean, or in laying and recovering deep cables, are not crucial concerns.

Unrepeatered Cables

Submarine cables shorter than 40–50 km don't require repeaters, which simplifies design.

Submarine cables much shorter than UK-Belgium 5 don't need repeaters at all. This greatly simplifies design because it eliminates concerns about submerged repeater performance. The active electronic and optical components can be housed safely on shore.

The cables are similar to other shallow-water submarine cables, except that the fibers in some are in loose tubes filled with high-viscosity material. This gives these cables greater bend resistance than tightly buffered cables. Because the cables are not laid at great depths, they do not have to withstand high static pressures. The major dangers are from shipping and fishing operations. In some cases, as in a cable laid across the mouth of the Hudson River from downtown Manhattan to the Statue of Liberty and Ellis Island, the cable may be buried under the seabed to protect it from damage. The large threat of damage makes cable armor critical.

Like most other telecommunication systems, repeaterless submarine cables are made with step-index single-mode fibers designed for transmission at 1300 nm. With no submerged repeaters to cause heat dissipation problems and the electronics readily accessible, the cables can contain a dozen or more fibers, allowing capacity for expansion. Most systems carry at least 140 Mbit/s over one fiber pair, with upgrades possible to 565 Mbit/s.

FUTURE TECHNOLOGY

Developing technology will allow transmission faster and further between repeaters.

If papers presented at research conferences are any indication, future long-distance systems will transmit at higher bit rates and over longer distances between repeaters. We examined some components earlier, but now we'll look at systems and applications.

Many refinements won't change operation of fiber-optic systems visibly. Integrated optoelectronic transmitters and receivers would merely replace devices performing similar functions. Integrated-optic switches will impact local and regional networks more than long-distance transmission, which doesn't require much switching. However, some innovations could have major impact.

1550-nm Transmission

A shift to 1550-nm transmission would let repeater spacing be stretched to 80 km on land and to 120 km at sea.

The first possible change in architecture is a shift to 1550 nm, where loss is lower and repeater spacings might be doubled or tripled. Success hinges on overcoming the chromatic dispersion problem.

The rewards could be significant. For example, Japanese engineers believe repeater spacings in 400-Mbit/s land systems might be doubled to 80 km from the present 40 km. Researchers at the Nippon Telegraph and Telephone Corp. have designed a 400-Mbit/s repeaterless 1550-nm submarine cable system that could transmit signals 120 km without a repeater, triple the distance possible at 1300 nm. Developers also are working on 1550-nm transmission for ocean-spanning cables, and planners of cables to be laid in the 1990s are counting on operating at that wavelength.

The most-detailed plans, NTT's design for a repeaterless 400-Mbit/s system, call for distributed-feedback laser diodes, which emit light at a very narrow range of wavelengths, and an InGaAs avalanche photodiode. Their 1550-nm lasers have a narrow enough spectral linewidth that dispersion effects are not significant. The loss budget for their system is shown in *Table 15-7*.

Table 15-7. Loss Budget for Unrepeatered 400-MBit/s, 1550-nm Undersea Cable System*

Power coupled into fiber from transmitter	+1 dBm
Receiver sensitivity at 10^{-11} BER	−39 dBm
120 km of fiber @ 0.27 dB/km	32.4 dB
System margin	7.6 dB

*Designed by Nippon Telegraph and Telephone.

An alternative approach, described in Chapter 4, is to use fibers with zero dispersion at 1550 nm. Each approach has advantages. Dispersion-shifted fibers are on the market, but few have been installed. Narrow-line 1550-nm lasers could work with already-installed fibers, but they are hard to make, and the first commercial models were not announced until late 1986.

The best performance may come by combining narrow-line lasers and dispersion-shifted fibers.

In fact, neither approach may emerge as the true winner. The best way to get the ultimate in performance from direct-detection fiber-optic systems may be to use narrow-linewidth transmitters with dispersion-shifted fibers. That would allow operation at the highest possible speeds over the longest possible distances.

Coherent Transmission Systems

Coherent detection systems promise much greater sensitivity than present types.

The next step, already visible in research papers, is to make fiber-optic systems coherent, as shown in *Figure 15-10*. The basic idea is similar to that used in heterodyne radio receivers. The light signal reaching the receiver is mixed with light from a local-oscillator laser that is close in frequency (i.e., in wavelength) to the transmitter. The two signals are mixed in a detector that detects the difference frequency. If ν_t is the transmitter frequency and ν_o is the oscillator frequency, the resulting signal is at a difference or intermediate frequency $\nu_d = |\nu_t - \nu_o|$ (the absolute value sign lets ν_o be greater than or less than the laser frequency).

Figure 15-10.
Elements of a Coherent
Transmission System

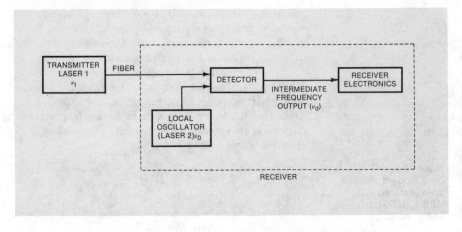

In experiments, this intermediate frequency is a few gigahertz or less, so conventional electronics can enhance the signal. Coherent detection avoids noise inherent in direct detection of amplitude-modulated light by an avalanche photodiode and allows use of other modulation schemes. The result is that lower signal levels are needed to get the same level of performance. Today's avalanche photodiodes need about 1000 photons/bit to operate at a bit-error rate of 10^{-9}. An ideal coherent system would only need 10 or 20 photons/bit to achieve that error rate. Nobody's done that well in the lab—yet—but they have gotten down to about 70 photons/bit.

The greater sensitivity of coherent detection could allow longer repeater spacings or higher transmission speeds.

That greater sensitivity can be translated into two advantages—one obvious and one subtle. If less signal is needed, the signal could travel farther through fibers between repeaters. If a practical coherent system had 10 dB greater sensitivity (e.g., needing 100 photons/bit rather than 1000), it could travel 40 km further through 0.25-dB/km fiber.

Alternatively, that higher sensitivity could be turned into faster speed. Each pulse contains a certain number of photons (i.e., a given amount of energy). Higher sensitivity would mean that less time is needed to accumulate the number of photons needed to detect a bit, if the power level stays the same. If it took 1 ns to detect a pulse with direct detection, that 10-dB improvement would allow detection in just 0.1 ns—opening the possibility of a tenfold improvement in speed. Another possibility is using the higher sensitivity to improve both speed and distance.

Use of coherent transmission and wavelength-division multiplexing might greatly expand fiber system capacity.

Coherent transmission could greatly expand system capacity when combined with wavelength-division multiplexing. Although AT&T used wavelength-division multiplexing in one major multimode-fiber telephone system, that technology was shunted to the background when single-mode fibers came into use.

The attraction of coherent transmission is that it would allow channels to be closer in wavelength. Coherent transmission enthusiasts say their techniques might allow channels to be only about 100 MHz apart, rather than the roughly 100 GHz now required. Electronics could separate channels in the intermediate-frequency signal, much as they separate radio stations in a heterodyne radio receiver.

It will not be easy to get all these advantages in the real world. Coherent detection requires lasers that emit an extremely narrow range of wavelengths, which must be matched to equally narrow-line local oscillators. Precise tuning of those lasers to the desired frequencies requires what is best labelled "tender loving care," which is hard to achieve in the field—especially under a few kilometers of sea water. The technology unquestionably has promise, but that promise is not going to be easy to realize.

Important problems remain in making coherent detection practical.

Mid-Infrared Fibers

Coherent detection may not be the ultimate improvement in fiber transmission. Further toward the technological horizon is the possibility of new fibers with attenuation as low as 0.001 dB/km, mentioned briefly in Chapter 4. If you run such a fiber 6000 km across the Atlantic Ocean, you've used only 6 dB of attenuation!

New types of non-oxide glass fibers in development might have loss as low as 0.001 dB/km.

These fibers are made of non-oxide glasses with much lower intrinsic losses than silicate glasses at longer mid-infrared wavelengths. Major problems remain in taming that technology. Researchers have only recently pushed loss of such fibers below 1 dB/km. The materials are difficult to handle and to make into fibers. Some may be extremely vulnerable to moisture. For example, alkali halides such as sodium chloride have extremely low attenuation but actually absorb water from the environment. Leave pure sodium chloride crystals out in a moist atmosphere too long and you'll find a puddle of salty water when you return. Imagine a cable containing such salt fibers immersed in water!

Many questions remain unanswered. Are suitable light sources and detectors practical? Do the fibers have insoluble problems in areas like dispersion? Can fibers be made rugged enough? If all those problems can be solved, it's possible to envision cables with repeaters thousands of kilometers apart—perhaps even on opposite shores of the Atlantic Ocean. But don't hold your breath.

Who Needs All the Capacity?

You don't have to be a complete cynic to wonder if anyone really needs all the capacity offered by future fiber systems. If there are plenty of spare fibers in today's backbone systems, why bother with coherent fiber systems?

Demands for bandwidth seem certain to continue to increase.

In part, it's the search for new ideas that drives many people who work on the cutting edges of technology. But leaving that aside, the need for transmission bandwidth has increased steadily as time passes. It's not as simple as saying "Communications expands to fill the bandwidth available." New services require more bandwidth. Data transmission has been growing rapidly. Image transmission is bandwidth-hungry. Someday those trends, like any other, may change, but we don't seem to be near that point yet.

WHAT HAVE WE LEARNED?

1. Telecommunications is a broad term, encompassing telephony, telegraphy, and other forms of electronic (and optical) communications. It includes voice, video, and data communications over long distances. The global telephone network interconnects with many other systems and carries non-telephone traffic as well.

2. Many separate telecommunication networks exist; some (but not all) interconnect with the telephone system.

3. Many signals are merged together for transmission over long distances and separated at the other end. The telephone industry has a hierarchy of transmission rates, with lower-speed systems feeding into faster ones. The higher-speed systems generally run for longer distances.

4. Specifications for the North American telephone hierarchy become hazy at high speeds because AT&T had not envisioned transmission above 274 Mbit/s when it defined its T carrier system. Long-distance phone companies now use 400 Mbit/s for most of their nationwide networks, and higher speeds are coming.

5. Other transmission media used in the global telephone network include twisted wire pairs, coaxial cable, microwave ground links, and microwave satellite links.

6. The telephone network is assembled from many point-to-point transmission links. Cables are good for point-to-point transmission, but satellites are better for reaching spread-out points with little traffic.

7. Long-distance backbone systems connect major population centers. Regional networks must interface with systems transmitting to distant areas and with local services that operate on smaller scales.

8. A 400-Mbit/s system can transmit signals 40–50 km between repeaters using single-mode fiber, an InGaAsP laser coupling -2 dBm into the fiber, and an InGaAsP PIN photodiode with -32 dBm sensitivity.

9. Telephone networks are designed for future expansion. Fiber systems allow for upgrades by including spare fibers in cables and by upgrading transmission speed on operating fiber pairs.

10. Telephone companies plan to upgrade capacity by increasing transmission speeds to 1.7 Gbit/s on existing fibers and by using spare fibers already installed in cables.

11. Fiber optics is replacing coaxial cables for submarine transmission because they allow transmission at higher speeds over longer distances.

12. There are three types of submarine fiber systems: those with no repeaters spanning 50 km or less, moderate-distance systems with a few repeaters, and ocean-spanning systems with dozens of repeaters.

13. The TAT-8 transatlantic fiber cable will transmit 278 Mbit/s through two pairs of single-mode fibers at 1300 nm. Speech compression will raise capacity to 40,000 voice circuits.

14. Submarine cables in shallow waters must be heavily armored to prevent damage from ships and fishing.

15. New technology should permit even longer repeater spacings and higher transmission speeds (e.g., 1550-nm transmission should double repeater spacings).

16. Coherent transmission systems demonstrated in the laboratory are more sensitive than direct detection, which could allow greater repeater spacings or higher-speed transmission. They might further expand capacity when used with wavelength-division multiplexing. However, major problems remain to be overcome before they can be used in practical systems.

17. The longest-term technology is mid-infrared fibers. It promises loss as low as 0.001 dB, which might allow a repeaterless transatlantic cable. However, attenuation has yet to reach that level, and major problems remain in making the fibers.

WHAT'S NEXT?

In Chapter 16, we'll look at the rest of the telecommunications network, including the part that contains the most cable: the connections between subscribers and switching facilities.

Quiz for Chapter 15

1. Which transmission speed is used both in Europe and the United States?
   a. 64,000 bit/s.
   b. 1.5 Mbit/s.
   c. 45 Mbit/s.
   d. 405 Mbit/s.
   e. 565 Mbit/s.

2. A 400-Mbit/s system operating in a backbone telecommunication network with repeater spacing of 40 km would use which type of light source?
   a. 1300-nm InGaAsP laser.
   b. 1300-nm InGaAsP LED.
   c. High-power 850-nm GaAlAs laser.
   d. 1550-nm InGaAsP laser.
   e. 1100-nm InGaAs laser.

3. How do systems operating at 400 Mbit/s and 1.7 Gbit/s for telephone backbone systems differ?
   a. Use different types of fiber.
   b. Lower-speed systems use LED transmitters.
   c. Higher-speed systems require shorter repeater spacings.
   d. The spectral width of the laser source is different.
   e. None of the above.

4. What are 400-Mbit/s fiber-optic systems used for?
   a. Transmission between central offices and subscribers.
   b. Long-distance transmission in nationwide backbone systems.
   c. Local-area networks.
   d. Temporary connections within telephone facilities.

5. Land-based long-distance phone cables include extra fibers to:
   a. provide room to expand services.
   b. replace broken fibers.
   c. allow fibers to be changed to take advantages of future improvements in fiber technology.
   d. use up surplus fibers.

6. What advantages do optical fibers have over coaxial cables for transoceanic submarine cables?
   a. Longer repeater spacings.
   b. Smaller cable size and greater flexibility.
   c. Higher transmission capacity.
   d. All of the above.

7. A 400-km long submarine fiber cable has repeater spacing of 50 km. How many undersea repeaters must it include?
   a. 7.
   b. 8.
   c. 9.
   d. 10.
   e. None of the above.

8. What part of a submarine cable is armored?
   a. The deep-sea part subject to high static pressure.
   b. Any part that must be recovered from the sea bed.
   c. The shallow-water part vulnerable to shipping damage.
   d. The whole thing.

9. Which of the following requirements are more stringent for transoceanic cables than for moderate-distance cables such as UK-Belgium 5?

   a. Armoring.
   b. Repeater lifetime.
   c. Temperature dissipation.
   d. System loss margin.

10. Which technology is expected to be used in the next generation of transoceanic cables?

   a. Coherent transmission.
   b. 1550-nm transmission.
   c. Non-oxide (fluoride) glass fibers.
   d. Graded-index fibers.

Short-Haul and Subscriber Telecommunications

ABOUT THIS CHAPTER

The international telecommunications network is more than just an array of long-distance transmission links. It is the interconnection of many networks that work on a smaller scale, regional and local systems that reach all the way to individual telecommunication subscribers. Optical fibers are used throughout that network. Fibers are most common for multiplexed transmission of many circuits, but they also are used in other parts of the telecommunications network. They are least common for transmission to individual subscribers.

BASIC CONCEPTS

We can look at uses of fiber optics in local and regional telecommunications in several ways. These multiple viewpoints will help you understand how telecommunication signals are distributed regionally and locally. Some different-sounding concepts are similar ideas hiding behind different names.

Telephone Networks

The telephone network interconnects all individual phones by running lines through specialized routing and switching equipment.

The most ubiquitous telecommunication network, particularly regionally and locally, is the telephone network. It starts at your home or business with a single telephone line (a voice circuit). Signals from individual phone lines are multiplexed with others at either a local distribution point or a switching center, as shown in *Figure 16-1*. These multiplexed signals are transmitted to other telephone facilities at the hierarchy of speeds shown in *Table 15-1*. In the figure, a long-distance call (from subscriber A), goes through a local distribution point, a local switching office, a long-distance network (Carrier 2), and a long-distance switching office before reaching its destination (subscriber D).

**Figure 16-1.
Simplified View of a
Telephone Network**

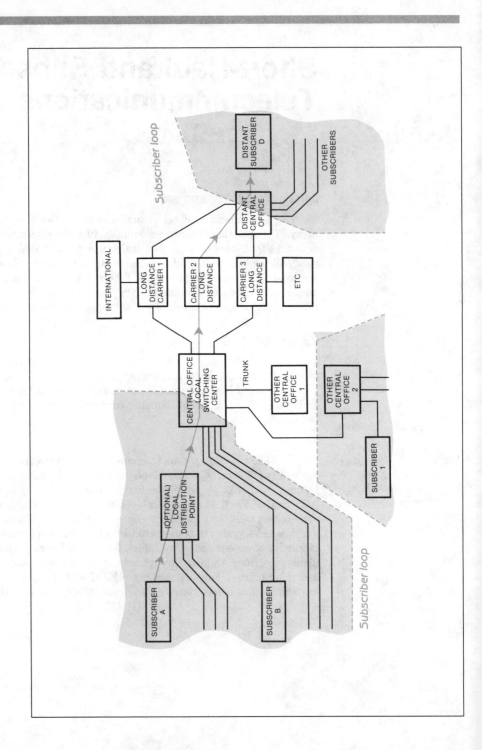

The telephone industry has its own terminology for this distribution system. Cables from the central office to the subscriber are part of the local loop or subscriber loop. Those carrying many voice circuits between telephone switching facilities are trunks.

Most signals start in analog form on conventional telephone lines. This is true even for computer data sent via modems, which modulate that data into analog signals that can be transmitted by the telephone network. (The word "modem" originated as modulator-demodulator.) The analog signals are converted to digital form before being processed by digital switches or transmitted over digital transmission lines. Eventually, at the other end of the transmission line, they are re-converted back to analog form for the receiving telephone.

Private Lines and Bypass

The telephone network described above is a public network, geared to serving individual telephones in separate homes or businesses. Larger businesses have different requirements. These might include:

- A transmission line directly from the central office to an internal telephone-switching system, called a PBX (for private branch exchange)
- A cable to carry signals directly from company offices somewhere else, without using the local telephone system (e.g., between two large plants separated by a few kilometers, from a plant to a satellite dish across town, or from a plant directly to a long-distance phone network)
- A transmission line able to carry digital data directly from telephone-switching facilities to a large computer at speeds higher than the 56-kbit/s equivalent of a phone line (The commonest such speed is the 1.5-Mbit/s T1 rate)
- A dedicated telephone circuit between two company facilities, typically far apart.

A transmission system leased from the phone company is called a private line. If somebody else owns and operates it (either the company or a third party), it's called bypass because it bypasses phone-company lines.

Why bother with private lines or bypass? They can be valuable either to save money or to get better transmission—or sometimes both. As with many other goods and services, buying telephone service in bulk can reduce the price. Bypass avoids telephone company overhead altogether.

Directly contracting with a carrier for digital services has important advantages when sending much computer data. Most computer modems can send only 1200 bit/s over an analog voice telephone circuit, which when converted to digital form is 56,000 bit/s. Modems also are expensive. Hooking the digital transmission line directly to the computer makes available the full 56,000-bit/s capacity of a voice-grade digital circuit. Thus, you can transmit data much faster than otherwise possible, on fewer voice circuits. It also avoids errors in analog-to-digital conversion and quality problems in analog voice circuits. Local and long-distance carriers offer digital fiber-optic transmission as a premium service for customers who need fast, error-free transmission. Speeds are 1.5 Mbit/s and up.

Regional-Area Networks

In Chapter 18, we will examine local-area networks (LANs), which transmit data within a small area (typically a building or campus) faster than a telephone network. LANs may carry voice and video signals as well as data and other information. Linking many LANs together gives a regional-area network (RAN). Forget its origins, back far enough away, and what you see looks suspiciously like the local telephone network linking together many business subscribers who require high-speed, multiline services.

Integrated Service Digital Networks (ISDN)

An Integrated Services Digital Network will carry many digitized tele-communication services.

As we saw in Chapter 15, much of the telephone industry is moving toward an Integrated Service Digital Network, or ISDN. The idea is to turn all telecommunications into digital strings of bits and carry them all through the same network. The network wouldn't care if bits represent computer data, voice, or video. The receiver on the other end would decode the bits and convert them into the desired signals. Fibers are a natural choice as a transmission medium because of their high-speed digital capacity.

The idea has both theoretical and commercial appeal. The theoretical appeal is the tremendous flexibility of a single network that can route any signal between any pair of subscribers. It's the sort of concept that makes the eyes of visionaries sparkle with dreams of universal communications. It also lights up the eyes of telephone marketing managers who want to carry as much traffic as they can. It is undoubtedly a logical direction for telephone networks to evolve, but it will take time and there will be competition from other telecommunication carriers, notably cable-television networks.

Broadband Services

The concept of broadband services is similar to ISDN, but it does not explicitly demand digital services.

It wasn't too long ago that telecommunication visionaries talked about broadband networks that would carry all sorts of signals to subscribers. If you think that idea sounds suspiciously familiar, you're right. The concept of broadband services is essentially the same as that of the Integrated Services Digital Network, only it does not rely so explicitly on digital technology. Nevertheless, "broadband" appears to be trailing "ISDN" in the futuristic buzzword race.

Special-Purpose Telecommunications

Telecommunications can exist outside the telephone network. One advantage telephone companies have is that they own rights of way allowing them to run cables along streets and other routes. However, utilities and railroads also own rights of way and need telecommunications to monitor their own operations (e.g., to make sure a power substation delivers the correct voltage and to follow trains on their routes). Some of these organizations are planning or offering to sell or lease excess communications capacity to other companies that need that capacity, either for their own use or to resell as telecommunications common carriers.

Hardware Realities

Many different telecommunication networks can use the same fiber-optic hardware.

One miracle of modern merchandizing is the ability to put the same product into several different packages and sell it to different customers looking for different things. If you've been reading carefully, you may begin to suspect that this is happening in fiber-optic communications. Do telephone companies and the companies offering bypass services buy the same type of hardware to carry signals across town? Of course they do. Is there really a difference between a 45-Mbit/s trunk line laid in telephone company ducts and one along an electric utility's high-voltage power lines? Yes, but the difference is mostly in the packaging. The same type of fiber might be used in different cables, and the same transmitters and receivers might be used, with different levels of shielding against electromagnetic interference.

In this book, our concern is with the realities of hardware, and that's what we're going to talk about for the rest of this chapter. We'll look at three forms of what we consider short- and moderate-distance telecommunications: trunk telecommunication lines, subscriber loop systems in present use, and future fiber-optic connections to subscribers. With some subdivisions, these cover the major hardware realities.

TRUNK SYSTEMS

The first success of fiber optics was in trunk systems carrying 45-Mbit/s T3 signals.

The first big success of fiber-optic communications was in trunk systems operating at the 45-Mbit/s T3 rate. Fiber optics filled a void because neither microwave systems nor metallic cables were economical at that speed. Until fibers came along, some communication specialists thought that T3 transmission might find few uses because no suitable technology was available. Fiber optics proved them wrong, and T3 lines—and the 90-Mbit/s T3C rate—have come into widespread use.

System Requirements

Trunk systems operate at 45–400 Mbit/s over distances to tens of kilometers.

The basic requirements for trunk transmission are to carry signals at 45–400 Mbit/s between central offices, economically and without repeaters if possible. However, there can be considerable variation in what "between central offices" means. In urban and suburban areas, central offices are typically several kilometers apart, while in rural areas distances between offices can be tens of kilometers. In most areas with reasonable population densities, fiber optics can transmit signals between switching offices without repeaters.

Four fiber cables can fit in an underground duct that could hold only one metal cable.

The nature of cable routes in metropolitan areas also helped push the early use of fiber optics. Most cables are routed under city streets through hollow ducts, such as that shown in *Figure 16-2*. Duct installation is expensive and inconvenient because it requires digging up city streets. Once the ducts are in place, they become a fixed asset—as long as room remains. The problem is that ducts fill up as transmission needs increase. Once no more duct space is left, the phone company must dig up the streets again. Digging up the streets can cost much more than new transmission hardware, so phone companies will pay well to increase duct capacity.

Figure 16-2.
Hollow Plastic Ducts
House Cables under
City Streets

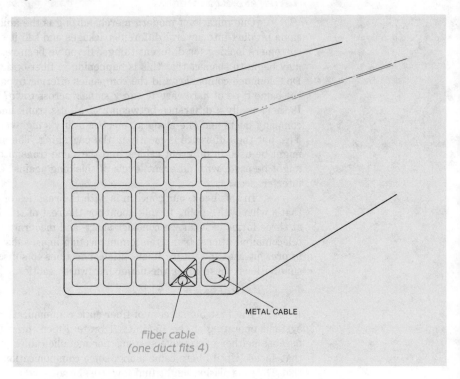

METAL CABLE

Fiber cable
(one duct fits 4)

That is a major attraction of fiber-optic cables. Four standard fiber-optic cables can fit in a duct that would hold a single (lower-capacity) metal cable. Extra fibers in the fiber cable can leave room for future expansion. The advantages are large enough to justify pulling out old metal cables to make room for new fiber cables.

Rural cables normally are installed overhead on utility poles or underground by direct burial. (A tractor with a plow opens a trench into which the cable is lowered, then the trench is filled.) Because cable size is not as crucial a concern as with cable pulled through ducts, it took a little longer for fibers to become accepted for such applications. However, fibers now are the standard transmission medium for virtually all such trunk transmission systems.

System Architecture

Fiber trunk technology
has evolved rapidly.

The first fiber-optic trunk systems used 850-nm GaAlAs lasers to transmit several kilometers through graded-index multimode fibers. By about 1980, phone companies began installing graded-index multimode fiber systems with 1300-nm InGaAsP lasers. Now virtually all new trunk systems contain step-index single-mode fiber and 1300-nm lasers. The reasons behind those changes are indicated in *Table 16-1*, which compares representative first-, second-, and third-generation fiber systems. These figures indicate performance that can be attained with each generation of fiber technology. That maximum data rate or repeater spacing is not required for all systems.

Table 16-1.
Typical Performance of Fiber-Optic Trunk Systems with Laser Sources

Type	First-generation 850-nm graded-index multimode	Second-generation 1300-nm graded-index multimode	Third-generation 1300-nm step-index single-mode
Data rate	45 Mbit/s	90 Mbit/s	400 Mbit/s
Fiber loss	3–4 dB/km	1 dB/km	0.5 dB/km
Repeater spacing	8 km	20 km	40 km
Limiting factors	Loss/bandwidth	Loss/bandwidth	Loss

It might seem that telecommunications companies would pick the technology with the lowest performance able to meet their needs (e.g., first-generation systems for a 5-km T3 link). However, the economics do not favor that choice. Single-mode fiber costs less to manufacture, and most telecommunication companies prefer to leave an upgrade route open. As mentioned in Chapter 15, cable-installation economics favor including extra fibers in a cable and upgrading capacity of installed fibers, rather than installing new cables. Thus, single-mode fiber and 1300-nm InGaAsP lasers are used in virtually all new cable trunk installations.

Telephone companies normally install aerial, direct-burial, or ducted cables, which were described in Chapter 5. The degree of armor and other protection used depends on system requirements. In lightning-prone areas, for instance, totally non-metallic aerial cable is preferred. Armored cable is needed to protect against gnawing rodents. Pressurized cable may be used in areas with high water tables or flooded manholes.

Phone companies install aerial, ducted, and direct-burial cables.

Special cables are made for electric utilities, as shown in *Figure 16-3*. These serve two functions: as ground wires for high-voltage power lines and as fiber-optic cables for telecommunication. The outer metal part of the cable serves as a ground wire. Because optical signals are not affected by electric currents, power transmission does not affect the optical signals carried by fibers in the cable core. Here, too, the major advantage is savings in installation cost, in this case by adding fiber-optic transmission capacity to a wire that would otherwise be required. It is even possible to combine electric power and fiber-optic signal transmission in an underwater cable.

**Figure 16-3.
Cable Transmits Both
Electrical Power and
Optical Signals**

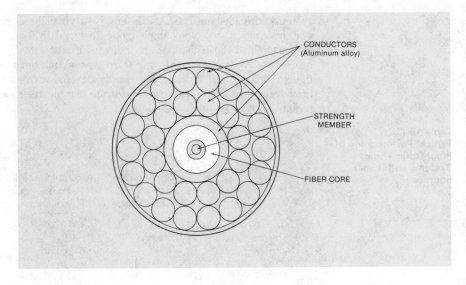

CONDUCTORS
(Aluminum alloy)

STRENGTH
MEMBER

FIBER CORE

Older generations of technology remain in use today.

The rapid progress of fiber-optic trunk technology does have some important practical consequences in working with existing rather than new systems. Although very few new multimode trunks are being installed many remain in the field, operating at 850 or 1300 nm. Those systems may no longer be cost-effective to install, but they are not being pulled out and thrown away. As long as those cables work, telephone companies will get rid of old analog wire cables first. The fraction of installed fiber-optic systems using third-generation technology will continue to increase, but do not blithely assume all systems are single-mode.

Despite the rapid changes of recent years, third-generation fiber technology seems likely to retain its domination of telecommunication trunking. Except when trying to stretch repeater spacing beyond about 40 km, there is no reason to shift to more-costly fourth-generation 1550-nm single-frequency lasers or dispersion-shifted fibers. There isn't much call for that in trunk systems. As we saw in Chapter 15, the bandwidth of conventional single-mode fibers at 1300 nm is high enough to allow 1.7-Gbit/s transmission over the same distances 400-Mbit/s systems operate.

TODAY'S SUBSCRIBER LOOP

Fiber optics are little used in the subscriber loop, which contains most cable in the telecommunications network.

We have seen why fibers have come to account for most new trunk and long-haul systems. However, fibers account for only a very small fraction of the part of the telecommunication network that contains most cable—the subscriber loop. Despite the tremendous advances in fiber-optic technology, pairs of copper wires remain more cost-effective for low-speed transmission over modest distances of a few kilometers or less, such as between your home and the local switching office.

Operating Requirements

Much new equipment has been added to the telephone network.

The basic elements of the subscriber loop are shown in *Figure 16-4*. The classical vision of the subscriber loop was a lot of heavy black dial phones hooked by wires to huge banks of electromechanical relays in a big brick building with "Bell" somewhere on the outside.

**Figure 16-4.
The Telephone
Subscriber Loop**

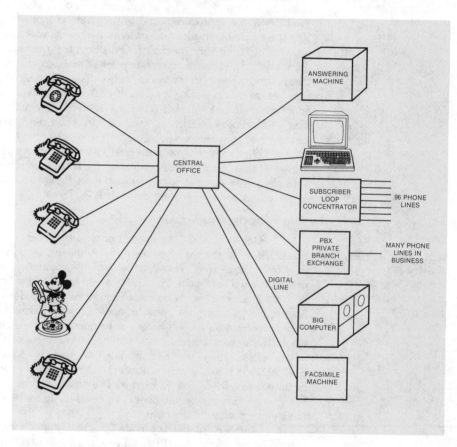

Lots of new equipment has been added to both ends of the subscriber loop since then. The big brick building still may say "Bell" on the outside, but on the inside matters are more complicated. Ownership of the building—and the internal equipment—is often divided between AT&T and the regional Bell operating company. In some, equipment on one side of a line taped down the middle of the floor belongs to AT&T's long-distance network; that on the other side is part of the regional phone company's network. Other long-distance carriers may have their equipment in that building or elsewhere nearby. More and more switching is done electronically, by special-purpose computers. Your phone probably is neither heavy nor black and has pushbuttons instead of a dial. However, pairs of copper wires still run between your house and the central office, and they still provide what the telephone industry calls POTS—plain old telephone service.

Plain old telephone service requires analog bandwidth of 4 kHz to give what the telephone company considers intelligible speech. Intelligible is not high fidelity. Telephones transmit sounds at about 200–3400 Hz; a good stereo can reproduce sound to the limits of human hearing, 20–20,000 Hz. The 4-kHz bandwidth also includes control signals and the sounds of rotary and Touch-Tone (pushbutton) dialling. (By the way, pushbutton dialling is analog, based on unique pairs of audio-frequency tones transmitted when each button is pushed. The old-fashioned rotary dial signals that it is replacing a series of clicks—ironically a more digital approach than tone signals.) Wires are perfectly adequate for this job. In fact, with proper conditioning, twisted wire pairs can transmit 1.5 Mbit/s a kilometer or so without repeaters.

Why, then, do you need fiber optics for the subscriber loop? The answer is that you don't for traditional POTS to existing residential areas with one phone line per house. However, as *Figure 16-4* shows, many parts of the subscriber loop require more than POTS.

Fiber-Optic Applications

One example of the use of fibers in the local loop is areas of new construction. In the old days, the phone company would string new phone lines from the nearest central office until the population crossed some predetermined threshold. Then they would build a new brick building and fill it with banks of new electromechanical relays. Modern electronics and fiber optics give the new phone company a new alternative. They can run wire pairs from a group of homes or businesses to a remote concentrator, where the signals are multiplexed and sent via fibers or wires back to the central office. One example of this is a unit made by AT&T called the SLC-96 (subscriber loop concentrator), which combines signals from 96 voice circuits onto a 6.3-Mbit/s T2 carrier. The data rate and distance covered are low enough that multimode fibers can be used, although a shift to single-mode fiber may be in the offing.

Another example is private lines to customers with high usage or private branch exchanges (PBXs). These are businesses with tens, hundreds, or even thousands of phones. Because many phone users are concentrated in one spot, it is much simpler to run a single cable there than to use many separate wire pairs. Both single- and multimode fiber optics are used.

Some customers want high-quality digital transmission direct to their premises. Typically, they need to transmit computer data at much higher speeds and without the transmission errors that can sometimes occur on analog phone lines. Telephone companies now offer fiber-optic transmission for this purpose explicitly because of its high quality. Data rates of 1.5–45 Mbit/s are sent over single- or multimode fibers.

These applications remain limited, but they are important because they represent the first of what are likely to be many uses of fiber in the subscriber loop.

Pairs of copper wires are perfectly adequate for 4-kHz analog POTS.

Fibers may be used in the local loop to multiplex voice circuits to many subscribers or deliver high-speed digital signals to subscribers.

System Architecture

Both single- and multimode fiber can be used in the subscriber loop.

From a functional standpoint, multimode fibers are well matched to short-distance transmission at moderate speeds using LED sources. Single-mode fibers also are possibilities, even with LEDs. Representative loss budgets are shown in *Table 16-2*. The system loss margin is large enough to allow multimode transmission even at the higher-loss 850-nm wavelength. However, chromatic dispersion is too high for an LED source to transmit 45 Mbit/s at 850 nm.

Table 16-2. Loss Budgets for Single- and Multimode Fiber in Subscriber Loop (3-km Transmission) Using 1300-nm LEDs or Lasers

Item	Multimode, 62.5/125 (conventional LED)	Single Mode (with edge-emitting LED)	Single Mode (with low-cost laser)
Fiber input	−13 dBm	−20 dBm	−10 dBm
Fiber loss	3 dB	1.5 dB	1.5 dB
Connector loss	2 dB	3 dB	3 dB
Receiver sensitivity	−35 dBm	−35 dBm	−30 dB
System margin	18 dB	10.5 dB	15.5 dB

Light from edge-emitting diodes is directional enough that a reasonable part of it can be collected by a single-mode fiber. Thus, as also shown in Table 16-2, single-mode fibers could transmit LED output over short distances in the subscriber loop, but the required edge-emitting diodes cost significantly more than surface-emitters. Another alternative is the use of lasers, packaged inexpensively by automated techniques, coupling about 0.1 mW (−10 dBm) into a single-mode fiber. Note that the higher laser power allows use of less sensitive, lower-cost receivers. Although the general trend in the telephone industry is toward single-mode fibers, the much higher light-collection efficiency of multimode fibers could tip the balance in their direction for the short distances of the subscriber loop.

TOMORROW'S SUBSCRIBER LOOP: FIBERS TO THE HOME

Optical fibers are not needed to carry POTS to homes. Installing the vast bandwidth of optical fibers for that task is like building a freeway for oxcarts. Suppose, however, that we turn the problem around and ask what services we could bring to home and business users if we had the vast bandwidth of optical fibers.

The huge capacity of fibers opens the possibility of new services to homes, breaking the bandwidth bottleneck.

This puts us into the exciting realm of possibilities. Fiber optics essentially changes the rules for telecommunications. Engineers have long had to squeeze services into a limited bandwidth. The huge capacity of fibers breaks the bandwidth bottleneck.

What new services would you like? You can let your imagination loose. Optical fibers do not have unlimited capacity, but single-mode fibers could bring at least hundreds of megabits per second to homes. They could carry services such as alarms, video telephones, high-speed computer data, videotex and information services, high-resolution television signals, and what have you. *Table 16-3* lists some possibilities proposed by communication planners, with rough indications of the capacity they would require. The requirements listed as "Low" probably could operate with less than a single 56-kbit/s digitized voice channel. Many services are listed as requiring the equivalent of one digital voice channel, although future developments could push those requirements up (to offer higher-quality transmission) or down (by compressing multiple services on a channel).

**Table 16-3.
Possible Services on
Fiber-Optic Networks**

Service	Bandwidth Requirement
Telephone	56 kbit/s or up
Computer data	56 kbit/s or up
Switched video (conventional TV quality, per channel)	45–90 Mbit/s (depends on compression)
High-resolution TV (per channel)	200 Mbit/s? (depends on compression)
Video telephone/teleconferencing	1–10 Mbit/s?
Simultaneous voice/data transmission	Circa 100 kbit/s
Electronic mail	Low
Automatic meter reading	Low
Fire/burglar alarms	Low
Home monitoring systems	Low
Home remote control systems	Low
Computer conferencing	56 kbit/s or up
Videotex	56 kbit/s or up
Information Services	56 kbit/s or up
Home banking	Low
Home-shopping (teleshopping)	1–10 Mbit/s? (one-way still video)
Educational services	56 kbit/s or up
Advanced voice capabilities	56 kbit/s or up
Facsimile transmission	56 kbit/s or up
High-quality audio	1.5 Mbit/s?
Energy management	Low
Video surveillance for home security	10–45 Mbit/s?

This wish list may look suspiciously like one for an Integrated Services Digital Network. Indeed, the people working toward an ISDN or broadband communications are talking increasingly about using fibers.

You can probably add some ideas to that list. If integrated broadband networks or ISDN do work, many new services will appear. The question of what services and who wants them will be a key one in deciding when and if optical fibers will reach homes. First, however, we should look at system requirements and concepts.

System Requirements

An integrated network must provide high transmission capacity at reasonable cost.

The basic requirements for a successful integrated broadband fiber-optic network are high-capacity transmission at reasonable cost. Most services require switching within the network, so signals can be routed to specific subscribers. Some special equipment will be needed at the transmission end to generate signals and to respond to requests from users. And some terminal equipment will be needed in the home (although what and who will pay for it remain to be seen).

The need for high-speed transmission clearly indicates that single-mode fiber is the best choice. There is always uncertainty about what "reasonable cost" means. One definition, borrowed from planners at Southern Bell Telephone, is shown in *Table 16-4*.

**Table 16-4.
Hardware Costs and Amortization (per Subscriber) for an All-Fiber-Optic Network, Envisioned by Southern Bell**

Item	Capital Cost	Monthly Writeoff
Optical switch	$135	$ 4.50
Remote equipment	420	14
Transmission equipment	334.50	11.15
Home terminal/information center	613.50	20.45
Totals	$1500	$50

The roundness of the totals makes them look suspiciously like the starting point, from which the breakdown came later. Nonetheless, cost per subscriber must be controlled if the idea is to be feasible.

There are some other subtle practical requirements that have as much to do with marketing, regulation, and organizational psychology as they do with technology.

Any new network must be compatible with the old one.

- The new network will have to be compatible with the old. For example, old and new telephones must be able to talk with each other. The new system should be compatible with existing audio and video programs and equipment. Customers may be willing to throw out their old phones if they can get better ones with the new system (as long as they can talk to someone over it). They are not likely to throw away thousands of dollars of audio and video electronics, although they might be willing to buy new high-resolution televisions if they could take advantage of a new signal transmitted into their homes.

- If customers have to pay extra for the system, they must get services worth the extra cost. Some people will buy the promise of services they can get someday. But not many people are going to pay extra for high-resolution television signals if they don't have a high-resolution monitor to see the pictures.
- Customers must have options. Just as with cable television, everybody won't want the same package of services. Computer-phobes won't want to pay for computer data links that they won't use, but computer-philes will demand them. And doubtless wars will be fought over pornographic video communications. (Some very interesting battles already are being fought over telephone sex services.)

New services will require changes in regulations.

- Existing telecommunications regulations must be altered. Despite all the hype about deregulation, telecommunications is still a regulated industry—it just isn't as heavily regulated as it used to be. Existing regulations now discourage combining video and telephone transmission on one network, but you must combine them to have an Integrated Services Digital Network.
- Organizations offering integrated services will have to fight vested interests who fear being put out of business. We're not talking about the meter readers' union, but about conflicts between separate companies offering telephone and cable-television services in the same town.
- Somebody somewhere in an organization with the technical, financial, and organizational means to offer integrated services will have to sell the idea successfully to other people in the organization.

Such concerns may seem remote from the hardware orientation in the rest of this book. However, it is those problems, together with the high cost of fiber-optic equipment, that keep plans for integrated service fiber-optic networks in the dream stage. We will return to such questions after we look at the realities of system architecture, field trials, and new concepts.

System Architecture

Network design must consider both hardware and service needs.

Design of an integrated fiber-optic network involves two levels of architecture: hardware and services (or, if you prefer, software). Although this section concentrates on hardware, the choice of services determines the hardware requirements. For instance, video services need high transmission capacity. If signals are to be directed to individual homes, rather than broadcast to all subscribers, the network must be switched.

Figure 16-5 shows a conceptual architecture for a switched video network. Because any such network would be complex, we show only services reaching an individual home subscriber. Services to businesses, government offices, and other organizations would be somewhat different.

**Figure 16-5.
A Switched Video
Network**

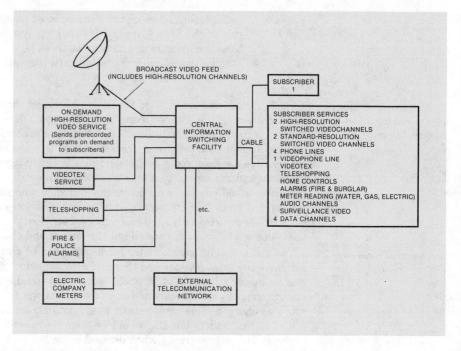

Figure 16-5.
A Switched Video Network

Switching Center

A central information facility would have functions similar to a telephone switching office. It would collect and distribute signals carrying services from information providers. In the example, these services include broadcast video (like today's cable television), switched video (programs from an automated video library), videotex and other computerized information services, and teleshopping. It also would be a switching center for audio and video telephones, electronic mail, data-transmission services, and data automatically collected from homes (e.g., meter readings and alarm signals).

Some switching might be handled outside the information center. For example, multiplexed signals destined for a group of subscribers might be sent through single-mode fibers, then divided among fibers going to individual subscribers. The choice between that approach and centralized switching will depend on system economics including:

- Relative costs of the extra fiber needed to transmit signals directly from the information center versus those of the extra transmitters and switching gear needed for remote switches.
- Percent utilization of system capacity—For example, if at most 20% of transmission capacity was used at once, you would not need as much capacity from information center to remote switch as from switch to homes. However, if nearly all the capacity was used (e.g., everyone had two high-resolution television sets on in the evening— not an unrealistic scenario), then you would need roughly the same capacity all the way from home to information center. (Telephone

multiplexing saves money because most phones are used only a small part of the time. Capacity of switching systems and trunk lines is based on average use and can at times be overwhelmed.)

- Distribution of subscribers and expansion potential.

Home Services

Home subscribers would receive all services through a single cable, containing one, two, or more fibers. They would select the services through terminals in their houses. For example, a subscriber could pick two high-resolution video programs, from broadcast services or an on-line video library. One program could go to televisions in the living room and master bedroom, and the second to a television in a child's room. Similarly, the subscriber could request a particular audio channel and make videophone calls or dial up to four computerized information services at once. The network shown would give the individual subscriber up to four telephone lines—enough to accommodate all but the largest and chattiest of families.

Some signals (e.g., meter readings) would be transmitted automatically or at prespecified times. If a household was charged for electric use as a function of time, the meter could send readings to the electric company at the start and end of each interval for which a different rate was charged. This arrangement would not only let utilities save the cost of meter readers but also would let them set rates to encourage customers to use electricity when demand is low.

Other signals might be sent automatically at the customer's option (e.g., burglar and smoke alarms). Your fire department would get very annoyed at being summoned every time your smoke detector went off while you were cooking, but you would want them to know if smoke was in the house when no one was at home. Video transmitters might let security-conscious (or paranoid) people monitor their homes while they were away. Being able to check with a video camera watching a sleeping baby might ease fears of first-time parents chatting at the neighbors. These video monitoring services might be sent over a videophone channel at much lower speed than needed for ordinary television.

All signals to and from the home would be carried in a modulated digital bit stream. A demodulator in the home would sort out individual signals and route them to the proper control systems and devices. A control center could route signals to individual devices, although such operations as picking audio, video, and data channels and dialling telephones could be done elsewhere. Signals sent back to the information center would pass through the home control center, where it would be multiplexed.

Transmission Requirements

Add up the transmission requirements and you will note a striking disparity. Services going into the home will require hundreds of megabits per second, particularly for high-resolution television. However, the only bandwidth-hungry channels leaving the home are the videophone and output

Home subscribers would receive many services through one cable.

Some signals could be sent automatically.

Signals would be sent to and from homes in a digitized bit stream.

More capacity is needed to send signals to homes than in the other direction.

from surveillance cameras, which probably could be sent via the videophone channel. Thus, a system design would need high bandwidth downstream to the home, but much lower bandwidth in the opposite direction.

Uncertainties

Most details of future networks remain to be defined.

Table 16-3 shows considerable uncertainty in bandwidth for certain services, particularly for video. That reflects the large uncertainties remaining about the form of those services. International standards groups are debating formats for high-resolution television. It probably will have about twice the resolution of present television transmission—giving it over 1000 lines per screen—but details remain to be defined. Bandwidth compression probably will be used, despite the large capacity of single-mode optical fibers, because high-resolution video needs much bandwidth. Without compression, present video channels require 90 Mbit/s, and doubling both horizontal and vertical resolution would up the requirement to 360 Mbit/s. However, some compression should be possible without noticeably degrading signal quality.

Videophone service likewise remains to be defined. Frames could be transmitted slower than in conventional television, and compression could reduce bandwidth by large factors (although at a noticeable cost in signal quality). What is not clear is what will finally emerge from development efforts. A serious but yet untested question is how potential customers will react to the trade-offs inherent with compression.

Hardware Requirements

System hardware will have to transmit hundreds of megabits per second.

I have been deliberately vague about the hardware needed for the multiservice network of *Figure 16-5*. That's because as long as the services themselves are undefined only very general specifications can be written for hardware.

LED transmitters could operate over a few kilometers of single-mode fiber.

Transmission will be at hundreds of megabits per second—probably several hundred. If the network takes a shape like today's telephone network, the distances to individual subscribers will be a few kilometers. Laboratory experiments have shown that semiconductor lasers are not necessary for such distances—1300-nm LEDs can send up to a few hundred megabits per second a few kilometers through single-mode fibers with zero dispersion at that wavelength. However, inexpensively packaged lasers may be economical and allow use of lower-cost receivers than special high-speed LEDs. The jury is still out.

The simplest, but most fiber-intensive, way to bring signals back from the home to the information center is with a second fiber. The high bandwidth of single-mode fibers is not essential, although using only single-mode fibers could simplify logistics. The economics might favor a multimode fiber and LED, perhaps at 850 nm where sources and detectors are less costly than at 1300 nm. However, such a system would need new cable to upgrade for bidirectional transmission at high speeds. A third alternative would be bidirectional transmission at two wavelengths through one single-mode fiber. Transmission to the subscriber might be with a laser or LED at the zero-dispersion wavelength; transmission back to the information center would be with an LED at a wavelength where fiber dispersion is higher.

All these conceptual designs remain in the realm of speculation; their future will depend on the demand for services and the evolution of the technology.

Business Service Requirements

Businesses generally pay for more non-entertainment telecommunications than homes.

Traditionally businesses have demanded more telecommunication capacity than home users—and have paid more for it. The first phones were installed in offices, not homes, as were the first systems for computer data transmission. Residential services are fascinating because of their impact on society, but by demanding more services, might business have more impact on the telecommunications industry?

In some ways, the answer is yes. Businesses will require much more data-transmission capacity. If videophone and videoconferencing ever are commercially successful, businesses will be the first to pay the bills. Businesses also will demand more channels and better transmission than home subscribers. News reporters do not want to conduct interviews over noisy phone lines, but they might tolerate some static to save a few dollars on a personal call. If the Integrated Services Digital Network becomes a reality, businesses will be the first "fibered" up. On the other hand, businesses won't want some entertainment services in *Table 16-3* (e.g., high-resolution television).

Evolution of Integrated Services

The vision of the wired city is evolving into the fibered society.

Integrated service networks do not exist, but the idea already has a long history. For years, telecommunication visionaries have been enchanted by the idea of the wired city or information society where vast amounts of information are exchanged by telecommunications. Visions of videophones date back to the Buck Rogers era of pulp science fiction over 50 years ago, as shown in *Figure 16-6*. Futurists speculated about transmission of signals over wires or radio waves.

The first to suggest optical fibers was John Fulenwider, speaking at the International Wire & Cable Symposium in 1972. In the years that followed, other people had similar ideas. Some of them even got money to build their dream systems. Japan's Ministry for International Trade and Industry spent an estimated $40 million on a prototype fiber-optic network called HI-OVIS that served 150 homes in the town of Higashi-Ikoma. A Canadian government-industry consortium tested fiber-optic links to about 150 homes in the tiny town of Elie, Manitoba. I lumped the concepts together as the fibered society.

Fiber services to homes work, but they are costly.

The technology worked, and the residents liked the services. HI-OVIS had two-way video transmission, so subscribers could do such things as have demonstrations of cooking in their own kitchens transmitted over the entire network. Young subscribers liked the on-request video service so much that they wore out tapes of children's programs. (The cold-hearted adults

Figure 16-6.
Videophones Were Part
of Hugo Gernsback's
RALPH 124C41+
(Courtesy Fantasy Books)

dealt with that problem by stopping the most popular children's programs.) Elie residents loved the new television channels they received, and their children enjoyed playing with the computerized information service.

The bad news was cost—tens or (for HI-OVIS) hundreds of thousands of dollars per subscriber. The developers had known costs would be high for the prototype systems, but they had hoped that new fiber technology would bring them down. They did, but not enough. Field tests continue in Japan and West Germany. Only in France have a substantial number of people had fiber optics strung to their homes, starting with residents of the coastal resort city of Biarritz. However, France is clearly an exception in the world of telecommunications. Its government-operated telecommunication system has invested heavily in modernization and has scored some unique successes. The most notable is videotex—a commercial home information service transmitted over conventional phone lines. The service bombed in the United States and Britain but gained acceptance in

France when the telephone system gave away inexpensive Minitel terminals. And even in France the spread of home fiber connections has been much slower than planned.

Plans for Subscriber Networks

Many organizations elsewhere in the world continue long-term efforts to bring fibers all the way to subscribers. Japan's Nippon Telegraph and Telephone Corp. is particularly active. NTT is evaluating two approaches, a near-term one to combine analog video and data transmission and a longer-term all-digital system. These approaches are summarized in *Table 16-5.*

**Table 16-5.
Fiber-Optic Subscriber
Systems Developed by
NTT for Near- and Long-
Term Use**

System Component	Near Term	Long Term
Service		
Video distribution	2 6.5-MHz analog	Digital
Telephone	2 64-kbit/s digital	64-kbit/s digital
Data	2-way 4 kbit/s	High-speed
Videophone	None	Digital
Digital rate	192 kbit/s	100–400 Mbit/s
Topology	Star	Star
Length	2 km	10 km
Wavelengths	780, 880, 1300 nm (multiplexed)	1300 nm
Light sources	LEDs	Laser diodes
Detectors	APDs @ 780, 880 and PIN @ 1300	APDs or PINs with optoelectronic integrated circuits
Fiber	50/125 graded-index	Single-mode

NTT planners expect the long-term approach to be more cost-effective by the mid-1990s. As is evident in *Table 16-5*, many details remain vague, and some changes are possible.

Somewhat similar concepts are envisioned by Southern Bell, which has taken the lead among American regional telephone companies looking at fiber-optic transmission to homes.

The most immediate prospects for bringing fibers all the way to the subscriber are in serving the needs of business customers. As we saw earlier, a few businesses already have fiber-optic connections direct to the central office. The numbers of these users will grow with business information needs. A big push comes from increasing data-transmission demands. Other pushes could come from growth in videoconferencing and video training and information systems.

Long-term efforts continue to bring fibers to homes.

Business subscribers will be the first connected to fibers.

Some planners talk about fiber-optic networks for smart buildings. In such a building, the communication network would be a fourth utility, analogous to electricity, water, and air (heating, ventilation, and air conditioning). The network would be installed during construction to provide many digital telephone channels and one video channel for each desk. The digital channels would carry voice and data; the video channel would be available for teleconferencing or video training or information systems.

Fibers might even play a role in future smart homes, which would be built with an automated system to control appliances, transmit audio and video signals, manage heating and cooling, and perform other functions.

Prospects for Fiber Networks

The biggest uncertainty about integrated fiber networks is what services could justify their high cost.

How realistic are all these futuristic plans? An objective answer would be "somewhat." Fiber-optic technology has an impressive development track record. The theoretical possibility of two decades ago has become vital for telecommunications. The technology still is advancing at a rapid rate, and prices have come way down.

On the other hand, the vision that started as the wired city, and evolved into what I call the fibered society, has been maddeningly elusive. The problem has been finding services worth what the customer would have to pay. That is a more severe problem for home subscribers than for businesses, though even business users balk at high-priced services they don't need.

Consumers have shown little interest in videophone and home videotex services.

The first generation of videophones was a market flop. AT&T demonstrated Picturephone at the 1964 New York World's Fair but couldn't sell customers on the idea. Two decades later, despite improvements that reduced its thirst for bandwidth and improved its picture quality, what became the Picturephone Meeting Service was little used by business and virtually invisible to the public at large.

A more recent failure on the consumer market was videotex, a computerized information service that transmitted signals over telephone lines to create graphic displays on home television screens. The idea sounded great, but there was so little content to the graphic-oriented service that subscribers paying for on-line time quickly lost interest.

Southern Bell estimates subscribers would have to buy over $50/month in services to pay for a fiber network.

The key problem remains finding services to entice customers to pay for a fiber-optic network. *Table 16-6* shows Southern Bell's estimates of what customers would have to pay for the company to make money with a fiber-optic network.

Table 16-6.
Monthly Revenues
Southern Bell Envisions
from a Fiber-Optic
Subscriber Network

Service	Monthly fee
Basic local phone	$15
Videotex	12
Cable-television transport	7
Home security	6
Energy management	4
Meter reading	4
Common channel signalling (data)	8
Total	$56

Note carefully what's in second place on that list—videotex, which has yet to find a viable home market in North America.

This is not to say that there are no realistic prospects for fiber-optic networks to homes. The point is that such a network must offer something subscribers will pay for. Other services are conceivable. One example is selection of video programs from a large on-line library for transmission to the home whenever the subscriber wants. Programs could be played on jukebox-like videodisc players, then transmitted over single-mode fibers to homes. The idea is similar to renting videotapes, but the on-line library could have a broader selection than a local dealer and could be reached from your living room any hour of the day or night. The convenience and selection (and, we hope, the quality of transmission) could help lure customers for a fiber-optic network.

WHAT HAVE WE LEARNED?

1. Most telephone cable is in the subscriber loop, the part of the network linking subscribers to switching offices. As yet this part of the network contains little fiber, except for special high-speed systems to a few businesses and for cables carrying multiplexed signals to regional dropoff points.
2. ISDN is a concept that calls for transmitting many digitized services over a single network.
3. The first commercial fiber-optic systems were 45-Mbit/s trunk lines over graded-index multimode fibers at 850 nm. Now digital fiber trunks operate at 45–400 Mbit/s, mostly at 1300 nm over single-mode fiber.
4. Modernization of the telephone network and deregulation of the telephone industry has led to the connection of much new equipment and increasing demands for sophisticated services. However, these services do not require the capacity of fiber optics.
5. 1300-nm LEDs could transmit signals through single- or multimode fiber the few kilometers between switching offices and subscribers, although more costly edge-emitting types would be needed for use with single-mode fibers. Inexpensively packaged semiconductor-lasers may allow lower-cost designs.

6. Many new telecommunication services have been proposed and are being tested. Some (e.g., high-resolution television transmission, switched video systems, and videophones) would benefit from or even require fiber-optic transmission. Some of these services have been demonstrated in experimental systems in Japan, Canada, and France. Subscribers have shown interest in the services, but the demonstrations have been expensive. It is not certain subscribers will be willing to pay enough to justify installing an expensive new fiber-optic system.

7. Cost extrapolations suggest telecommunication companies would want subscribers to pay over $50 a month for services before they would be willing to invest the $1500 worth of fiber-optic equipment needed per subscriber.

WHAT'S NEXT?

In Chapter 17, we will see how fiber optics are used for video transmission and why the cable television industry has made little use of fiber.

Quiz for Chapter 16

1. The Integrated Services Digital Network is:
   a. a meaningless marketing buzzword.
   b. high-speed digital transmission between two switching offices.
   c. a network that carries many services in digital form.
   d. incompatible with fiber-optic transmission.

2. Which of the following speeds is used in fiber-optic telecommunication trunks?
   a. 45 Mbit/s.
   b. 90 Mbit/s.
   c. 405 Mbit/s.
   d. All of the above.

3. What technology is used in a second-generation fiber-optic system?
   a. Multimode fibers transmitting 850 nm several kilometers between repeaters at 45 Mbit/s.
   b. Multimode fibers to transmit 1300 nm up to 20 km between repeaters at 90 Mbit/s.
   c. Single-mode fibers to transmit 1300 nm for 40 km between repeaters at 45–417 Mbit/s.
   d. Plastic fibers to transmit 665 ns up to 100 m at speeds below 1 Mbit/s.

4. Which of the following might you still find operating in the telephone network? (Multiple answers are okay.)
   a. Multimode fibers transmitting 850 nm several kilometers between repeaters at 45 Mbit/s.
   b. Multimode fibers to transmit 1300 nm up to 20 km between repeaters at 90 Mbit/s.
   c. Single-mode fibers to transmit 1300 nm for 40 km between repeaters at 45–417 Mbit/s.
   d. Plastic fibers to transmit 665 ns up to 100 m at speeds below 1 Mbit/s.

5. Optical fibers are used in the subscriber loop to:
   a. multiplex several dozen home phone lines together for transmission to the central office.
   b. transmit signals on trunk lines between central offices at 45–417 Mbit/s.
   c. carry individual phone circuits to home subscribers.
   d. replace damaged wires in existing cables.

6. What services could optical fibers deliver to individual (home or business) subscribers?
   a. High-speed digital data.
   b. High-resolution television.
   c. Video telephone.
   d. Multiline telephone service.
   e. All of the above.

7. What is the most important requirement that must be met before fiber-optic cables can be installed to home subscribers?
   a. Raising data-transmission speeds on optical fibers.
   b. Developing lower-cost fiber technology.
   c. Developing services that customers will pay for.
   d. Making enough fiber.

8. Which technology is most likely to be used for home fiber-optic subscriber links?
   a. 665-nm LEDs sending 1 Mbit/s through plastic fibers.
   b. 850-nm LEDs sending 45 Mbit/s through step-index multimode fibers.
   c. 1300-nm LEDs or low-cost lasers sending hundreds of megabits per second through step-index single-mode fibers.
   d. 1550-nm lasers sending gigabits per second through dispersion-shifted single-mode fibers.
   e. Impossible to tell.

9. Which of the following services that could be provided over a fiber-optic subscriber network would businesses be likely to use? (Multiple answers are okay.)
   a. Videoconferencing.
   b. High-speed data transmission.
   c. Multiple telephone circuits to a PBX.
   d. High-resolution television transmission.
   e. Video surveillance transmission.

10. Business telecommunications subscribers:
    a. already use some fiber-optic connections to the central office.
    b. are willing to pay more for premium-quality services than home subscribers.
    c. will use fiber services before residential customers.
    d. all of the above.
    e. none of the above.

Video Transmission

ABOUT THIS CHAPTER

Video transmission, which we discussed in earlier chapters, is an important part of telecommunications. In many places, cable television (sometimes called CATV for Community Antenna TeleVision) networks parallel the local phone network. Video signals are transmitted for many other purposes, including electronic news gathering, high-resolution video displays, signal distribution in broadcasting studios, and surveillance.

In this chapter, we will examine fiber-optic video transmission. First, we will look at video-transmission requirements and how well—or in some cases, how poorly—they match capabilities of optical fibers. Then we will look at applications where fibers are successful or unsuccessful. We will concentrate on video transmission per se rather than as an element of an Integrated Services Digital Network (ISDN) where video signals are only bits added to the general flow of telecommunications in digital form.

VIDEO-TRANSMISSION CONCEPTS

Video transmission re-
quires high bandwidth.

Video transmission is by its nature bandwidth intensive. The old saying may hold that a picture is worth a thousand words, but it takes about 50 times more capacity to transmit than 1000 words, as shown in *Table 17-1*. Video telephones have had the same basic problem: it takes roughly a thousand voice telephone circuits to transmit a standard television picture, using either analog or digital technology.

Video signals can be com-
pressed in bandwidth, but
high compression reduces
quality visibly and re-
quires costly electronics.

Special techniques can compress digital video signals so they do not demand such inordinate bandwidth. Good results have been demonstrated in the laboratory, but high compression factors are difficult to achieve. Sophisticated methods can identify what parts of the picture remain unchanged between two successive frames. However, the more effective and sophisticated the process, the more costly it is to implement electronically. In addition, compression eventually begins to degrade the picture, causing effects such as blurring of moving objects and fuzzy edges. Thus, both cost and performance are trade-offs with the degree of compression.

**Table 17-1.
Comparison of Video,
Voice, and Text
Transmission
Requirements***

Transmission	Analog Equivalent	Digital Equivalent
Standard television channel	6.3 MHz	100 Mbit/s**
Telephone circuit	4 kHz	56 kbit/s
TV-quality video frame	—	3.3 Mbit**
1000 words of written text	—	60,000 bits
High-resolution TV†	25 MHz	400 Mbit/s**

*Using North American formats. All digital values except for telephone transmission are approximate.
†No standard yet developed, but proposed high-resolution (or high-definition) systems would have about twice as many vertical lines and about twice as many picture elements on a line as conventional television. Thus, quadrupling standard video approximate requirements.
**Approximate—no firm standard.

Although compression is used in some practical video systems, there are no generally accepted standards for degrees or levels of compression. The picture is further muddied by cost and performance trade-offs. Determining an acceptable level of cost or performance is inherently a subjective judgment, based on many external considerations. (When the people who decide what to pay don't have to use the equipment, the results can be eyestrain-inducing displays and closets full of unusable electronic dust-collectors.) The resulting uncertainty about compression causes some fuzziness in discussing video-transmission speeds and requirements.

Types of Video Transmission

Like telephone transmission, video transmission is not simply a matter of hooking up cables between two points and turning on the transmitter and receiver. Video signals are transmitted in many circumstances for many different applications. The most important present examples (excluding videophone and videoconferencing) are listed in *Table 17-2.*

Requirements for video transmission vary widely.

As the table shows, transmission requirements differ widely. In some cases (e.g., electronic news gathering), a single channel is sent between two points, and the transmission medium must be portable and not subject to interference. In others (e.g., remote pickup), point-to-point transmission can go over fixed commercial telecommunication lines, including satellite channels and cables. Other applications, including cable and broadcast television, require simultaneous transmission to many points.

**Table 17-2.
Current Video
Transmission
Applications and
Requirements**

Application	Requirements
Broadcast television (ground)	Radiate signals on one channel to many antennas in local area
Broadcast-signal delivery to transmission equipment	High-quality transmission of one channel between two points
Direct broadcast satellite	Transmit signals to satellite dishes over wide area on ground
Cable-television distribution	Transmit many channels one way through cable to many local subscribers
Cable-television trunks	Transmit multiple channels between cable-company facilities
Closed-circuit television	Point-to-point transmission between fixed TV camera and monitor
Electronic news gathering	Point-to-point transmission from portable camera to fixed equipment
Remote pickup	Transmission from a fixed remote site (e.g., a stadium) to a broadcast studio
High-resolution displays	Point-to-point transmission from central computer to remote display
Studio transmission	High-quality transmission between points in television studio
Network feeds	Transmission from a central network facility to many remote stations or cable television systems

Fibers are suitable for some video applications but not for all.

Fiber optics obviously is unsuitable for some of the applications listed in *Table 17-2* (e.g., broadcast television and satellite transmission). Much long-distance video transmission (e.g., network feeds) already is handled efficiently by satellites. However, fibers are candidates for many transmission jobs that can be done by satellites.

As we saw earlier, there are important distinctions between point-to-point and multipoint transmission. Point-to-point transmission is easy over optical fibers; multipoint distribution is not. Later in this chapter, you will learn how this problem has handcuffed efforts to distribute cable television signals to individual subscribers over fiber.

Analog Versus Digital Video

Most video transmission remains analog, and little video equipment is digital.

The video world, unlike the telephone industry, has yet to jump on the digital bandwagon. There is plenty of talk about digital television, but not much more. Like home telephones, home video equipment is analog, requiring analog input and using analog techniques to generate analog output. However, digital technology has not made as deep an impact on video processing and transmission as it has on voice communications.

Much of the reason for that difference lies in the much greater bandwidth of video signals. Digitized video requires much higher bit rates and switching speeds than digitized audio. Digital video electronics must be much faster than those used for analog video signals. (Note that in *Table 17-1* the bit rate of raw digitized video signals is more than ten times their analog bandwidth.) That means higher costs and more stringent transmission requirements.

Different types of processing and distribution are needed for video signals and voice telecommunications. Telephone and other telecommunication signals must be switched in ways amenable to computer-controlled automation (e.g., multiplexing of signals and routing individual signals to their destinations). Video signals are switched much less often, and much of the processing they undergo is editing, which is only automatable indirectly. That is, editing must be done under human control, even if the human is using a computer to translate visible frames into bits on tape.

Likewise, transmission differs. In the present telephone network, multiplexed signals are sent at high speeds between switching facilities, and separate signals are sent at much lower speeds to individual homes. On the other hand, cable television systems transmit many video channels to all home subscribers, as shown in *Figure 17-1*. Different homes get separate telecommunication signals, but all get the same signal from the cable-television network, which is arranged like a tree with signals distributed from the head end at the base of the trunk out to individual subscribers on the leaves or branches. The figure inverts the picture so the flow of signals is down from the roots at the head end to the subscriber branches.

Cable-television systems are using fiber optics in trunk cables, which carry multiple video channels between distribution centers in a region or from satellite antennas to a control center. However, fibers are poorly matched to the needs of signal distribution to subscribers. We'll look at such supertrunks first, then highlight the differences between that application and the distribution of cable television signals to subscribers.

**Figure 17-1.
Function of Cable-
Television System Head
End**

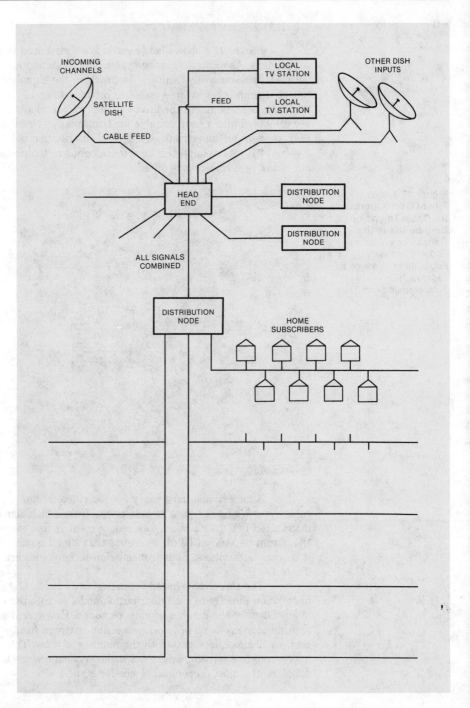

CABLE-TV SUPERTRUNKS

Supertrunks carry cable-television signals from a system control center to regional distribution points.

Figure 17-1 shows how signals are distributed to cable-television subscribers. As in the telephone system, these signals also must be sent between cable-company facilities. In practice, the signals must first be collected from other sources (satellite broadcasts and local television stations) and brought to the head end control center. At the head end, they are combined into a form suitable for transmission to subscribers. Then they are sent on supertrunks to regional distribution points, as shown in *Figure 17-2*. (These supertrunks are analogous to telephone trunk systems, but industry terminologies differ.)

**Figure 17-2.
Fiber-Optic Supertrunks
for Cable-Television
Distribution in the
Dallas Area** *(Courtesy
Robert Hoss, VP
Engineering, American
Express TRS
Telecommunications)*

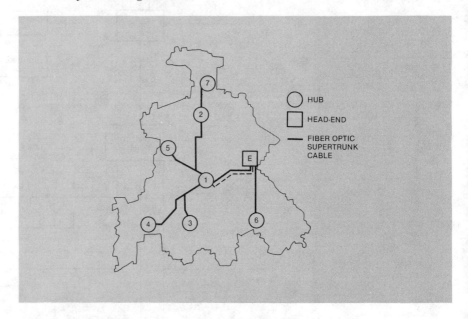

A supertrunk must carry 30–100 video channels 8–16 km from a cable system's head end to remote points from which signals are distributed to subscribers. A key requirement is that signal-to-noise ratio at the output be at least 53 dB to ensure that signal quality is adequate when it reaches subscribers. Four transmission technologies are compared in *Table 17-3*.

The three cable-based technologies in *Table 17-3* require installation directly to each distribution node, so cost per channel is largely independent of the number of hubs or nodes. However, because microwave transmission needs no physical connection between head end and node, the cost per channel decreases with the number of hubs. Thus, while microwave transmission might not make economic sense to two nodes, it might be the most economical choice for 20 nodes.

**Table 17-3.
Performance
Comparison for
Supertrunk
Transmission of 100
Channels over 16 km
(10 mi)**

	FM Coax	AM Coax	Microwave	Fiber
Number of repeaters	56	56	0	0
Channel capacity	25/cable	25/cable	50–70/band	6–10/fiber/wavelength
Number of cables	4	4	0	1
S/N ratio	≥ 53 dB	≤ 52 dB	53 dB	55–60 dB
Installed cost per channel	$11,000	$5000–$6000*		$9000–$11,000

*Cost depends on number of receivers; it was $22,000 for a baseline system.

Based on a compilation by Robert Hoss of American Express Telecommunications, *Table 17-3* shows why fiber is starting to make inroads in supertrunk systems. AM coaxial cable systems are the cheapest, but their signal-to-noise ratio is too low. Fiber offers performance well above the minimum level, giving added margin that can be used elsewhere in the system or improving signal quality received by subscribers. A single-fiber cable can include all the fibers needed, making installation easier than multicable coax systems.

Both analog and digital fiber-optic supertrunks are available. Transmission is at 1300 nm, generally through single-mode fibers, although a few older systems may operate at 800–900 nm. Analog and digital transmissions have different advantages. With frequency modulation and frequency-division multiplexing, analog systems can transmit eight channels per fiber with signal-to-noise ratio of 60 dB over 20–25 km, which is adequate for most supertrunks. Total bandwidth of those systems is around 320 MHz. Digital pulse-code modulation with time-division multiplexing can allow transmission over 30–40 km at 560 Mbit/s. However, the encoding scheme used (70 Mbit/s per channel) leads to a lower (but still adequate) signal-to-noise ratio (57 dB) and the need for analog to digital conversion increases cost.

Some analog systems use wavelength-division multiplexing to increase (typically double) the capacity of a single fiber.

BROADCAST SYSTEMS

Fibers can carry high-quality signals among broadcast equipment.

Although fiber optics cannot broadcast television signals, they can carry signals between equipment used in broadcasting (e.g., from a studio to a transmitting tower). Such transmission (weighted signal-to-noise ratio of 67 dB) requires a much more exacting specification than for supertrunks. Those tighter transmission standards are needed to provide an adequate signal-to-noise ratio at home television receivers because more noise is introduced into broadcast signals than into signals carried on cables.

Because the signal-to-noise ratios of fiber systems are high, they are attractive for broadcasting applications. To meet stringent broadcast standards, each fiber carries only one channel. That is not a serious limitation because each station broadcasts only one channel, unlike cable networks, which carry many channels. The transmission may be analog or digital.

VIDEO FEEDS

Table 17-2 includes several video-transmission jobs quite similar in nature. Electronic news gathering, closed-circuit television, signal distribution in studios, and transmission to high-resolution video displays all require carrying video signals between two points. In many cases, only one channel is being carried. This is well within the capabilities of fiber-optic transmission and has become the commonest video use of fibers.

Fibers are used for many single-channel video feeds because of their small size, high bandwidth, and low loss.

Fiber Versus Coaxial Cable

Coaxial cable can be used in many video feeds, particularly short ones (e.g., within a moderate-sized building). However, fiber transmission may be preferable:

- If the distance is long enough to require repeaters in coaxial cables
- If electromagnetic interference (e.g., from power lines) might affect coax transmission
- If ground loops or potential differences exist between the points connected by the cable
- If environmental conditions (e.g., moisture) degrade coax transmission
- To simplify installation and handling by using a lighter, more flexible cable
- To avoid the need to adjust transmission equipment to account for differences in cable length.

Most of those advantages of fiber optics are familiar ones mentioned earlier. However, the last one is a subtle concern that applies specifically to video transmission and deserves more explanation.

The video amplifiers that distribute signals in television facilities are designed to drive coaxial cables with nominal impedance of 75 Ω (ohms). However, actual impedance of coaxial cables is a function of length. As cable length increases, so does its capacitance, degrading high-frequency response if the cable is longer than 15–30 m (50–100 ft). Boosting the high-frequency signal, a process called equalization, can compensate for this degradation, but proper equalization requires knowing the cable's length and attenuation characteristics. Compensation also becomes harder with cable lengths over 300 m (1000 ft) and is impractical for cables longer than about 900 m (3000 ft). Because signals are transmitted as light waves through optical fibers, there is no analogous effect in fiber cables and their effective impedance (actually the impedance of the transmitters and receivers that convert electrical signals into optical form) is constant regardless of cable length.

Representative Applications

Fiber video applications include electronic news gathering and connections to high-resolution CAD/CAM terminals.

If the descriptions of fiber-optic video applications have seemed vague, it's because they are many and diverse. You may wonder where and why such systems actually are used. We can't give an exhaustive list, but a few examples might help.

News organizations covering space shuttle landings at Edwards Air Force Base in California had to run video-feed cables from a NASA van to their own vans for processing and selection. The fiber cables provided broadcast-quality video signals, eliminating ground-loop and hum problems with coaxial cables. The fiber cables also were easier to install.

Terminals used in computer-aided design and manufacturing (CAD/CAM) require high-resolution video signals to give large, high-quality displays. The signals are sent from a central computer to a remote terminal. Cables must be strung through existing ducts and walls and must be immune to noise that could disturb data transmission. Fiber optics has solved prior data-transmission problems.

Molinare, a London television producer, needed to send broadcast-quality video between two buildings 500 m apart. The cable had to pass under a busy street, alongside power lines. Fiber optics was chosen to avoid the electromagnetic interference that the power lines would cause with coax. It also avoided ground loops and hum that could arise when connecting the two buildings electrically.

Finally, television news camera operators must take cameras with them at the news scene—such as a fire—but often need a cable hook-up to a van or other vehicle. Light single-fiber cable can do the job better than bulkier, heavier, and more fragile coax.

CABLE-TELEVISION DISTRIBUTION

Fibers are poorly matched to needs of cable-television distribution.

The biggest use of video cable is one where optical fiber has found very little use—to carry cable television signals from the distribution center to homes. Distribution cable must have high bandwidth to carry many video channels. Although that might sound like a job for fiber optics, it distributes signals in only a tiny handful of cable-television systems. The most ambitious of them, in Alameda, Calif., was a financial disaster for the company that installed it, and that fiasco has warned off other potential developers.

What's the matter? Actually, the problems are fairly simple to understand when you look carefully. Major problems are:

Fibers cannot readily branch out as required for cable-TV distribution.

1. The standard architecture for cable-television systems is a tree, branching out from a base to individual subscribers, as shown in *Figure 17-1*. A coaxial cable runs down the street and drop cables to individual homes are tapped from it. This design is viable for electrical cables. However, as we saw in Chapter 10, it is hard to divide optical signals. There are no fiber-optic equivalents of the cheap passive taps for individual homes off a main coaxial cable.
2. Expensive fiber-optic receivers are needed to separate many channels carried on a single fiber and convert them into electronic form for home television inputs.

3. Expensive optical transmitters are needed to avoid non-linearities that could distort high-bandwidth analog signals. Even with such transmitters, multiple fibers might be needed for analog distribution. Digital transmitters might be easier to build but would require even more fibers to transmit signals to homes, as well as digital-to-analog conversion in homes. (Most experimental fiber video systems described in Chapter 16 included switches and sent only one or two user-selected video channels all the way to homes.)

4. Components needed for switched fiber-optic video transmission aren't yet available at a reasonable cost.

Switched video may eventually be included in an integrated fiber-optic network.

In the long term, switched video may be incorporated into an integrated fiber-optic network, as described in Chapter 16. Present tree-structured coaxial cable networks are fine for distributing many video channels to homes, as long as everyone receives the same signal. The selection you get when you pay for certain services is only apparent. All signals are transmitted to your home, even the ones you don't buy. However, the added services are scrambled. People who pay for the extra services get special decoders (built into channel selectors) so they can see those extra channels.

Any interactive video system requires selective two-way transmission not possible over a simple coax tree. A switched fiber-optic network also could deliver high-resolution (or high-definition) television. First, however, broadcasters hope to work out compromises to make the technology work with incompatible broadcast-television standards in North America, Europe, and Japan. Whether they will succeed in developing a universal standard is unclear, but it does seem that high-resolution television is on the way.

WHAT HAVE WE LEARNED?

1. Video transmission is bandwidth intensive in analog or digital format. Most current systems are analog. Digital video requires much more transmission capacity than analog unless special techniques are used to compress the digital signal. However, the more the compression, the higher the costs in signal quality and special electronics.

2. There are many different applications requiring video transmission. Fibers are good for some but not for others.

3. Major uses of fiber optics for video transmission are for point-to-point systems (e.g., cable-television supertrunks, electronic news gathering, and transmission of signals within and between broadcast facilities). Fibers are best for applications that require high-quality transmission over long distances or where light weight or ease of installation are paramount.

4. Fiber optics is not economically practical for distribution of cable-television signals to subscribers using present tree architectures.

WHAT'S NEXT?

In Chapter 18, you will learn how optical fibers are used in the digital world of data communications.

Quiz for Chapter 17

1. What is analog bandwidth for a single standard television channel?
   a. 56 kHz.
   b. 1 MHz.
   c. 6.3 MHz.
   d. 25 MHz.

2. What is the approximate bit rate needed to transmit a standard digitized video channel without data compression?
   a. 56 kbit/s.
   b. 1 Mbit/s.
   c. 10 Mbit/s.
   d. 100 Mbit/s.
   e. 1000 Mbit/s.

3. Video transmission is:
   a. largely digital.
   b. shifting rapidly from analog to digital.
   c. mostly analog.
   d. about equally divided between analog and digital.

4. Cable-television systems use optical fibers most often for:
   a. supertrunks between the head end and distribution nodes.
   b. signal distribution to subscribers.
   c. prevention of tapping into their networks.
   d. they do not use fiber optics at all.

5. Which is the most important attraction of fiber optics for cable-television trunk systems?
   a. Lowest cost per channel.
   b. Best signal-to-noise ratio.
   c. Most immune to interference.
   d. Best able to connect many nodes.

6. What is the preferred technology for fiber-optic supertrunks?
   a. 850-nm transmission through single-mode fiber.
   b. 850-nm transmission through graded-index multimode fiber.
   c. 1300-nm transmission through single-mode fiber.
   d. 1550-nm transmission through dispersion-shifted single-mode fiber.

7. Television broadcasters use optical fibers to:
   a. carry signals from studios to antennas.
   b. distribute signals to viewers.
   c. pick up transmission from other stations.
   d. compete with cable-television networks.

8. What is the most important reason why fiber-optic cable is used in electronic news gathering?
   a. The cables are cheap and non-breakable.
   b. The cables are small and lightweight.
   c. The cables can be made over 10 km long.
   d. Fiber cables have the highest transmission bandwidth.

9. What is the most important reason fibers have found little use in distributing cable-television signals to subscribers?
   a. Difficulty in dividing signals from a single cable among many subscribers.
   b. Inadequate transmission bandwidth.
   c. Fiber cables are too fragile.
   d. Incompatibility with analog equipment.
   e. Stupidity of cable-television industry.

10. Optical fibers ought to be
compatible with a cable-television
network under which of the
following circumstances?
- **a.** Transmission was digital with
  a tree architecture.
- **b.** Digital signals were switched
  to individual subscribers.
- **c.** Cable costs were reduced.
- **d.** Existing coaxial cable was
  scrapped but the network
  was otherwise unchanged.

Computers and Local-Area Networks

ABOUT THIS CHAPTER

A major strength of fiber optics is in digital communications. One of the fastest growth areas in digital communications is computer data communications. Fiber optics operates well at high speeds. Data communications is moving toward higher speeds. Put those facts together, and it seems clear that fiber optics should play a growing role in data communications and computer technology.

If you had put your money behind that logical assumption back in 1980, you probably would have lost it. Only recently has anyone started making money in fiber-optic data communications. What happened? Needs were much less urgent than market analysts had thought. Likewise, computer designers and marketers were far more cautious than expected. People outside the computer industry often underestimate the time it takes to design a new machine and write software for it (as opposed to churning out the 50th clone of the IBM PC).

Nonetheless, fiber-optic data communication is not like computer-industry vaporware—promised products that are always six months away from delivery. It already has special-purpose uses, ranging from secure data transmission to extending local-area networks to nodes outside their normal reach. More is coming. The first standard for 100 Mbit/s networks calls for fiber-optic transmission.

To understand the role of fiber optics in the world of data communications, we will first review basic concepts of computer communications and the attractions of fiber optics. Then we will look at the major types of computer communications and see how fibers meet system requirements.

BASIC CONCEPTS

Identical computers decode the same data in the same way, but different computers require a common format to interchange information.

Computers think and communicate in binary bits—ones and zeros. Identical computers—and circuit boards within the same computer—decode binary information in the same way. Different computers must convert information to a standard format before they can interchange it. This requires both hardware and software compatibility on both ends. It won't do any good if identical computers are running different programs that expect information in different formats. Communication must be in a standard format usable by the software and hardware on both ends.

Computers also must agree on the speed at which they're going to talk. Computer data rates normally are measured in baud, which means signal-level transitions per second. (In coding schemes where there is one transition per bit, baud equals bit rate. In schemes such as Manchester coding, where there are two transitions per bit, the baud rate is twice the bit rate.) The speeds are often slow by fiber-optic standards. Computers normally exchange data over telephone lines at 300 or 1200 baud. Even connections that are fast by computer standards are tortoise-like compared with telecommunication systems. A fast personal computer might transfer 500,000 bit/s between an external disk drive and its central processor. As long as distances are modest, electrical wires are adequate.

Larger computers and peripheral devices interchange data faster, and computer users want both more data storage and faster data exchange. As speeds increase, so do demands on the transmission medium. Wires work well over short distances, but transmission can suffer from electromagnetic interference. Bandwidths decrease with distance. As the degree of semiconductor circuit integration increases, it gets hard to get signals in and out of the chip because the number of circuit elements increases faster than the space on the chip's perimeter available for connections. Developers are seeking new technologies to help them with various data-transfer problems.

Point-to-Point Transmission

Much data is exchanged between pairs of devices.

Much data transmission is simply moving information from one device to another (e.g., from a computer to a printer or disk drive), as shown in *Figure 18-1*. Even if data goes from the external disk to the printer, it must pass through the computer, as shown in *Figure 18-1*. This is the simplest of tasks for fiber optics because it requires only a transmitter, a receiver, and some fiber. However, in most cases, it's even simpler to do with wires because the electrical output from one device can be carried directly to the other by wires without a transmitter or receiver.

**Figure 18-1.
Point-to-Point
Connections Between a
Computer and
Individual External
Devices**

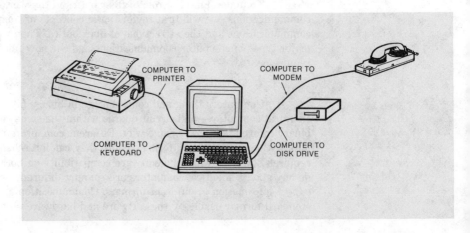

COMPUTER TO PRINTER

COMPUTER TO MODEM

COMPUTER TO KEYBOARD

COMPUTER TO DISK DRIVE

Optical fibers enter the picture when data rates become too high, distances too long, and the environment too noisy or when other factors make it hard for wires to work. Such problems are in the minority, but they can be a significant minority.

The boundaries between point-to-point transmission and local-area networks can be hazy. In some local-area networks, signals are transmitted by individual cables running between one central node and many peripheral devices. We'll consider those local-area networks because they interconnect many terminals. However, we do not consider cables all connected to the same central computer, as in *Figure 18-1*, to make a local-area network as long as the devices transfer data to and from the central computer, not directly between each other. Nor will we consider switched networks (e.g., the telephone system) to be local-area networks.

Local-Area Networks

A local-area network interconnects many nodes.

The local-area network (LAN) is best known as one of the hottest buzzwords in data communications. The basic idea is shown in *Figure 18-2*: a network that transmits data among a multitude of nodes. Details vary widely, but the key idea is that all nodes can interchange data with each other. Data packets carry header information that routes them to particular nodes on the network. For example, the users of computers 1 and 2 could each retrieve data from the shared data base, then prepare reports for output on the laser printer, with user 1 also getting data from the mainframe computer.

**Figure 18-2.
A LAN Interconnects
Many Nodes that Can
Send Messages to Any
Other Node**

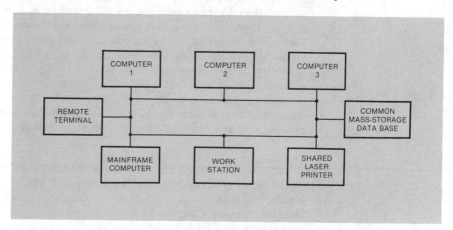

What advantages does a LAN have over the telephone network? Big advantages are ease of interconnection and speed. That allows computer users in large (or moderate-sized) organizations to share data files (and often software, much to the distress of software producers) much more readily than is possible over slower switched telephone circuits. If enough data is being transferred, that savings of time can mean a savings of money. However, if data transfer needs are small (or only involve a few terminals), users might do as well, at lower cost, with either phone lines or dedicated cables.

Many different types of local-area networks are possible, and that is one of the biggest practical problems developers have encountered. Effective, efficient communications require standards. Customizing network hardware and software to meet particular needs is very expensive. Because unit costs of both hardware and software drop sharply as the number of units increases, it is much more economical to buy standard, mass-produced versions.

Engineers know this but also have different ideas of the most efficient network architecture. Users also have differing requirements, so no single design will be the best for everyone. Marketing strategy is another complication because companies like to have proprietary products that their customers cannot buy elsewhere. (Of course, customers are wary of proprietary designs for exactly the same reason.) This situation makes network standardization hard, but some standards have emerged.

The three basic approaches to LANs are shown in *Figure 18-3*. In the star topology, all signals pass through a central node, which may be active or passive. (An active star, which switches signals to particular nodes, is functionally similar to a telephone switch.) In a ring network, the transmission medium passes through all nodes, and signals can be passed in one or both directions (sometimes over two parallel paths for redundancy). In the data-bus topology, a common transmission medium connects all the nodes, but is not closed to form a loop (i.e., the signal does not pass serially through all nodes). Variations on these approaches exist making classification more complex than it might sound, but we will ignore those here.

It was inevitable that some people would combine fiber optics and LANs, if only because the two are among the hottest communication technologies of the 1980s. One obvious problem in building fiber-optic LANs is couplers. Metal wires and coaxial cables are easy to tap at nodes; fibers are not. Another is the extra cost of fiber-optic terminal equipment, which is hard to justify in current 1- to 10-Mbit/s LANs unless node spacing is larger than coaxial cable can handle. The real promise of fiber optics is for next-generation LANs, operating at 100 Mbit/s and up.

Optical Interconnection

All data transfer is not between separate devices. Data must be moved between circuit boards in computers, between chips on circuit boards, and even within chips. In some cases, such as supercomputers, data must be transferred at very high speeds, and wires and other circuit components must be very tightly packed. Fibers can help both by allowing faster data transmission (one fiber connection might replace several wires) and by avoiding interference between adjacent wires.

Fibers and other optical techniques also may make connections directly to integrated-circuit chips. Such connections grow harder as the scale of integration increases because the number of circuit elements can increase much faster than the room to make electrical connections. One way to tackle that problem is to multiplex many signals and transmit them via a modest number of optical fibers.

**Figure 18-3.
Star, Bus, and Ring
LAN Architectures**

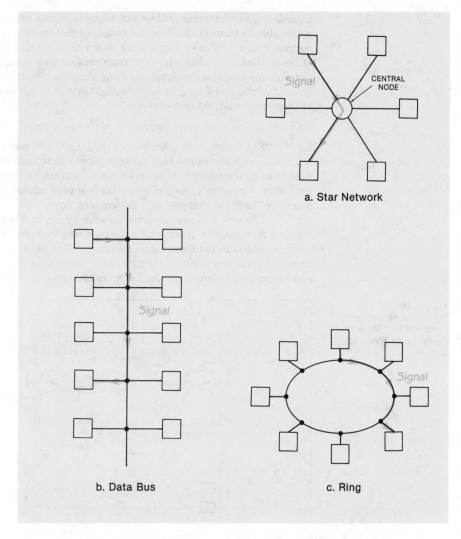

Signal

CENTRAL
NODE

a. Star Network

Signal

b. Data Bus

Signal

c. Ring

Other researchers are looking at ways that light could make connections without fibers. Holographic optical elements can direct light from a single source to several points, such as light detectors on a semiconductor chip. And optical devices themselves may learn to perform computing functions. Such technologies remain in the laboratory and are outside the scope of this book, but you should keep your eyes open for new developments.

REASONS TO USE FIBER OPTICS

The vast bulk of computer data communications is over wires.

Although fiber optics can solve data-communication problems, they are not needed everywhere. The vast bulk of computer data goes over ordinary wires. Most is sent over short distances and/or at low speeds. In an ordinary environment, it does not make economic sense to use a fiber-optic cable to transmit 9600 bit/s a couple of meters from a personal computer to a printer or to use it to hook a telephone to a wall jack. The value of fiber is its ability to solve problems such as getting high-speed signals over long distances, transmitting through electrically noisy environments, and fitting cables into tight spaces. Let's take a closer look at these major advantages of fibers.

Immunity to Electromagnetic Interference

Electromagnetic interference (EMI) is a common type of noise that originates with one of the basic properties of electromagnetism. Magnetic field lines generate (or induce) an electrical current as they cut across a conductor. Conversely, flow of electrons in a conductor generates a magnetic field that changes with the current flow.

The way this can generate noise is shown in *Figure 18-4*. The strong current in the upper wire induces a weak current in the lower wire, which is added to the input signal as noise. If the noise power is high enough, it can overwhelm the input signal. A noise spike, caused by a sudden surge of current, can add a spurious data bit to a digital signal.

Electromagnetic interference is caused by noise currents generated in a conductor, which adds noise to electrical signals.

**Figure 18-4.
Induction of Noise
Currents in a Wire
Carrying an Electrical
Signal**

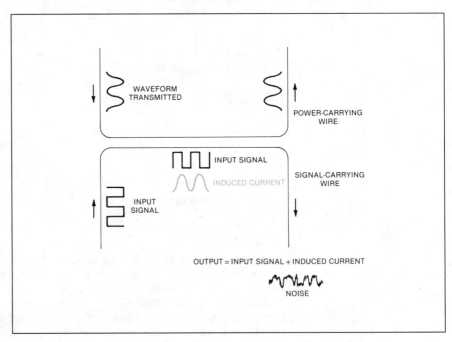

You don't need two wires next to each other to get EMI. All you need is passage of magnetic flux through a conductor. That's how a radio or television antenna picks up broadcast signals. Stray magnetic fields can induce a current in any wire exposed to them. Often those currents are simple noise (e.g., the crackle on AM radio when a nearby light is switched on). Sometimes the result is crosstalk (e.g., a radio program in the background on your telephone line).

Electromagnetic interference is less in coaxial cable, because the outer conductor can shield the inner one, thus reducing induced noise currents. However, even coax isn't immune to EMI. Optical fibers are immune because they transmit signals as light rather than current. Thus, they can carry signals through places where EMI would otherwise block transmission.

This is not to say that fiber-optic systems are perfectly immune to EMI. Electric currents drive fiber-optic transmitters and emerge from fiber-optic receivers. If transmitter and receiver circuits are not shielded, they can pick up EMI, and the system will suffer from noise, although it's not the fiber's fault.

Data Security

Magnetic fields and current induction work two ways. They don't just generate noise in a signal-carrying conductor; they also let the information the conductor is carrying leak out. Fluctuations in the induced magnetic field outside a conductor carry the same information as the current passing through the conductor. That's good news for potential spies because they can eavesdrop on these magnetic fields without cutting into the cable. Shielding the wire, as in coaxial cable, can alleviate the problem, but shielding imperfections can let enough signal leak out to allow tapping.

There are no radiated magnetic fields around an optical fiber—the electromagnetic fields are confined within the fiber. That makes it impossible to tap the signal being transmitted through a fiber without cutting into the fiber. That would increase fiber loss sharply, in a way easy for users of the communication channel to detect, making fibers a much more secure transmission medium.

Non-conductive Cables

Metal cables can encounter other signal-transmission problems because of subtle variations in electrical potential. Electronic designers assume that ground is a uniform potential. That is reasonable if ground is a single metal chassis, and it's not too bad if ground is a good conductor that extends through a small building (e.g., copper plumbing or a third wire in electrical wiring). However, the nominal ground potential can differ by a several volts if cables run between different buildings or sometimes even different parts of the same building.

That doesn't sound like much—and in the days of vacuum tube electronics, it wasn't worth worrying about. However, signal levels in semiconductor circuits are just a few volts, creating a problem called a ground loop, which isn't mentioned in many engineering textbooks. When the difference in ground potential at two ends of a wire gets comparable to the signal level, stray currents start to cause noise. If the differences grow

Fibers do not pick up EMI, so they can transmit signals through noisy environments.

Fibers do not emit electromagnetic fields that can be tapped by eavesdroppers.

Non-conductive fiber-optic cables are immune to ground loops and effects of lightning surges.

large enough, they can even damage components. Electric utilities have the biggest problems because their switching stations and power plants may have large potential differences.

A serious concern with outdoor cables is that they can be hit by lightning, causing power and voltage surges large enough to fry electronics on either end. The telephone industry is well aware of this problem, and many wire circuits include protective devices to block current and voltage surges.

Any conductive cable can carry power surges or ground loops. Fiber-optic cables can be made non-conductive by avoiding metal in their design. Such all-dielectric cables are economical and standard for many indoor applications. Outdoor versions are more expensive because they require special strength members, but they can still be valuable in eliminating ground loops and protecting electronic equipment from surge damage.

Eliminating Spark Hazards

Sparks produced by electrical wires can be dangerous in explosive atmospheres.

In some cases, transmitting signals electrically can be downright dangerous. Even modest electric potentials can generate small sparks. Those sparks ordinarily pose no hazard, but can be extremely dangerous in an oil refinery or chemical plant where the air is contaminated with potentially explosive vapors. One tiny spark could make one very large boom. Potential spark hazards seriously hinder data and telephone communication in such facilities.

Ease of Installation

Fiber cables can be much easier and less expensive to install than metal cables.

Increasing transmission capacity of wire cables generally makes them thicker and more rigid. Shielding against EMI or eavesdropping has similar effects. Such thick cables can be difficult to install in existing buildings where they must go through walls and cable ducts. Fiber cables are easier to install because they're smaller and more flexible. They also can run along the same routes as electric power cables without picking up excessive noise.

One way to simplify installation in existing buildings is to run cables through ventilation ducts. However, fire codes require that such plenum cables be made of costly fire-retardant materials that emit little smoke. You can buy metal as well as fiber plenum cables. The advantage of fiber types is that they are smaller and hence require less of the costly fire-retardant materials.

The small size, light weight, and flexibility of fiber-optic cables also make them easier to use in temporary or portable installations. In Chapter 17, we saw one example—electronic news gathering. The use of fiber-optic cables also can ease logistics in many other applications requiring portability (e.g., a portable battlefield communication system being developed for the Army).

High Bandwidth over Long Distance

We have talked enough about the capacity of fibers to carry high-speed signals over longer distances without repeaters than other types of cable. Nonetheless, we mention it here to be complete because it is an important factor that leads to the choice of fibers for data communication. Fiber can be added on to a wire network so it can reach terminals outside its normal range, as we will see later in this chapter.

PROBLEMS WITH FIBERS

Fibers do have some technical limitations. The most significant is the coupler problem described in Chapter 10, which makes it hard to split signals among many devices and complicates the architecture of fiber-optic LANs. Their inability to carry much power also causes some problems in remote powering of devices, which can be done with the signal power sent over electrical wires. However, ways generally can be found to make fibers do most data-transmission jobs.

The major practical problem with fiber-optic transmission is that it usually costs more than wires. The higher costs often are not caused by the fiber itself. Fiber-optic cable costs less than many types of coaxial cable. The difference comes when other components—transmitters, receivers, couplers, and connectors—are added. Fiber systems require separate transmitters and receivers because they cannot directly use the electrical output of computer devices; that signal must be converted into optical form and then converted back to electrical form. Fiber-optic connectors and couplers are more expensive than their electrical counterparts. These costs add up.

A related but often ignored issue is the cost of converting existing systems to use fibers. For example, it may be more efficient to transmit high-speed computer data serially (one bit after another) than sending several bits at a time in parallel over separate wires. However, that changeover requires modifications both to software and hardware. The need for such modifications can be minimized by designing fiber-optic systems with interfaces that look just like electrical ones, but that raises the cost and does not make the most efficient use of fiber's transmission capacity.

POINT-TO-POINT DATA COMMUNICATIONS

Point-to-point data communications is similar to point-to-point telecommunications. The data rates transmitted differ from those in the telephone hierarchy, reflecting the needs and standards of the computer industry, and are much less than the maximum used in telecommunications. Distances are limited, typically to within a building or campus, although at times up to a few kilometers. (Longer-distance data communication usually is considered to be telecommunications, but borderlines between the two categories can be hazy.)

Technology

The main technical concern in point-to-point fiber data communication is minimizing cost.

Point-to-point data communications rarely push the limits of fiber-optic technology. The main technological concerns are minimizing system costs. Although fiber-optic cable may cost less than coaxial cable, the extra cost of the optical transmitter and receiver may tip the economic scales in favor of coax. Typically, it is the advantages listed earlier in this chapter that shift the balance in favor of fiber.

Most commercial point-to-point fiber links accept input and deliver output in standard electrical formats.

Most commercial point-to-point fiber links accept input and deliver output in standard electrical formats. Some come with integral electrical connectors that plug into standard sockets on computers or peripheral devices. Some companies call such products modems or fiber-optic modems because their function is analogous to that of electronic modems, which convert digital data for transmission in different form over telephone lines. Normally, transmitter and receiver are packaged together. (Note, however, that such fiber-optic modems can't send data over wire telephone lines.)

The operation of such a point-to-point fiber-optic modem is shown schematically in *Figure 18-5.* The example shows signals going through one fiber from the computer to the disk drive and through the second fiber in the opposite direction. Each fiber-optic interface is a transceiver, containing a transmitter to send optical output and a receiver to accept optical input. Most fiber-optic modems are made to replace electrical cables and be functionally transparent or invisible to the user.

**Figure 18-5.
Operation of Point-to-
Point Fiber-Optic
Modem**

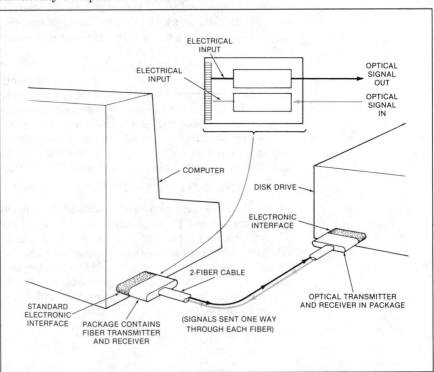

Many different fiber-optic systems are offered for short-distance communications, designed for different levels of price and performance. A small sampling of parameters is given in *Table 18-1*.

**Table 18-1.
Characteristics of Point-to-Point Fiber-Optic Systems**

Data Rate	Wavelength/Source	Transmission Loss Allowed	Fiber
0–2 Mbit/s	820-nm LED	15 dB	100/140
0–100 kbit/s	820-nm LED	40 dB	100/140
0–100 kbit/s	820-nm LED	8 dB	100/140
5 Mbit/s	820-nm LED*	4 dB	50/125
5 Mbit/s	820-nm LED*	9 dB	62.5/125
5 Mbit/s	820-nm LED†	6 dB	85/125
5 Mbit/s	820-nm LED†	9 dB	100/140
5 Mbit/s	820-nm LED†	14 dB	200/250
50 Mbit/s	850-nm LED	23 dB	50/125
50 Mbit/s	1300-nm LED	14 dB	50/125
50 Mbit/s	1300-nm LED (high-radiance)	23 dB	50/125
50 Mbit/s	850-nm laser	45 dB	50/125
50 Mbit/s	1300-nm laser	45 dB	50/125

*Transmitter designed for small-core fiber.
†Transmitter designed for large-core fiber.

Most systems are designed for distances no more than a few kilometers at 820 or 850 nm.

Table 18-1 shows that most systems are intended for distances no more than a few kilometers, assuming loss of 4 dB/km for multimode fibers at 820 nm. Because loss at 1300 nm is much lower than at 820 or 850 nm, a smaller margin in decibels at the longer wavelength allows transmission over longer distances. For example, a 14-dB margin at 1300 nm allows transmission farther than 10 km, versus only 5- to 6-km transmission for 23-dB margin at the shorter wavelength.

Higher system margins are possible with high-radiance 1300-nm LEDs or with diode lasers at either wavelength. However, that margin comes at a significant price penalty.

Fiber loss may be only a small part of total system loss.

As we saw in Chapter 14, fiber loss may account for only a small fraction of total loss in short-distance systems assembled from inexpensive components. This is illustrated forcefully in *Table 18-2*, which shows loss budget for a 100-m link of 200/250 fiber with a broad-area 900-nm LED source. The assumed fiber loss of 25 dB/km is unusually high; specifications for commercial 200/240 fiber give loss of no more than 10 dB/km, which would reduce the fiber's contribution to total loss from 2.5 dB to 1.0 dB.

Table 18-2.
Loss Budget for a Short Point-to-Point System with a Connector in the Middle of 100 m of 200/250 Fiber (*Example Courtesy Motorola Inc.*)

LED to fiber connector loss	2.7 dB
First fiber entry Fresnel loss	0.2 dB
LED to fiber NA loss	6.79 dB
Attenuation of 50 m of fiber	1.25 dB
Fiber exit Fresnel loss	0.2 dB
Connector loss	1.5 dB
Fiber entry Fresnel loss	0.2 dB
Attenuation of 50 m of fiber	1.25 dB
Fiber exit Fresnel loss	0.2 dB
Fiber-detector connector loss	1.5 dB
Detector entry Fresnel loss	0.3 dB
Total non-fiber loss	13.77 dB
Fiber attenuation	2.5 dB
Total system loss	16.27 dB
LED output	−9.03 dBm
Power into detector	−25.3 dBm

Table 18-2 also shows one other subtle but important point. Most of the loss—9.69 dB—comes in coupling light from the LED into the first length of fiber, including loss at the first connector. If the light source had come with a fiber pigtail terminated in a connector, most of that attenuation would have been invisible, hidden in the lower specified output power of the transmitter.

Multiplexed Transmission

Fibers can provide multiplexed data transmission between two points, replacing many cables.

So far, we have talked only about transmission of signals between two individual devices. However, multiplexed data transmission between two points can replace many cables. Suppose, for example, a central computer is in one building but a group of 20 user terminals are in a second building, 500 m away. You could run 20 separate cables between the two buildings. Or you could run 20 cables from the mainframe to a multiplexer, which combines all the signals onto a fiber-optic cable and transmits the signals to a receiver that demultiplexes the signal and distributes it to the separate terminals, as shown in *Figure 18-6*. This approach can save money, especially on cable-installation.

Applications in Point-to-Point Transmission

In most cases where fibers are used for point-to-point transmission, they are picked to overcome special problems or because they can cut costs.

Multiplexing signals onto a single-fiber cable to save the cost, installation, and handling problems of multiple cables has been one successful use of fibers for point-to-point data transmission. We can sample only a few other examples where fibers were picked because they offered important cost or performance benefits.

**Figure 18-6.
Multiplexing Signals
through a Fiber
Eliminates Many
Parallel Cables**

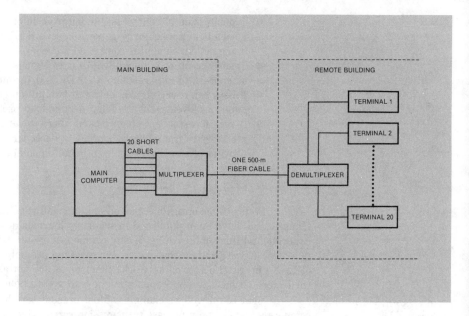

- Data transmission from a hospital computer to remote terminals. The radio-frequency paging system the hospital used to call physicians was interfering with data transmission.
- Transmission of data from a screen room used in electromagnetic testing to the outside world. Screening shields the room from intense electric and magnetic fields generated in the laboratory, which prevent electronic transmission. Optical fibers can carry signals through that hostile environment without picking up noise.
- Transmission of classified and/or sensitive information to, from, and within secure facilities, mostly operated by military and security agencies. One of the first fiber-optic systems for distribution of computer data was installed to meet security requirements at the Army's Harry Diamond Laboratories. Although shielded metal cables were available, fiber was less expensive to install in the existing building.
- Transmission of data from underground nuclear tests to the surface. Cables must hang down deep, narrow shafts to instruments near the bomb blast. Because each test costs tens of millions of dollars, it is crucial to collect as much data as possible—putting a premium on high-speed transmission. (The instruments near the bomb do not survive the tests; they collect and transmit data between the detonation and their destruction.)
- Transmission of control signals in experimental nuclear fusion reactors at government laboratories. The reactors include huge banks of capacitors that produce electrical pulses and high levels of noise that would otherwise obstruct data transmission.

- Transmission of control and monitoring information in factories where heavy machinery generates severe EMI. This is particularly important in computer-controlled factories.
- Transmission of high-bandwidth data signals from mainframe computers to high-resolution CAD/CAM terminals.
- Fibers can carry signals between two buildings with different ground potential levels, thus avoiding ground-loop problems.
- A major restaurant chain picked fibers to carry signals from cash registers to the manager's office in back, largely because the cables were easy and inexpensive to install in compliance with local fire codes.

FIBER OPTICS IN LOCAL-AREA NETWORKS

Fiber optics has been used in few practical LANs but has been tested in many laboratory LANs.

So far, fiber optics has played only a small role in practical LANs. Many papers have been published in scholarly journals describing experimental fiber-optic LANs. A few companies even offer commercial fiber LANs, but the few that have been installed are only a small fraction of LANs in use. Local-area networks themselves are not as widespread as their more enthusiastic proponents might lead you to believe.

Interest in higher-speed transmission could push uses of fibers in LANs.

That could well be changing. Most present terminals and personal computers require only modest transmission rates, which often can be sent over telephone lines. Current LANs operate at speeds to about 10 Mbit/s, at which there is little benefit from using optical fibers (except in the special cases described below). Future devices will require faster data transmission to allow better-quality graphics, faster access to data bases, and more efficient sharing of information. The need for faster transmission will accelerate the development of LANs—and their use of fiber optics.

To understand how fiber optics can be used in local-area networks, we will first look at how fibers can enhance capabilities of present LANs. Then we will turn to the next generation of LANs being developed around fiber optics.

Fiber Optics in Ethernet

The most accepted LAN is the 10-Mbit/s Ethernet, which uses coax.

The first LAN to gain much acceptance was the Ethernet standard developed by Digital Equipment, Intel, and Xerox. Ethernet distributes digital data packets of variable length at 10 Mbit/s to terminals dispersed along a coaxial cable bus, as shown in *Figure 18-7*. Separate cables up to 50 m long, containing four twisted wire pairs, run from individual devices (e.g., terminals or printers) to transceivers joined to the coaxial cable.

The network has no overall controller; control functions are handled by individual transceivers. If a terminal is ready to send a signal, its transceiver checks if another signal is going along the coaxial cable. Transmission is delayed if another signal is present. If not, the terminal begins transmitting and continues until it finishes or detects a collision, the transmission of data at the same time by a second terminal. Such collisions happen because it takes time—several nanoseconds a meter—for signals to travel along the coax. If the delay is 6 ns/m, a collision would occur if two

**Figure 18-7.
Basic Elements of
Ethernet**

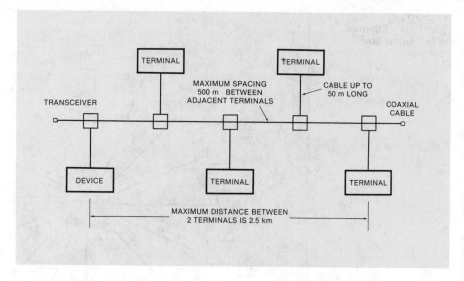

terminals 300 m apart on the coax started sending within 1.8 μs of each
other. The terminal stops transmitting if it detects a collision and waits a
random interval before trying again.

Data signals contain an address header specifying the terminal to
which they are directed. They pass through all transceivers but are sent
only to the terminal to which they are directed.

The choice of coaxial cable limits Ethernet performance. Maximum
terminal separation is 2.5 km, with no more than 500 m between terminals
and no more than 1024 terminals altogether. This is adequate for most, but
not all, applications. Optical fibers can stretch transmission distances
beyond the limit imposed by loss of the coaxial cable to limits imposed by
delays in signal propagation. (Round-trip time through the system must be
under 45 μs.)

One all-fiber-optic version of Ethernet is based on a central
transmissive star coupler. As shown in *Figure 18-8*, this system relies on
transmission directly between network nodes and the central passive star
coupler, which mixes and distributes signals throughout the network. This
network requires separate transmit and receive fibers to each node, which
has a separate fiber-optic transmitter and receiver. The output fiber goes to
the input of the transmissive star coupler; the input is connected to the star
coupler's output. The standard design uses a coupler with 32 input and
output ports, but the number of connections can be increased by adding
signal-amplifying repeaters and extra star couplers in place of nodes. With
a second level of identical star couplers, the number of terminals can be
raised to 32 $\times$ 32 or 1024, the upper limit for Ethernet. The standard
version of this system operates with short-wavelength LEDs, which couple
-4.6 dBm into 100/140 fiber; receiver sensitivity is -28.2 dBm.

*Optical fibers can stretch
Ethernet transmission be-
yond the 500-m limit of
coax.*

**Figure 18-8.
Fiber-Optic Ethernet
with Central Star
Coupler**

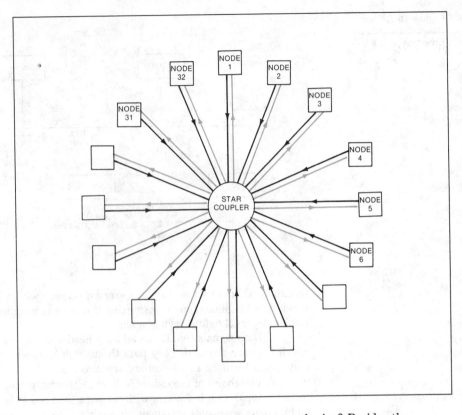

What advantages does the fiber-optic approach give? Besides the typical ones for all data transmission that we described above, it also can deliver signals to nodes over an area up to 5 km², larger than possible with a coax Ethernet.

That is not the only fiber-optic Ethernet variant. Other all-fiber versions include one with a ring topology and one that uses an active star coupler, which contains repeaters that detect input signals and amplify them for retransmission through the network. Fiber-optic components also can be added to coaxial-cable Ethernets. One standard product is an Ethernet extender, which uses fiber-optic cable to extend transmission distance between an Ethernet transceiver and station to a kilometer from the standard 50 m.

Fiber-optic LANs based on Ethernet-like concepts are used today. Some are small-scale networks, used where fibers are essential because of problems such as ground loops or severe EMI. Others are designed to allow future upgrading. Probably the largest-scale system is one Southwestern Bell installed in its 44-story headquarters in St. Louis. Over 2000 terminals in the 1.5-million-ft² building are connected by 144 km of 85/125 fiber. Part of the network is an Ethernet-like system carrying asynchronous data at 10 Mbit/s. The rest of the network allows 2.5-Mbit/s synchronous transmission

There are several fiber-optic Ethernet variants, and fibers can be added to coax Ethernets to extend transmission distances.

Fiber LANs allow upgrading of capacity.

between individual terminals and/or a central computer facility. Company officials looked at other possibilities but picked fiber largely because of the ease of upgrading. Communications manager David Stein was quoted as saying, "If we put in fiber, we put it in once, and it's in for the life of the building."

Fiber-Optic Distributed Data Interface

The 100-Mbit/s FDDI standard LAN is built around fiber optics.

Ethernet is designed for present computer data-transmission needs, but new technology will be needed as transmission requirements grow. Many more designs for fiber-optic LANs have been proposed than can be described here. We will concentrate on one that is being accepted as a standard for 100-Mbit/s transmission: the Fiber Distributed Data Interface (FDDI).

The FDDI standard calls for a ring topology, as shown in *Figure 18-9.* There actually are two fiber rings, transmitting signals in opposite directions, to give redundancy in case of component or cable failure. Data transmission in each direction is 100 Mbit/s.

**Figure 18-9.
FDDI's Dual Ring**

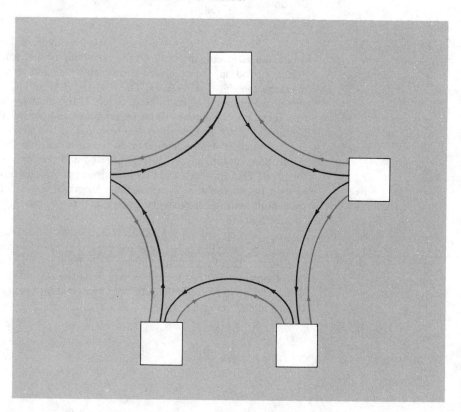

Tokens

The FDDI network picks
the transmitting node by
passing a token message
around the ring.

The FDDI network picks the transmitting node in a different way than Ethernet. Ethernet uses a contention scheme in which all nodes contend with each other. As long as a node does not detect another signal, it can seize the opportunity to transmit. In FDDI, only one node can transmit at a given time. The authorization to transmit is given by a message called a token, which is passed around the ring when no other message is being transmitted. If a node receives the token and is not ready to transmit, it sends the token message to the next node. However, if it is ready to transmit, it holds the token and sends the message. The message contains a code that identifies its destination. As it circles about the ring, it is ignored by every other node except the one marked as its destination. Once it returns to its origin, the message is cancelled and the token is passed along the ring again. This controlled access scheme is more efficient in a heavily used network than the Ethernet's contention scheme, and calculations show that such a network could transmit data up to 99.5% of the time.

Nodes

FDDI nodes include
transmitter, receiver, bypass switch, and a termi-
nal interface.

Each node includes a transmitter, receiver, and bypass switch, as well as an electronic interface to the terminal, as shown in *Figure 18-10*. In normal operation, the receiver—transmitter pair acts as a repeater that monitors the signal it receives. The receiver detects and amplifies the signal and passes it along to be decoded. If the message is not addressed to that terminal, the signal will be regenerated and passed along to the next terminal, but not processed by the terminal. If the signal is addressed to that terminal, it will be passed on to the terminal (via the electronic connections on top) as well as regenerated and transmitted to the next terminal. This approach allows transmission of signals through up to 2 km of fiber between nodes. The maximum length of the entire ring, constrained by default settings of recovery timers, is 200 km, passing through a thousand nodes.

The optical bypass switch is important because it lets the network work if the node has failed or is not receiving the electrical power it needs to operate. This keeps failure of one node from putting the whole network out of service. Further backup comes from the second ring in *Figure 18-9*, which can be used continually or only switched into service in case of a failure in the first ring.

**Figure 18-10.
A Node in an FDDI
Network**

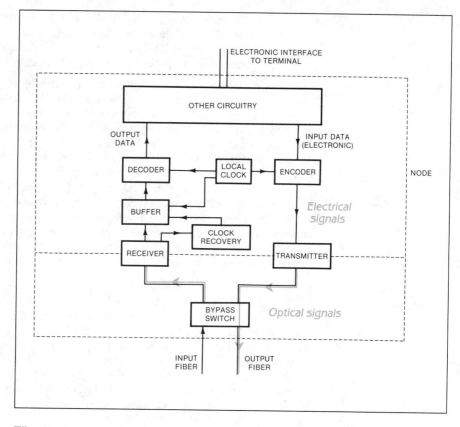

Fibers

FDDI uses multimode fiber and 1300-nm LEDs.

Developers of the FDDI standard balanced cost and performance to decide hardware specifications. Their choices were:

- Multimode graded-index fiber, either 62.5/125 or 85/125, as offering the best combination of light acceptance and bandwidth. Specifications call for fibers with modal bandwidth of at least 400 MHz-km and attenuation no more than 2.5 dB/km, both easily available in commercial cables.
- 1300-nm transmission because of the low loss and dispersion of fibers at that wavelength, allowing use of LED sources.
- InGaAsP (1300-nm) LED sources because of their much lower cost and better reliability than lasers. Peak output must be at least −16 dBm into either 62.5/125 or 85/125 fiber.
- PIN photodiode detectors because of their lower cost and better reliability than 1300-nm APDs. Minimum power needed at the detector for a bit-error rate of 1 in 2.5×10^{10} bits must be no higher than −27 dBm.

- The transmission code will be a 4 of 5 code, with one extra bit added for every four data bits. This means the actual data rate will be 125 Mbaud for 100 Mbit/s of user data. This coding scheme balances transmission between on and off bits to enhance operating efficiency.

Comparing transmitter output and receiver input power shows that the loss budget for one hop in the ring is 11 dB, with 5 dB allocated for the cable and 6 dB for other components, including splices, connectors, and switches. Although that seems tight, the allowed attenuation is generous for multimode fiber at 1300 nm. One leading fiber maker offers different grades of both 62.5/125 and 85/125 fiber with specified attenuation at 1300 nm of 0.7 to 1.5 dB/km, but nothing higher.

FDDI is intended to connect computer mainframes and lower-speed LANs.

What will the high data rate of FDDI be used for? Initially, it is intended to connect computer mainframes and to connect lower-speed LANs. An enhancement to FDDI allows circuit switching so the network could handle integrated streams of voice, video, and computer data. It is part of a developing set of related data-transmission standards and will find increasing use as data-transfer rates continue their inexorable rise.

Other Fiber Local-Area Networks

The Hubnet fiber LAN uses a rooted tree architecture.

The FDDI network is far from the only high-speed fiber-optic LAN. Another interesting example, developed at the University of Toronto and Canstar Communications, is the 50-Mbit/s Hubnet shown in *Figure 18-11.* The Hubnet LAN is a directional system, sending signals from the transmitters to the receivers. The transmit and receiver nodes, shown as separate, are attached to the same device. The architecture is called a rooted tree, but more properly it might be considered a pair of interconnected trees—one selecting and transmitting the signals and the other broadcasting and receiving them.

Like Ethernet, Hubnet is a contention system. A node starts transmitting a signal whenever it wants to. The signal enters a sub-hub, which is an active device that transmits it if it is not handling any other signal. However, if the sub-hub already is transmitting another signal, it ignores the second input. This process goes on through other sub-hubs to the central hub, which selects one signal for transfer to the broadcast network, which diffuses signals out to all the receiving sub-hubs and receiving sides of nodes. When a node receives a signal addressed to it, it will retransmit the signal back to the source node. Successful receipt of the return transmission indicates that the signal reached its destination. If the return transmission is not received in a given time, the original node will try transmitting the signal again.

Active hubs are used in Hubnet; they discriminate among signals and regenerate only input.

The hubs in Hubnet are not passive star connectors but are active devices that detect signals and retransmit them (i.e., they function as repeaters but add a discrimination function so they process only one input at a time). The design calls for separate transmit and receive fibers, multimode types up to 2 km long with minimum bandwidth of 500 MHz-km. Although transmit and receive nodes are shown separate for convenience, they actually are packaged together.

**Figure 18-11.
The Hubnet LAN**

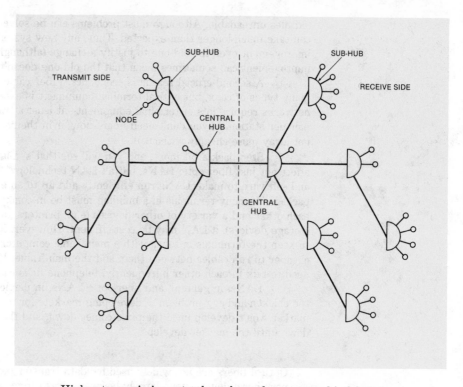

Higher transmission rates
have been demonstrated
in laboratory fiber LANs.

Higher transmission rates have been demonstrated in laboratory fiber LANs, including 200 Mbit/s in a dual-ring system similar in outline to the FDDI standard. Experimental transmitters and receivers have operated in the burst mode required for LANs at speeds to 500 Mbit/s. Increasing transmission speed will increase demands on decision, logic, and switching components as well as on transmitters and receivers. It also may begin to approach modal bandwidth limits of multimode fibers. Some laboratories have shown interest in single-mode fiber LANs, but that interest may be hard to translate into economical systems because of the high cost of terminal components needed to couple enough light into single-mode fibers.

Limitations of Fiber LANs

Now that you have learned how fiber optics can work in LANs, we should reiterate a point made earlier in this chapter: fiber optics are not used widely in LANs. You may wonder why.

Technological inertia and
economics are holding
back fiber LANs.

One reason is the inevitable inertia that holds back any changing technology. People are comfortable with what they have and don't want to change. This is not irrational. Anyone who has gone through the trauma of shifting between incompatible computer systems can testify why systems that work should not be changed lightly. It is not merely that Murphy's Law—"anything that can go wrong, will"—holds, but that Murphy was an optimist. Minor differences cause old programs to crash and make data in old files unreadable. Although most problems can be solved, the solutions can take much longer than expected. Thus, any new system has to be a big improvement over the old one to justify a change (although the big improvement can sometimes mean that the old one doesn't work).

A second crucial reason is economics. Fiber cable is cheaper than many types of coax, but optical terminal equipment is expensive. Local-area networks require lots of terminal equipment—at least a transmitter/receiver pair per station. If you don't need fiber—now or in the near future—you can get away more cheaply without it.

Step back a bit more, and you will see that similar considerations affect not just fiber-optic LANs, but all LAN technology. The hardware and software to make LANs run efficiently add up to an expensive package. If many terminals in a building must be in constant touch with each other and a variety of other devices (e.g., printers and archival storage devices), a LAN may be cost-efficient. However, if the real need is to keep the terminals in touch with a mainframe computer, it would be cheaper to run cables between them and the mainframe. If the terminals need to talk to each other infrequently, telephone lines can do the job.

LANs, in general, and fiber-optic LANs, in particular, suffer from the chicken-and-egg problem of developing markets for new technology. The market won't develop until the price comes down, and the price won't come down until the market develops.

> The factors limiting the use of fibers in LANs are similar to those limiting the use of any LANs.

WHAT HAVE WE LEARNED?

1. Optical fibers are not widely used for data transfer between computers and peripheral devices now, largely because wires can generally do the job cheaper, but increasing demands for speed and distance will push the use of fibers.
2. Local-area networks interconnect many terminals, allowing them to communicate directly with each other. Point-to-point transmission carries data between pairs of devices.
3. Fiber optics is a candidate for point-to-point data transmission, LANs, and interconnection of circuit boards and chips.
4. Local-area networks come in star, ring, and data-bus configurations.
5. The main uses of fibers for data transmission are in applications where wires won't work. This includes avoiding EMI, enhancing data security, avoiding conductive cables, stretching transmission distance at high speeds, avoiding spark hazards, and simplifying installation.
6. Because data rates and distances are modest, computer systems tend to use multimode fiber rather than single-mode types. Many new computer systems still use short-wavelength LEDs. In general, the loss budget of computer data links and LANs is dominated by coupling losses rather than fiber attenuation.
7. Standards for the Ethernet LAN call for coaxial-cable transmission, but fibers can be used to extend the network over larger areas than otherwise possible.

8. Multimode fibers and 1300-nm LEDs are required in the 100-Mbit/s FDDI LAN standard. The FDDI is a ring network. Fibers also can be used in star-architecture LANs.

WHAT'S NEXT?

In Chapter 19, we will look at some interesting fiber-optic applications in communications outside of the normal world of telecommunications and data transmission—automotive, aircraft, and military systems.

Quiz for Chapter 18

1. Point-to-point data transmission involves:
   a. transmission of signals from a central computer to remote nodes.
   b. interchange of data between pairs of devices.
   c. communication between any two devices connected by a common transmission medium.
   d. the connection of two devices through a switched network like the telephone system.

2. A LAN is:
   a. a system that interconnects many nodes by making all signals pass through a central node.
   b. a ring network with a transmission medium that passes through all nodes.
   c. a common transmission medium or data bus to which all nodes are connected but that does not form a complete ring.
   d. all of the above.
   e. none of the above.

3. What makes optical fibers immune to EMI?
   a. They transmit signals in as light rather than electric current.
   b. They are too small for magnetic fields to induce currents in them.
   c. Magnetic fields cannot penetrate the glass of the fiber.
   d. They are readily shielded by outer conductors in cable.

4. Which of the following is not a reason to use fiber-optic cables for point-to-point data transmission?
   a. Need to assure data security.
   b. Avoidance of ground loops.
   c. Data-transfer rates too low to use metal cables.
   d. Ease of installing smaller fiber-optic cables.
   e. Elimination of spark hazards.

5. Which of the following types of fiber would deliver the most power to a receiver from an inexpensive 850-nm large-area LED source? Assume that the transmission distance is short, about 100 m.
   a. Single-mode.
   b. All plastic.
   c. 100/140 step-index multimode.
   d. 85/125.
   e. 50/125.

6. In a system such as the one in question 5, what would contribute most to total system loss?
   a. Fiber attenuation.
   b. Connectors.
   c. Splices.
   d. Coupling light from LED into fiber.
   e. Coupling light from fiber to receiver.

7. When would optical fibers be used in an Ethernet-type LAN?
   a. Never, the standard calls for coaxial cable.
   b. To extend transmission distance to reach remote terminals.
   c. Routinely, the standard allows for optical fiber.
   d. When transmission speeds exceed 50 Mbit/s.

8. The FDDI standard calls for nodes to be:
   a. attached to a data bus in the form of a transmissive star coupler.
   b. attached to a pair of fiber rings carrying signals in opposite directions.
   c. attached to a network of star couplers that detect collisions and transmit only one signal to the next level.
   d. attached to fibers by passive T couplers.

9. The type of light source and fiber chosen for FDDI networks are:
   a. single-mode fiber and 1550-nm lasers.
   b. single-mode fiber and 1300-nm lasers.
   c. multimode fiber and 1300-nm lasers.
   d. multimode fiber and 1300-nm LEDs.
   e. multimode fiber and 850-nm lasers or LEDs.

10. Speeds of laboratory fiber-optic LANs have reached:
   a. 1 Mbit/s.
   b. 10 Mbit/s.
   c. 100 Mbit/s.
   d. hundreds of negabits per second.
   e. gigabits per second.

Auto, Aircraft, and Military Communications

In the past four chapters, we have examined a wide range of fiber-optic voice, data, and video communications. Some other uses of fiber-optic communications don't properly fall under those headings. One is signal transmission in moving vehicles—cars, planes, ships, and even spacecraft. Why use optical fibers when anything mankind could pilot in the near future is small compared with fiber transmission ranges? As we will see, reasons include EMI immunity, small size, and light weight. Similar reasons underlie use of fibers by military organizations, particularly for portable battlefield communication systems. (We will cover military sensing systems in Chapter 20).

In this chapter, you will see why optical fibers are used in vehicles and military systems and how the technologies used in such systems differ from those in more conventional telecommunication and data-transmission systems.

When the automobile industry started using electronic controls, it found it had an EMI problem.

A few decades ago, automotive engineering took a brief nap reminiscent of Rip Van Winkle's, and automotive innovation was confined mostly to sculpting tail fins. The industry finally woke in the early 1970s, kicked awake, perhaps a bit brutally, by government and public concern over pollution, safety, and fuel economy. Engineers discovered that electronic control systems might solve some of their problems.

They also discovered that EMI could do nasty things to those electronic controls because semiconductor electronics need only a few volts to switch states. Electromagnetic interference made strange things happen on the road. Take the case of an electronic system to prevent brakes from locking in a skid, which was tested on buses in the Chicago area. When the buses crossed a certain bridge, the electronic brakes could seize up. Engineers were puzzled until they found that radio signals from a transmitter 0.5 mi away were bouncing up from the bridge's metal deck and inducing spurious signals that confused the brakes.

The proliferation of automotive electronics made the EMI problem critical.

Other problems also appeared as automotive electronics proliferated. *Table 19-1* lists some electrical and electronic systems in 1986 cars. Each car may contain a few microprocessors. Wiring harnesses, always a problem, have evolved into a nightmare of complexity. Because many electronic systems are optional, automakers must stock many different harnesses in their plants and make sure the proper one is installed in each car. If you have ever tried to find and fix a defective electrical connection somewhere in an automobile, you can appreciate how the problem of servicing wiring harnesses evolved from hard toward virtually impossible.

Table 19-1.
Some Electrical and
Electronic Systems
Available in 1986 Cars

Speedometer	Odometer	Chime
Battery guard	Interior lights	Illuminated entry
Dimming control	Exterior lights	Fuel monitor
Transmission control	Clock/calendar	Driver diagnostics
Climate diagnostic panel	Heating/ventilation	Air Conditioning
Rear-window defogger	Radio	Tape player
CB radio	Cruise control	Windshield wipers
Instrument panel	Cigarette lighter	Horn
Electronic fuel injection	Coolant fan control	Hazard flasher
Seat-belt warning buzzer	Windshield washer	Power windows
Power door locks		

Controls operated from the driver's seat pose another problem. In conventional designs, the power to drive an electrical accessory (e.g., power windows or windshield wipers) must pass through a switch on the instrument panel. The more accessories, the harder it is to find a place to mount the switches and the thicker the tangle of wires behind the instrument panel.

Automotive engineers want to separate control signals from power. One way to do so is to use fiber optics.

Automotive engineers began seeking ways to separate control signals from power lines. That would let them send from the dashboard a signal to roll down the window, without passing current to drive the motor through the dashboard switch. Instead, a power bus would distribute electrical power through the car, with the power switched on and off by remote controls. They could simplify signal wiring by multiplexing many control signals together. The first step was to develop multiplexed metal-wire systems, but the next—already demonstrated in experimental cars—will be to send signals through fibers to avoid the not-inconsiderable EMI problems.

Bulb-Outage Indicators

Plastic fibers are used for illumination and bulb-outage indicators in present cars.

Fiber optics is not totally new to the automotive industry. Some cars already contain fibers, but they do not use them for communications. Their function is to alert the driver that a bulb has burned out. As shown in *Figure 19-1*, a cheap plastic fiber is put near a bulb (e.g., a headlamp) to collect some light for delivery to a dashboard display. If the light is on, there is a bright spot; if not, there is no light. Fibers also can be used as light pipes to carry light from bulbs hidden behind the dashboard to a display visible to the driver.

**Figure 19-1.
A Fiber-Optic Bulb-
Outage Indicator**

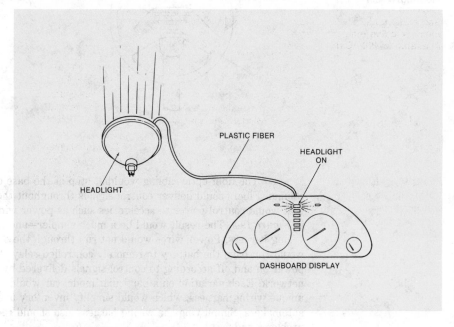

PLASTIC FIBER

HEADLIGHT
ON

HEADLIGHT

DASHBOARD DISPLAY

Multiplexed Fiber-Optic Transmission

Fiber's EMI immunity makes it attractive for control-signal transmission.

What's the matter with wires for transmitting multiplexed control signals? EMI is a big problem because cars are very noisy. Spark plugs and electric motors generate EMI normally, and adding a Citizen's Band radio or mobile phone can make matters much worse. Wires carrying multiplexed signals require special EMI filters to keep the noise from confusing electronic instruments and controls. Costs of those EMI filters add up. Worse yet, as transmission speeds increase, EMI filters start to remove signals as well as noise. At the 1-Mbit/s data rate General Motors is considering to handle the many electronic systems in its cars, the limited bandpass of EMI filters becomes a problem. Thus, GM as well as Japanese automakers are working on fiber-optic multiplex systems for cars.

Multiplexed fiber-optic transmission will impact what the driver sees as well as the hardware behind the dashboard. From a human-engineering viewpoint, the center of the steering wheel is a logical place to put control switches. However, there are large hardware-engineering problems in fitting anything more on the steering wheel. Fiber-optic multiplexing could carry signals down the steering column in a system such as the one shown in *Figure 19-2*.

**Figure 19-2.
Multiplexing Control
Signals Down Optical
Fibers Allow Controls at
the Center of the
Steering Wheel**

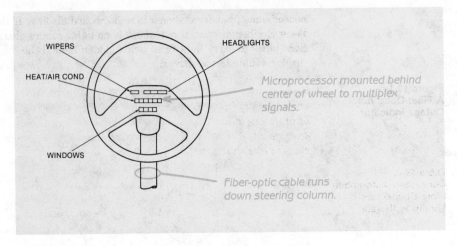

WIPERS

HEADLIGHTS

HEAT/AIR COND

*Microprocessor mounted behind
center of wheel to multiplex
signals.*

WINDOWS

*Fiber-optic cable runs
down steering column.*

Fibers could deliver control signals throughout the car, simplifying and lightening the wiring harness.

The fiber-optic cabling wouldn't stop at the base of the steering column. Fibers could deliver control signals throughout the car to relays that would control power to accessories such as power windows, as shown in *Figure 19-3*. The result would be a much simpler—and much lighter—wiring harness. Power wires would not run through the dashboard, but would run from the battery to remotely controlled relays that would switch power on and off according to control signals delivered by the fiber network. Each variation on a particular model car would not require a unique wiring harness, which would simplify inventory and assembly for automakers. Simplifying the wiring harness also should reduce maintenance problems and costs.

**Figure 19-3.
Optical Fiber Delivers
Control Signals to a
Relay that Switches
Power to Power Window
Motor**

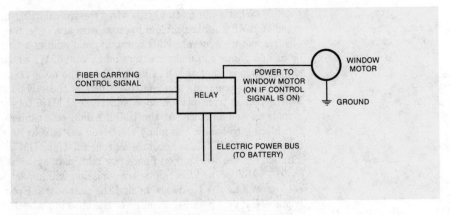

WINDOW
MOTOR

FIBER CARRYING
CONTROL SIGNAL

POWER TO
WINDOW MOTOR
(ON IF CONTROL
SIGNAL IS ON)

RELAY

GROUND

ELECTRIC POWER BUS
(TO BATTERY)

Fiber-Optic Hardware for Cars

Automotive fiber optics need not transmit signals far, but they must be cheap and durable.

Fiber-optic hardware must meet very different requirements for automotive use than for the applications described earlier. Distances are much shorter because standard cars are under 10 m long. True mass production is a must since domestic automakers produce several million cars a year. Components must be easy to install on a highly automated assembly line and cheap, with light source, connector, and detector well under a dollar each.

That isn't all that the auto industry asks. Any equipment installed in a car must be able to withstand the daily abuse of driving as well as exposure to extreme temperatures, foul weather, road salt, and other environmental hazards. It also must withstand the best (and worst) efforts of professional and amateur mechanics. To see what the auto industry worries about, read the partial list of conditions that connectors should survive in *Table 19-2*.

Table 19-2.
Conditions Automobile
Fiber Connectors
Should Withstand

Temperature	−40 to +100°C
Contaminants	Ambient light
	Antifreeze
	Axle grease
	Battery fumes
	Brake fluid
	Degreasing chemicals
	Engine oil
	Exhaust gas
	Gasoline
	Gravel bombardment
	Road salt
	Sand and dust
	Soap
	Transmission oil
	Water

These conditions can be relaxed somewhat if fibers are used only in the passenger compartment, where temperatures are less extreme than next to the engine and many environmental contaminants are not present.

Automakers plan to use all-plastic fiber with 1-mm core.

To meet those specifications, automakers have picked all-plastic fiber with a 1-mm core. Its advantages include ease of coupling light into the large core, ease in termination (simply slice it with a razor blade), flexibility, and ease of handling. The ease of coupling greatly relaxes connector tolerances and costs. Tolerances are so loose that some demonstration automotive fiber links can operate even when the fiber is pulled partially out of the connector. The plastic fiber also allows use of cheap, large-area red LEDs, compatible with inexpensive detectors.

Promises and Problems of Automotive Fiber Optics

General Motors may have started earlier, but as is distressingly typical in the automotive world, Japanese companies were the first to produce cars using fiber-optic communications. However, even in Japan, such cars are rare. Costs remain high, and the auto industry still takes years to adapt new technology to mass production.

One critical technical problem is the lack of fibers usable at the high temperatures in the engine compartment. Glass fibers can withstand the heat, but large-core types are inflexible and hard to handle. Present plastic fibers cannot tolerate the high temperatures, but new fibers may be able to operate at the required temperatures.

FIBERS IN AIRCRAFT

The design of aircraft communication systems is somewhat similar to that of automotive systems, although aviation requirements are often more demanding. Planes are more complex than cars, and pilots require more information than drivers. The toughest requirements are for military aircraft, which must withstand hostile conditions and include much high-performance electronic equipment.

It was not too long ago that most aircraft were controlled by hydraulic systems. When the pilot moved a lever, it would cause hydraulic fluid to move a control surface (e.g., a wing flap), much as hydraulic brakes work in an automobile. Newer planes have fly-by-wire electronic controls that send electronic signals to motors that move control surfaces. In addition, aircraft—particularly military craft—use many electronic systems and sensors. In addition to radars, military aircraft may have targeting systems and electronic countermeasure equipment to knock out the radars, targeting systems, and internal electronics of enemy aircraft. Conversely, military aircraft must be protected against enemy countermeasures.

A decade ago, the Department of Defense was studying fiber-optic transmission in aircraft. One motivation was development of composite non-metallic materials for aircraft skins. Such materials are stronger per pound than metals, but unlike a metal fuselage they cannot shield the inside of the plane from electromagnetic interference and countermeasures. Using fiber could overcome such problems as well as reduce cable weight.

Uses in Military Planes

It takes years for new technology to work its way into military hardware. Limited amounts of optical fiber are used in the B-1 bomber and the MX missile, weapons originally designed in the 1970s when fiber was new and largely untested. Other planes also use some short lengths of fiber. A 6-m (20-ft) length of fiber cable carries sensor data in the Marine Corps' AV-8B Harrier attack aircraft. Future aircraft might benefit more from more extensive use of fiber. Calculations indicate that the B-1's weight could have been reduced by as much as a ton if all its wire cables had been replaced by fiber. Such weight reductions could mean greater range, lower fuel requirements, or higher load capacity.

It's a good bet that new-generation planes using super-secret Stealth technology will use optical fibers extensively. Stealth technology is supposed to make planes nearly invisible to radar, a task which probably will mean drastically reducing the amount of metal on board. Such craft probably will have composite fuselages and much electronic equipment—and thus would benefit from fiber transmission. Fiber transmission also would help keep the wiring in Stealth planes from radiating its own EMI, which an enemy might spot.

Helicopters Fly by Light

An Army program is developing fiber controls for helicopters.

The U.S. Army is working on a fiber-optic Advanced Digital Optical Control System to control the flight of helicopters. In this system, the pilot's actions are converted into optical signals and transmitted through fiber to an internal flight computer, which analyzes this information together with control settings and data on flight-path stability. Fibers then send signals from the computer to control position of the copter's rotors. The system also includes some optical sensors.

The Army system uses some 30 m (100 ft) of fiber cable per helicopter. That lowers the copter's mass by about 18 kg (40 lb), but the Army's main goal is to cut maintenance and training costs.

Commercial Airplanes

Boeing is looking to optical fibers to carry control signals in its next generation of commercial planes, planned for the early 1990s. The new plane will extensively integrate flight-control electronics, using only two or three flight-control computers instead of 21 in a 757 or 767. Fiber cables from these computers to control surfaces would avoid the need for bulky metal cables shielded against EMI and lightning strikes.

Spacecraft

NASA is considering fiber-optic communications for its manned space station.

With light weight a crucial concern in spacecraft, you would expect fibers to be considered for use in large satellites. One candidate for which fibers are already being considered is the space station that NASA hopes to put in orbit in the 1990s. The agency has been working with fibers for years; it has been operating underground fiber-optic systems at the Kennedy Space Flight Center in Florida for over a decade.

Technology Choices

Most aircraft systems will use multimode fiber, and military planes are likely to use radiation-hardened fibers.

Not much detail has been announced about the fiber technology in military aircraft. Because transmission is over short distances, multimode or large-core fiber is the clear choice. This also is clear from military requirements for other components that put a premium on durability rather than tight tolerances and high performance. Over short distances, most data-transmission requirements can be met by LEDs, which presumably would be chosen because of their lifetimes. Similar considerations will affect choices in civilian planes.

Military systems presumably will use radiation-hardened fibers designed to allow system operation even during and after exposure to nuclear radiation. (Virtually all standard radiation-hardened fibers are multimode.) Note that interest in radiation hardening is not confined to fiber optics; most new military specifications call for it.

MILITARY SYSTEMS

One hundred fifty military programs are using, testing, or planning to use fiber-optic communications.

Fiber optics is used in many other types of military systems besides aircraft, including surface ships and submarines. The list of military programs and systems involving fibers is long and impressive. A tabulation compiled by the Defense Electronics Supply Center in late 1985 listed 37 systems using fiber that are installed or in production, 45 systems using fiber in development, and 84 programs evaluating or planning to use fiber.

The list includes many cryptic code numbers and acronyms, such as AN/TPN-19, AN/UYK-20, and ROTHR. Some systems are fixed strategic-communication networks that are similar in function to civilian telecommunication lines. Others are large programs that use a small amount of fiber, such as the command, control, and communication system for the MX missile. Some are classified. I can't describe all 150 systems here, so I will briefly describe three of the more interesting uses of fiber for which information is available.

Radar Remoting

Fibers can carry signals from remote radars to control centers.

Radar is invaluable in spotting and tracking enemy aircraft, but it also has an important drawback. Microwave emissions from radar dishes are a target for enemy radiation-seeking missiles. Military planners figure they are bound to lose a few radar dishes so they want to put the dishes far from ground control centers. That lets the control center—and the soldiers operating it—survive if a radiation-seeking missile takes out a radar dish.

That's fine in theory, but in practice there are problems. Military radar generates an analog intermediate-frequency output at 70 MHz. That frequency is high enough to carry the signal but too high to go far through metal cable. Thus, the radar dish can't be located a comfortable distance away from the control center without repeaters. However, fiber can transmit 70 MHz a few kilometers without repeaters. That's why existing metal cables are being removed from many military radar dishes and replaced by fibers.

Battlefield Communication Systems

Fibers are attractive for portable battlefield communication systems.

Military field command centers require extensive communication capabilities to do their job properly. That means that the Army must have portable communication systems ready to lay down quickly at temporary field headquarters. The need for portability means the systems should be small and lightweight. For battlefield use, they also should be rugged. The cables may be laid in the dirt, or strung from trees or buildings. Soldiers might step on them, or vehicles might drive over them. One of a number of proposed military systems—for long-haul interconnection of tactical shelters (truck-mounted communication centers)—is shown in *Figure 19-4*.

**Figure 19-4.
Fiber-Optic Cables Can Connect Truck-Mounted Military Communication Centers Several Kilometers Apart without Repeaters**
(©1985 IEEE, from Donald H. Rice and Gerd E. Keiser, "Application of Fiber Optics to Tactical Communication Systems," IEEE Communications Magazine 23, 46–56 [May 1985])

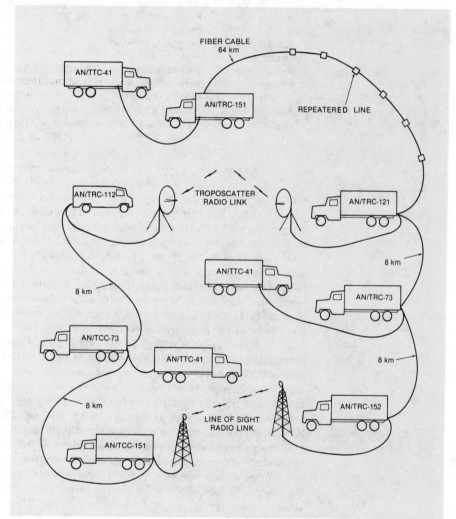

The Army's standard choice before fiber optics was 26-pair metal cable for distances under 1 km, and twin coaxial cable for 1–60 km. These cables work, but they have plenty of problems. They are heavy and bulky, take a long time to install, and are subject to lightning and EMI. Thus, they aren't very portable. Worse yet, the bulky cables are easy to damage. Wires in the 26-pair cable tend to break as it is installed and picked up, rendering many circuits inoperable. The characteristics of metal cables are far from the Army's wish list in *Table 19-3*.

Table 19-3.
Desired Characteristics
of Battlefield Cables

Characteristic	Goal
Mobility	Rapid deployment (half the time the systems are moved in 10 hours)
Installation	On ground or poles
Temperature	−55° to +160°F
Nuclear effects	Non-metallic cables with radiation-hardened fiber
Strength	Very durable cable (must withstand rough handling and being run over by vehicles)
Size and weight	As small as possible
Junctions	Connectorized, with connectors designed to keep out dirt; include provisions for field repair

By switching to fiber, the Army could reduce the number of trucks needed to carry a system and speed its installation.

What does this mean in a practical sense? It takes a few trucks to carry enough 26-pair cable for a regional command center. A couple of hefty soldiers are needed to carry a reel full of cable, which is not a very long cable because of its thickness. By switching to fiber, the Army could cut back to a single truck and lay cable much faster, either by hand or from a helicopter. One early film demonstrating the system highlighted the ease of laying fiber cable by showing two hefty male soldiers struggling with a reel of metal cable while a lightly built woman soldier strolled by unreeling a much longer fiber cable from a reel.

Fiber is also replacing 26-pair cable in other portable communication systems. Another example is the U.S. Communications Grid Network, designed to augment American communications in an emergency in western Europe. It lets military support groups use civilian phone networks of friendly countries rather than building separate military networks. U.S. allies such as West Germany will install junction boxes to let U.S. support groups communicate directly with local switching facilities. U.S. troops would string two-fiber cables from the junction points to temporary headquarters, which could move every few days. The cable must run 1–5 km without a repeater. Field tests in West Germany showed that the fiber cable took less than an hour to install and could survive most abuse—except getting caught by a street sweeper. Those tests convinced the Army to shift its design from 26-pair cable to fiber.

The system—like many others—uses 50/125 fiber to transmit light from 850-nm LEDs with silicon PIN photodiode detectors. It uses a two-fiber cable with Kevlar strength members and a polyethylene jacket the ruggedness of which was demonstrated by laying it across the entrance to a parking lot, where it survived being run over by 30,000 cars. It comes on 1-km reels that weigh 55 lb.

Fiber-Optic Guided Missile

A fiber could transmit images from a television camera in a missile back to a soldier guiding the missile to its target.

One of the more unique military systems is the fiber-optic guided missile, FOG-M. The idea of FOG-M, shown in *Figure 19-5*, is to send images from a video camera in the missile to a soldier on the ground, who could guide the missile to its target. The images—and control signals in the reverse direction—would be sent through a single optical fiber.

**Figure 19-5.
A Soldier Guiding FOG-M Missile to Its Target**

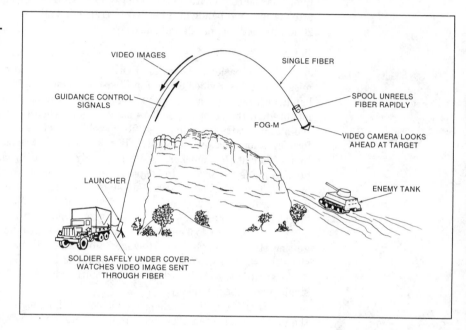

Guidance Mechanisms

Precision-guided munitions have been hot for the past two decades because they can deliver more bang for the buck. Or, to be more precise, they can deliver more bang to the target for the buck. Conventional missiles, guns, and bombs are fired or dropped, then left to hit the target on their own. If the original aim was off, if the target moves, or if the weapon drifts off course, there is no way to correct its path and it misses. Precision-guided munitions actively home in on targets as they travel.

The first major success was the smart bomb, which homed in on a spot on the target designated by a soldier armed with a laser. The laser beam projected a characteristic pattern on the target, which was seen by a four-element detector in the bomb. The bomb's control system automatically corrected its path so the spot was in the center of its field of view (i.e., so the bomb was on target).

The soldier guiding a FOG-M can be safely under cover.

The problem with laser guidance, as well as with some other precision-guided weapons, is that the soldier guiding the weapon must be in a line of sight with the target. For example, the soldier holding the laser must be able to see the target (e.g., a tank) if he is to "mark" it with a laser spot. If the tank sees the soldier, he's in big trouble. The advantage of the fiber-guided missile is that the soldier guiding it doesn't have to be in the line of sight with the target—but rather can be hiding safely behind a hill, as in *Figure 19-5*.

Missiles can be guided by wire, but wire can't carry video images and, thus, can't show the soldier where the missile is going. Only an optical fiber has the combination of small size, light weight, strength, low attenuation, and bandwidth needed to transmit video signals over the 10 or 20 km needed for missile guidance.

Basic Hardware

FOG-M contains a video camera, a fiber-optic video transmitter, a low-bandwidth receiver, and a special reel of fiber.

The basic hardware for FOG-M is shown in *Figure 19-6*. The missile contains a video camera, a fiber-optic video transmitter, a low-bandwidth receiver, and a special reel of fiber. One end of the fiber is mounted on the launcher, so it remains behind when the missile is fired. As the missile starts travelling, the fiber rapidly unwinds from the reel, forming a long arc over the battlefield. The camera transmits a video image to a soldier at the launcher, who sends control signals back through the fiber to guide the missile to keep the target in the proper place in his field of view. Thus, the soldier remotely steers the missile to its target. In one demonstration, an Army spokesperson said the gunner put the missile "right through the pilot's seat" of a parked helicopter 10 km away.

At first, it looked like one of the trickiest problems would be getting strong enough fiber. However, fiber strength has improved so much since the start of the program that commercial fibers are used in demonstrations. Initial work was with multimode fibers, but a shift to single-mode fibers is likely.

Wavelength-division multiplexing separates video and control signals at 1300 and 1550 nm.

FOG-M uses wavelength-division multiplexing. Video signals from the missile are sent at 1300 nm through fibers with low dispersion at that wavelength. Low-bandwidth control signals are sent at 1550 nm in the opposite direction. The present range is 10 km, but advocates say that could be doubled to 20 km.

Uses for FOG-M

For essentially political reasons, the first FOG-M systems probably will be designed for use against helicopters out of the line of sight. (That was one mission of the ill-fated DIVAD weapons system cancelled by the Department of Defense because of poor performance.) Later FOG-Ms could be designed to hit tanks from above, where they usually have little armor.

The remote-control technology used in FOG-M could be used in robots.

FOG-M is one military program that could have real civilian payoffs. Similar remote-control technology could be applied to robots fighting fires or working inside nuclear reactors. A human operator in a safe place could guide a robot in a hazardous environment to do tasks that require human judgment and skills.

**Figure 19-6.
FOG-M Components**

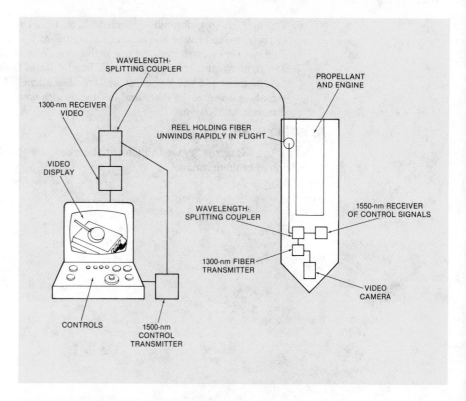

WHAT HAVE WE LEARNED?

1. Large-core plastic fibers now are used as bulb-outage indicators in automobiles.
2. The automobile industry is working on systems to transmit multiplexed control signals and other data through large-core all-plastic fibers. These are needed because of the proliferation of automotive electronics. The key attractions of fibers for this application are EMI immunity, small size, light weight, and ability to avoid problems of electrical wiring harnesses.
3. The technology needed for automotive fiber optics is very different than for telecommunications. Emphasis is on low cost, mass-produced components to transmit 1 Mbit/s several meters and withstand normal car operation and repair.
4. Multimode glass fibers are being developed for signal transmission in aircraft because of EMI immunity, small size, and light weight. They are likely to be used in super-secret Stealth fighters and bombers. Fiber controls also are being developed for military helicopters and civilian airliners.
5. Because fibers can transmit signals farther without repeaters than other cables, they can be used to put radar dishes far from military battlefield control centers.

6. The light weight, durability, bandwidth, and EMI immunity of fibers have led to their use to replace 26-pair cable and dual-core coax in portable military field networks. Such systems use multimode fiber at 850 nm.

7. The light weight, strength, and high bandwidth of fibers have led to the development of FOG-Ms. These systems use a single fiber transmitting video signals one way at 1300 nm and low-bandwidth control signals in the other direction at 1550 nm.

WHAT'S NEXT?

In Chapter 20, we will look at the many uses of fiber optics outside of communications.

Quiz for Chapter 19

1. How are optical fibers now used in cars?
   a. To transmit control signals to power-window motors.
   b. To indicate whether bulbs are operating.
   c. To transmit power to windshield washers.
   d. To provide noise-free transmission of stereo programs.

2. Which of the following is not a reason for using optical fibers to transmit control signals in cars?
   a. Optical fibers are immune to EMI.
   b. Optical fibers are lighter than wires.
   c. Fiber optics can eliminate the need for bulbs.
   d. Fiber-optic cabling harnesses should be simpler than wires.
   e. Fiber-optic cables should be more reliable than electrical harnesses.

3. What type of fiber and light source will be used in automotive signal-transmission systems?
   a. Bulbs and large-core plastic fibers.
   b. Red LEDs and large-core plastic fibers.
   c. GaAs LEDs and large-core plastic fibers.
   d. Red LEDs and plastic-clad silica fibers.
   e. Infrared LEDs and plastic-clad silica fibers.

4. The most likely data rate in multiplexed automotive fiber systems is:
   a. 56 kbit/s.
   b. 64 kbit/s.
   c. 100 kbit/s.
   d. 1 Mbit/s.
   e. 10 Mbit/s.

5. Which of the following reasons do not influence the use of fibers for signal transmission in aircraft?
   a. Optical fibers are immune to EMI.
   b. Optical fibers are lighter than wires.
   c. Aircraft don't carry enough power to drive wires.
   d. Military aircraft must be hardened against enemy electronic countermeasures.
   e. Fiber optics can help reduce aircraft visibility to radar.

6. Why are fiber-optic cables used to connect radar dishes to battlefield control centers?
   a. Because they can transmit the 70-MHz signal generated by the radars a few kilometers without repeaters.
   b. Because they are immune to electromagnetic eavesdropping.
   c. Because they are non-conductive.
   d. Because the control centers already use fiber optics.
   e. Because a top military official's brother-in-law sells fiber-optic cable.

7. What will fiber-optic cable replace in portable battlefield communication systems?
   a. Obsolete plastic fibers that have become brittle with age.
   b. 26-pair cable and coax.
   c. Telephone cables.
   d. Microwave transmission.
   e. Nothing, without fibers such systems were not practical.

8. What type of technology is used in portable battlefield communication systems being developed by the army?
   a. Plastic fibers transmitting in the red.
   b. Multimode fibers transmitting 850 nm.
   c. Multimode fibers transmitting 1300 nm.
   d. Single-mode fibers transmitting 1300 nm.
   e. Single-mode dispersion-shifted fibers transmitting 1550 nm.

9. What are the major attractions of fiber-optic cables for portable battlefield communications?
   a. Small cables require fewer trucks to transport.
   b. Fiber cable is more durable than metal cables.
   c. Light weight eases the logistics of laying cable.
   d. Cable is radiation-hardened and immune to EMI.
   e. All the above.

10. What advanced technology is used in the FOG-M?
    a. Wavelength-division multiplexing for bidirectional transmission.
    b. Dispersion-shifted fiber for video transmission at 1550 nm.
    c. Mid-infrared fiber for ultra-long-distance transmission.
    d. Star couplers to allow one soldier to control many missiles.

Non-Communication Fiber Optics

ABOUT THIS CHAPTER

Communications was a latecomer in the world of fiber optics; the early developers of optical fibers had other things in mind. I have mentioned some of those other uses of fibers in passing, but now it's time to take a closer look at applications including light piping, imaging, inspection, sensing, and medical treatment. Many require fibers different from those used for communications. Such fibers may be bundled together in a particular way (e.g., to make a sign) or may be designed to be very sensitive to external effects (e.g., pressure or temperature).

If you are going to work regularly with fiber optics, you should know about such applications. You may not encounter some of them outside of narrow fields such as medicine. However, fiber used in some applications —notably sensors—could soon be interfacing with communication fibers (e.g., in control systems). The major applications—classical fiber-optic light piping and imaging, medicine, and sensing—differ in important ways, so we will look at each separately. First, however, we will look at light guiding in short, bundled fibers.

BASICS OF FIBER BUNDLES

Some simplifying assumptions valid for single communication fibers are not valid for bundles.

The basic principles of fiber optics are the same if the fibers are separate or bundled. However, some implicit assumptions always hide behind any discussion of basics. Those assumptions are valid for the communication fibers we have described so far, but they may not be for all fibers. Thus, it's time to go back and face some complications we earlier simplified away.

Light Rays in Optical Fibers

One important simplification was treating light in fibers collectively rather than as individual rays. We assumed a cone of light entered a fiber and found that a cone of light emerged, dependent on the fiber's numerical aperture.

Light rays entering a fiber at one angle emerge in a diverging ring.

Looking at individual light rays, as in *Figure 20-1*, gives a slightly different view consistent with what we learned earlier. If a light ray enters the fiber at an angle θ within the fiber's acceptance angle, it will emerge at roughly the same angle to the fiber axis, although not necessarily in the same direction. "Roughly" is the operative word since the ray will emerge in a ring of angles centered on θ because of imperfections in the fiber, effects of fiber length, and other factors.

This does not conflict with what we learned earlier because then we looked at light guiding only collectively. If we had broken down the cone of light entering the fiber into individual rays, we would have seen the same effect. At that point, there was no need to do so.

**Figure 20-1.
Light Rays Emerge from a Fiber in a Diverging Ring**

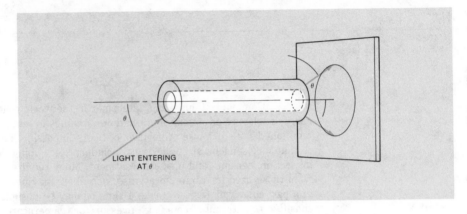

LIGHT ENTERING
AT θ

Step-index fibers with constant-size cores do not focus light.

One other thing should be pointed out: step-index fibers with constant-diameter cores do not focus light. (As we will see later, graded-index fibers can focus light.) All light entering the fiber emerges at roughly the same angle it entered, not at a changed angle, as would be required to focus light. As long as the fiber's sides and ends are straight and perpendicular to each other, a fiber or a fiber bundle can no more focus light than a pane of flat window glass can.

If the fiber's output end is cut at an angle not perpendicular to its axis, light entering at an angle θ still emerges in a cone, but the center of the cone is at an angle to the fiber axis. If the slant angle (from the perpendicular) is a small value ϕ, the angle β by which the rays are offset is approximately:

$$\beta = \phi(n - 1)$$

where n is the refractive index of the fiber core, as long as ϕ is small.

Tapered Fibers

Tapered fibers magnify or demagnify objects seen through them.

We assumed that fiber cores are straight and uniform, but they also could be tapered (although obviously not over long distances). *Figure 20-2* shows what happens to a light ray entering such a fiber at an angle θ_1. If the ray meets criteria for total internal reflection, it is confined in the core. However, it meets the core-cladding boundary at different angles on each bounce so each total internal reflection is at different angles. The result is that it emerges from the fiber at a different angle, θ_2. If input core diameter is d_1 and output core diameter is d_2, the relationship between input and output angles is:

$$d_1 \sin \theta_1 = d_2 \sin \theta_2$$

The same relationship holds for the fiber's outer diameter as long as core and outer diameter change by the same factor d_2/d_1.

Figure 20-2.
Light Passing from the Narrow to the Broad End of a Tapered Fiber

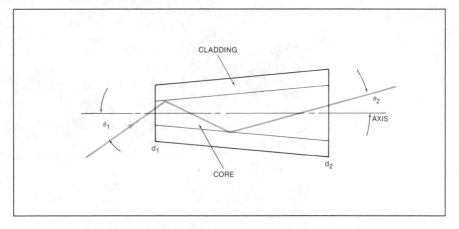

As a numerical example, suppose input angle was 30° and the taper expanded diameter by a factor of 2. The output angle would be about 14.5° (the inverse sine of 0.25). Thus, light exiting the broad end of a taper would emerge at a smaller angle to the axis than at which it entered (and, conversely, light entering the broad end would leave the narrow end at a broader angle). This effect lets fiber tapers magnify or demagnify objects.

Focusing with Graded-Index Fibers

Graded-index fibers can focus light in certain cases.

Although step-index fibers cannot focus light, graded-index fibers can focus it in certain cases. They are not used for image transmission or other fiber-bundle applications, but as described later in this chapter, graded-index fibers segments can function as components in some optical systems.

In Chapter 4 you saw that light follows a sinusoidal path through graded-index fiber. Looking at the transmission of a cone of light through a long fiber, we saw output as a cone of the same angle. Now look instead at the path of an individual ray through a short segment of graded-index fiber, shown in *Figure 20-3*, and compare that with the path of a light ray in step-index fiber.

There is an important but subtle difference. Total internal reflection from a step-index boundary keeps light rays at the same angle to the fiber axis all along the fiber. However, graded-index fibers refract light rays, so the angle of the ray to the axis is constantly changing as the ray follows a sinusoidal path. If you cut the fiber after the light ray has gone through 180° or 360° of the sinusoid, the light emerges at the same angle that it entered. However, if the distance the light ray travels is not an integral multiple of 180° of the sinusoid, it emerges at a different angle. Thus, segments of graded-index fiber can focus light.

**Figure 20-3.
Ray in Graded-Index
and Step-Index Fibers**

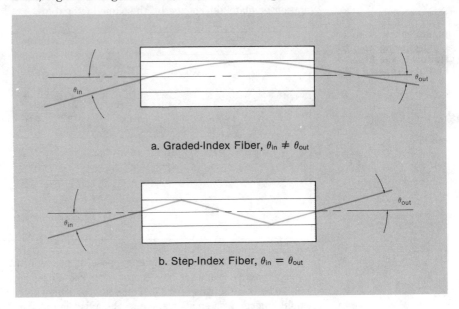

a. Graded-Index Fiber, $\theta_{in} \neq \theta_{out}$

b. Step-Index Fiber, $\theta_{in} = \theta_{out}$

Pitch is a critical parameter of graded-index lenses.

In design of graded-index lenses (sold under the tradename Selfoc), the key parameter is the fraction of a full sinusoidal cycle that light goes through before emerging. That fraction is called the pitch. A 0.23-pitch lens, for instance, has gone through 0.23 of a cycle, or $0.23 \times 360° = 82.8°$. The value of the pitch depends on factors including refractive-index gradient, index of the fiber, core diameter, and wavelength of light.

Although the lenses are segments of fiber, they are short by fiber-optic standards, just a few millimeters long. Thus, they can be considered as rod lenses as well as fiber lenses.

Imaging and Resolution

Bundle resolution depends on the size of the fiber cores it contains.

If step-index fibers cannot form images unless they are tapered, how can we talk about imaging fibers? *Figure 20-4* indicates how. If an image is formed at one end of a bundle, each fiber core will carry its segment of the image to the other end of the bundle. As long as the fibers are aligned in the same way on both ends, this will recreate an approximation of the image.

To visualize what happens, imagine that each fiber core captures a chunk of the image and delivers it to the other end of the bundle. The transmission process averages out any details that fall within a single core. For example, if the input to a single core is half black and half white, the output will be gray. Thus, the fiber cores must be small to see much detail. For a static fiber bundle, the resolution is about half a line pair per fiber core, meaning two fiber core widths are needed to measure a line pair. Numerically, that means 50-μm fiber cores could resolve 10 line pairs per millimeter (1 line pair per 100 μm).

**Figure 20-4.
Image Transmission
through a Fiber Bundle**

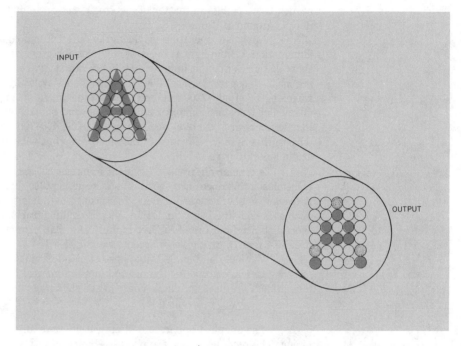

Resolution is significantly higher—about 0.8 line pair per fiber core diameter—if the fiber bundle is moving with respect to the object.

Cladding Effects

Light that falls into fiber claddings in bundles is lost.

Optical fibers are not 100% core. Even in bundles, the cores must be surrounded by claddings. The fate of light that falls onto the claddings rather than the cores or leaks from the cores into the claddings depends on bundle design. Some light may leak from the cladding and be absorbed in the jackets of flexible bundles made up of discrete fibers. Some fused fiber bundles have two claddings—the usual transparent inner cladding and an outer opaque cladding to keep light from leaking between fibers. In others, some light may be transferred between fiber cores.

Usually, most light entering the cladding is lost, which can limit transmission efficiency. This causes a loss not present in single fibers and makes one factor in a bundle's light-collection efficiency the fraction of its surface made up of fiber cores. That is, the collection efficiency depends (in part) on the packing fraction, defined as:

$$\text{PACKING FRACTION} = \frac{\text{TOTAL CORE AREA}}{\text{TOTAL SURFACE AREA}}$$

Transmission Characteristics

Typical attenuation of
bundled fiber is around 1
dB/m.

Short-distance fibers, particularly those used in bundles, do not have as low attenuation as communication fibers. Such low loss is not needed to carry light a matter of meters. Typical attenuation of bundled fiber is around 1 dB/m.

Likewise, operating wavelengths differ. For imaging and illumination, visible light is needed, and even for other applications the short distances make it unnecessary to operate at wavelengths where fibers are most transparent. Short-distance glass fibers typically are usable at wavelengths of 400–2200 nm, and special types made from glass with good ultraviolet transmission are usable at somewhat shorter wavelengths. Plastic fibers are usable at visible wavelengths, 400–700 nm. Some special-purpose fibers may be made of other materials, but their only present uses are in laboratory research.

Bundled fibers are step-index multimode types with large NA.

A similar difference is evident in numerical aperture. Because transmission distances are short, effects such as pulse dispersion are irrelevant in short-distance fibers. Increasing the numerical aperture decreases coupling losses and, thus, increases light transmission. Thus, short-distance fibers tend to have higher NAs than communication fibers (from 0.4 to 1.1), giving large acceptance angles.

Virtually all bundled fibers are step-index multimode types, which are easy to make and offer the desired large numerical aperture. Graded-index fibers are used in bundles only in special arrays used to focus light.

LIGHT PIPING

Light piping transfers light from one place to another.

The simplest application of optical fibers of any type is light piping. The term describes the process—piping light from one place to another. The light can be carried by one or many fibers without regard to how the fibers are arranged. Thus, alignment of individual fibers need not be the same at the two ends of a fiber bundle. An individual fiber may be in the center of the input end and at the outer edge of the output end.

Illumination

A fiber bundle can deliver light to small, hard-to-reach areas.

Most light-piping is for illumination, the delivery of light to some desired location. Why bother with optical fibers to do a light bulb's job? A flexible bundle of optical fibers can efficiently concentrate light in a small area or deliver light around corners to places it could not otherwise reach (e.g., inside machinery). They can illuminate places where light bulbs cannot be used (e.g., areas where explosive vapors are being used). And they can deliver light from one bulb to many places, desirable in some cases where low illumination is needed (e.g., for backlighting labels).

Such illumination fibers also are used in indicators (e.g., the automobile bulb-outage indicator discussed in Chapter 19). And illumination fibers—in the form of fiber-optic lamps and displays—gave many of us our first real view of fiber technology.

Signs

If all the fibers in an illumination bundle wind up in the same place, they illuminate a single spot. If they are routed to different places, they can form an image, such as the fiber-optic display shown in *Figure 20-5*. All the fibers are brought together in one place to collect light from a bulb, then they are splayed out to create the desired pattern. Diffusing lenses at the ends of the fiber can spread out light to make large spots. Or smaller lenses can be used to create smaller spots.

Figure 20-5.
A Fiber-Optic Sign

FUSED AND IMAGING FIBERS

In many cases, fibers must keep their alignment in a bundle. Most involve image transmission, where it is necessary to keep parts of the image in the right positions relative to each other.

Image Transmission

For image transmission, fibers in a bundle must have the same relative positions on input and output faces.

The basic concept behind image transmission by a fiber-optic bundle was shown in *Figure 20-4*. The basic requirement is that the fibers must maintain identical relative positions on input and output faces. These are called coherent bundles because of how the fiber ends are aligned. (Light travelling through them is not coherent in the same sense as laser light.)

One way to make coherent bundles is to pack many fibers together in parallel and heat them so they fuse together. The resulting bundle can be drawn to greater lengths, shrinking the cross-section of individual fiber cores and, thus, increasing resolution. The fused bundle, sometimes called image conduit, can be heated and bent to the desired shape, which it retains when cooled. A 0.5-in square bundle can be bent to a radius of curvature as small as 1 in.

Images can be viewed through coherent fiber bundles in two ways. One is by placing the bundle's input end close to the object. It can even be put directly on the object, if the bundle can carry enough light to illuminate the object for viewing. (Otherwise, the object would be too dark to see because it would have no light to reflect.) The other approach is to project an image optically onto the input end. In either case, light from the object or the image is transmitted along the bundle, and the input image is reconstructed on the output face.

Many factors influence performance of fiber bundles. Earlier we discussed how resolution depends on fiber core size. Packing fraction strongly influences the fraction of light transmitted; typically 8–30% of input surface is cladding. Image quality also depends on effective isolation of light in fiber cores from other light, including light transmitted through the cladding. Absorptive cladding can increase the apparent brightness— even though actually reducing the total amount of light transmitted— because it cuts down on background illumination without reducing intensity of the image transmitted through the fiber cores.

Like incoherent bundles for illumination, coherent fiber bundles are not the standard means of image transmission. Their rigidity, high cost, and resolution limit their applications. However, they can be very valuable in probing otherwise inaccessible areas (e.g., inside machinery or inside the human body). (Flexible coherent fiber bundles can be made using different technology.)

Faceplates

Image transmission does not have to be over a long distance. Indeed, one of the most common uses of fiber-optic image transmission uses coherent bundles less than 1 in long. That is the fiber-optic faceplate, a rigid array of short fibers fused together into a vacuum-tight sheet. Typical core diameters range between 6 and 25 μm.

The main role of a faceplate is to transmit an image from inside a vacuum envelope (e.g., a cathode ray tube) to the outside, with as little light loss as possible. We are not talking about replacing the glass front of an ordinary television picture tube. Most fiber-optic faceplates are much smaller and are used in small, high-resolution displays or in image-intensification tubes, demanding applications that can justify the added expense. They can help flatten the curved field of a cathode-ray tube, correct for distortion, and enhance the display's effective brightness by suppressing ambient light. They also concentrate output toward the front of the screen, minimizing losses in other directions. Faceplates may not sound as exciting as other uses of coherent fiber bundles, but they probably are the most widespread application of the technology.

Image Manipulation, Splitting, and Combining

Coherent fiber bundles can manipulate images.

Coherent fiber bundles can do more than just transmit images; they also can manipulate them. Twisting a coherent bundle by 180° inverts the image. You can do the same with lenses, but a fiber-optic image inverter does not require as long a working distance, which is of critical importance in some military systems. (I have a sample image inverter less than 1 in long.)

Another type of image manipulation possible with fused fiber optics is the image combiner/splitter shown in *Figure 20-6*. This is made by laying down a series of fiber-optic ribbons, alternating them as if shuffling a deck of cards. One ribbon goes from the single input to output 1, the next from the input to output 2, the next to output 1, and so on. Put a single image into the input, and you get two identical output images. Put separate images into the two outputs, and you get one combined image.

**Figure 20-6.
A Fiber-Optic Image
Combiner and
Duplicator** *(Courtesy
Galileo Electro-Optics
Corp.)*

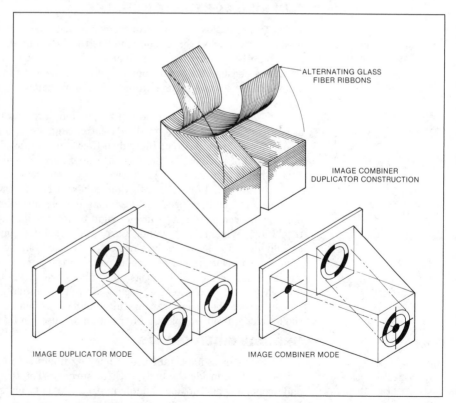

ALTERNATING GLASS
FIBER RIBBONS

IMAGE COMBINER
DUPLICATOR CONSTRUCTION

IMAGE DUPLICATOR MODE

IMAGE COMBINER MODE

Similar ideas could be used in other image manipulators or in devices to perform operations on optical signals. However, before you rush out for a patent application on your own bright idea, you must face the ugly reality of cost. Manufacture of the fiber-optic image combiner in *Figure 20-6* requires time and exacting precision, making it too expensive for practical use. Image inverters are used in some systems, but only where less-costly lens systems won't do the job. Conventional lenses also do not suffer the same resolution limits as fiber bundles.

Tapers

Earlier we saw how tapered fiber cores could bend light rays. If a bundle was tapered, the same principle could be used for imaging. Controlling the degree of tapering across the area of the taper could vary the degree of magnification. For example, the center of the image could be magnified more than the outer part, or vice versa. However, the practical problems are similar to those with other fiber-optic image manipulators— lenses are less costly and have higher resolution so they are preferred for most applications.

FLEXIBLE COHERENT BUNDLES

Flexible coherent bundles are made from many separate fibers which are aligned at their ends.

It is easiest to make coherent fiber bundles if the fibers are permanently aligned throughout the length of the bundle (i.e., if they are fused together). However, such fiber bundles are not usable in many situations because they don't bend. You can't make a rigid fiber bundle follow the twists and turns of a drainpipe, for example. You need one that is flexible.

A flexible fiber bundle can be made by bonding the fibers together at two ends, so they maintain their relative alignment, but leaving them free to move in the middle. Individual fibers, unlike fused bundles, can take a good deal of bending. In practice, the fibers are housed in a sheath, generally of stainless steel, to protect them from damage.

Individual fibers in a flexible coherent bundle can be small, but not quite as small as in a fused bundle. Some performance limits of flexible bundles are comparable to those of rigid bundles (e.g., packing fraction and resolution). When flexible bundles are used, an added concern is breakage of individual fibers, which does not occur in fused bundles. This can prevent light transmission from a particular point on the input face. The loss of a single point is not critical, but as more fibers break, the transmitted light level drops and resolution can decline as well. Eventually breakage reaches a point where the image-transmitting bundle is no longer usable. Because of the breakage problem, plastic fibers often are used in flexible bundles.

MEDICAL FIBER OPTICS

Imaging fiber bundles let physicians look inside body cavities without surgery.

The most important use of imaging fibers is in medicine, to let physicians look inside the body without surgery. The tool they use is an endoscope, a coherent fiber bundle up to a couple of meters long. Endoscopes may be rigid or flexible; the choice depends on the procedure.

Suppose a physician wants to examine a patient's bronchial tubes. He passes an endoscope down the patient's throat toward the lungs. Some fibers in the bundle transmit light from a lamp to illuminate the airway. Others return light so the physician can see conditions inside.

In some cases, the physician can do more than just look. With a laser that emits a wavelength transmitted by the fiber, the physician can perform laser surgery through the endoscope. For example, an endoscope could be aimed at bronchial cancer, then laser pulses fired to vaporize the cancer cells. The physician could look through the endoscope to check progress after each pulse. (Care must be taken to keep the bright laser pulses out of the physician's eye.) Only a small fraction of endoscopes are used with laser surgery, but the possibilities of avoiding some major surgery are exciting. That can save the patients not only money but the risk of hospitalization and major trauma.

In the longer term, some medical researchers are testing the use of optical fibers and lasers to remove plaque that clogs arteries. The idea is to thread a single thin fiber through an artery to deliver laser energy to a blocked region. Some promising results have been reported, but there are some serious problems. You can't just blast away with a fiber-optic version of a plumber's snake. The techniques that remove arterial plaque can also remove arterial walls, posing serious control problems for physicians.

GRADED-INDEX FIBER LENSES

The graded-index fiber segments described earlier in this chapter have uses quite different from imaging fiber bundles and faceplates. Their applications are as lenses.

Some uses are in fiber optics. A graded-index fiber microlens might focus output from an LED or diode laser so it could be coupled efficiently into a fiber. However, most transmitters have light sources butted up against the collecting fiber, without intermediate components. In addition, many fiber pigtails could serve the same function without the need of another discrete component.

Most applications of graded-index fiber microlenses are in optical systems such as photocopiers. One example is the use of a linear array of fiber-optic microlenses to focus light reflected from a small area of a page being copied onto individual sensors in a linear array that detects reflected light.

FIBER SENSORS

Sensing is the detection of external events by the way they influence a sensor. After learning how insensitive optical fibers are to outside influences (e.g., EMI), you might think fiber-optic sensing a strange idea. It isn't. Transmission of light in optical fibers can be influenced by the outside world. Factors such as temperature, pressure, and magnetic fields can alter light transmission in fibers, although not always in ways that affect ordinary communication signals. Many changes are subtle, but subtle

Some endoscopes can be used for internal laser surgery.

Lasers and optical fibers might some day help clean out clogged arteries.

Graded-index fiber segments can be used as optical components.

Fiber sensors rely on changes in the way fibers transmit light.

changes can be detected when you transmit light through a kilometer of fiber. Optical fibers also can collect information in the form of light from other sensors—sometimes delivering light to the sensors as well.

There are several types of fiber-optic sensors. They range in complexity from simple light pipes looking for the presence or absence of light to sophisticated instruments looking for changes in length smaller than a wavelength of light.

Types of Fiber Sensors

There are many types of fiber sensors.

We can break the broad field of fiber-optic sensors down into a few major subcategories, which makes it easier to discuss the basic principles of each. These are:

- Fiber-optic probes, which only look for the presence or absence of light
- Remote optical sensors, which are not fibers themselves but which work with light received or sent through fibers
- Fiber intensity sensors in which a fiber's light transmission changes in response to an external influence (e.g., pressure)
- Color sensors, which detect changes in light transmission at different wavelengths
- Interferometric sensors in which changes in the effective path length of light through a fiber are monitored by comparing light transmission through the sensing fiber with light in another fiber
- Polarization sensors, which detect externally induced changes in the polarization of light travelling through a fiber.

We will discuss each of these and indicate some applications.

Fiber-Optic Probes

Fiber-optic probes look for presence or absence of light.

A simple fiber-optic probe is shown in *Figure 20-7*. In this example, one optical fiber delivers light from an external source, and a second fiber collects light emerging from the first, as long as nothing gets between the two. If something does pass between the two fibers (e.g., blocks the light signal passing between them) the signal is turned off; otherwise it is on.

Figure 20-7.
A Fiber-Optic Probe

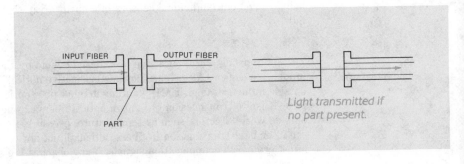

INPUT FIBER   OUTPUT FIBER

PART

Light transmitted if no part present.

This basic concept can be used in many applications. One example is reading holes in a punched card. The card passes an array of fibers at a fixed speed, and detectors monitor light transmission as a function of time. The timing cycle is chosen to match the speed of the passing. Passage of a hole lets light reach the detector, and the light is turned off at the end of the hole, when part of the card blocks it. It's a simple and effective technique, but unfortunately for the fiber industry the punched computer card is well on the way to the oblivion of obsolescence.

A more refined example is monitoring parts in a jig on a production line. An array of fibers could be mounted near places where the edge of the part should lie. If all the places that should be covered are and all those which should not be covered are not, the machine can go ahead. If any fibers see light that should not see it—or don't see light that should see it—the anomaly would halt the machine and trigger an alarm to tell a worker to check what was wrong.

Optical Remote Sensing

Fibers can deliver light to and from remote optical sensors.

More sophisticated sensors use fibers to carry light to and from optical sensors that are not themselves fibers. These sensors change optical properties in response to external influences. The fiber's role is similar to that of the wires attached to an electronic sensor.

One example is the liquid-level sensor shown in *Figure 20-8*, which senses when the gasoline in tank trucks reaches a certain level. Many of these trucks are filled from the bottom so vapor left in the tank can be collected to control pollution, and the liquid level must be sensed to prevent overfilling. One fiber delivers light to a prism mounted at the proper level. If there is no liquid in the tank, the light from the fiber experiences total internal reflection at the base of the prism and is directed back into the collecting fiber. If the bottom of the prism is covered with gasoline, total internal reflection cannot occur at the angle light strikes the prism's bottom face and no more light is reflected back into the fiber. When the light signal stops, the control system shuts off the gas pump.

A more sophisticated example is a remote temperature sensor that detects changes in fluorescence of a phosphor in a glass blob at the end of a fiber. Ultraviolet light transmitted by the fiber stimulates fluorescence from the phosphor at several wavelengths, with the amounts at different wavelengths dependent on temperature. The same fiber collects the fluorescence for delivery to an optical analyzer that measures intensities at different wavelengths and, thus, measures the temperature.

Fiber Intensity Sensor

Some effects can change the amount of light transmitted by a fiber.

Outside influences can directly affect transmission characteristics of optical fibers. Many of these effects are weak but become significant when averaged over long lengths of fiber. You may not want to measure average pressure on a straight 100-m fiber, but you can wind the 100-m fiber around a reel and put that where you want to measure pressure.

Alternatively, fibers can be designed or mounted in special ways that enhance their sensitivity to effects that communication fibers don't notice—or are designed not to respond to.

Figure 20-8.
Operation of a Liquid-Level Sensor

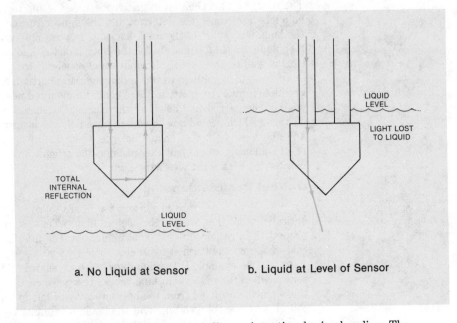

a. No Liquid at Sensor   b. Liquid at Level of Sensor

One fiber intensity sensor relies on intentional microbending. The fiber is mounted between a pair of plates containing parallel grooves. Pressure on the plates pushed them together, causing microbends in the fiber. The microbends increase loss, which is monitored by measuring light transmission through the fiber. This allows measurement not just of static pressure but also of instantaneous changes in pressure (i.e., sound waves). The goal is not better microphones for ordinary use but very special microphones, acoustic sensors to rest on the ocean bottom keeping watch for submarines.

A more subtle approach is temperature sensing with a fiber in which the refractive indexes of core and cladding vary with temperature in different ways. Suppose that at 0°C the core index is 1.50, the cladding index is 1.49, and the core index decreases by 0.0005 per degree while the cladding index decreases by 0.0004 per degree. At 100°C, the two refractive indexes would be equal (1.45). At that point, the fiber would stop guiding light, so output light intensity would drop to essentially zero. The example is an artificial one, but the concept has been demonstrated in a sensor that can measure temperature within a few degrees.

Color Sensors

A variation on simple intensity measurement is to observe what happens at multiple wavelengths. The phosphor temperature probe described above is one example, although the fiber per se is not a sensor. Even though this is more complex than simple measurement of total intensity, it can allow more sophisticated measurements.

Interferometric (Phase) Sensing

Changes in effective transmission distance through a fiber can be measured by observing the phase of light.

One of the most general phenomena usable in sensing—although one of the more complex to measure—is a change in effective transmission distance through a fiber. That depends on the fiber's physical length, its refractive index, and its cross-sectional dimensions. Pressure, strain, and temperature can affect all those characteristics. The result is a change in the phase of the light emerging from the fiber.

To understand what a phase change means and how it can be measured, consider a simplified example in which a fiber transmits coherent light waves. If the fiber is stretched by half a wavelength of light, the light intensity from that fiber alone will not change. To detect the change, light from the stretched fiber must be mixed with light in phase with the light from the original unstretched fiber (in a device called an interferometer). Those two light waves will be 180° out of phase and, thus, will interfere and add to zero. Further stretching will cause intensity to increase until the two waves are in phase (meaning the stretching has shifted phase by 360°).

Interferometric measurements are complex but very sensitive.

Although interferometric measurements can be cumbersome, they are very sensitive and can be used in many different sensing problems. For example, if a fiber is coated with metal or wrapped around a special cylinder, a magnetic field applies pressure to it in a way proportional to field strength, making a magnetic-field sensor. Tying a weight to the end of a fiber applies a strain to the fiber when the object accelerates, making an accelerometer. The stretching of the fibers is detected by interferometry.

A fiber-optic gyroscope measures rotation interferometrically.

One interferometric fiber sensor likely to find applications soon relies on principles somewhat different than those explained so far to detect rotation. This is the Sagnac interferometer—often called a fiber-optic gyroscope—shown in *Figure 20-9*. Light from a single source is split into two beams directed into opposite ends of a loop of single-mode fiber. In actual sensors, the fiber is wound many times around a cylinder, but the drawing shows only one turn. Light takes a finite time to travel around a fiber loop with radius r, and in that time the loop can rotate an angle θ, which in practice is very small. That means the starting point will have moved a distance Δ. Light going in the direction of the rotation must travel a distance $2\pi r + \Delta$ to get back to its starting point, but light travelling in the opposite direction will only have to travel $2\pi r - \Delta$. That slight difference in distance will mean a difference in phase when the two beams are superimposed. If the beams are coherent, that difference can be detected interferometrically.

The result is a simple, compact rotation sensor with no moving parts. Military planners are enthusiastic and plan several applications for fiber gyros.

**Figure 20-9.
Workings of a Fiber-Optic Gyroscope**

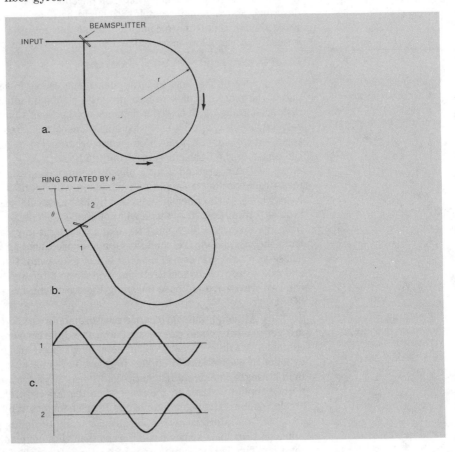

Polarization Sensors

Changes in the polarization of light in fibers can be used for sensing.

Single-mode fibers normally transmit two orthogonal polarization modes, but as we saw in Chapter 4, some can control polarization or limit transmission to one polarization. Outside phenomena can affect light polarization in a fiber, and the resulting changes can be used in sensing applications.

One example is measurement of current by Faraday rotation, which rotates the plane of polarized light by an amount proportional to the magnetic field. This phenomenon can be used to measure current by winding a fiber around a conductor, which generates a magnetic field surrounding it, and seeing how this magnetic field affects the polarization of light travelling through the fiber.

System Implications of Sensors

Fiber-optic communication networks could include fiber sensors for controls.

We have seen how optical fibers can carry light to and from sensors and how fibers themselves can be sensing elements. This can let fiber-optic networks go beyond just communicating. Consider, for instance, a fiber-optic factory control system. Optical fibers carry data to and from process control computers. Now some data originates in central computers, and some, in electronic sensors. But eventually some might come directly from optical sensors, such as the one we examined earlier to check for position of a part on an assembly line.

Fiber intensity sensors and probes are the simplest to implement, but other types can be developed. Suppose, for instance, a sensor compares intensity of three fluorescence wavelengths to measure temperature. Light at the three wavelengths could be collected by a single fiber, separated at a wavelength-division demultiplexer, and fed to three separate detectors. Output from the three detectors could be compared electronically to generate a digitized temperature reading. Further in the future, output of the wavelength-division demultiplexer might instead go to bistable optical devices, designed to switch to certain logical states (and turn on other controls) if the temperature went above or below the desired range. The technology is young, and further possibilities are many.

WHAT HAVE WE LEARNED?

1. Rigid or flexible bundles of optical fibers can transmit images if the fibers that make them up are properly aligned at the ends (coherent). Rigid bundles are made of fibers fused together; flexible bundles contain separate fibers bonded at the ends. Resolution is limited by the size of the fiber cores.
2. Bundles of fibers in which the ends are not aligned with one another illuminate hard-to-reach places. The bundle can be broken up on one end to form an image or display (e.g., a WALK sign).
3. Step-index multimode fibers do not focus light; however, tapered step-index fibers can. Segments of graded-index fiber can serve as lenses.
4. Imaging and other short-distance fibers generally have much higher attenuation than communication fibers.
5. Thin fiber-optic faceplates are used in high-resolution and image-intensifying tubes, but not for ordinary cathode-ray tubes (e.g., television picture tubes).
6. Endoscopy is the use of coherent fiber bundles to view inside the body without surgery.
7. Fiber-optic sensors include probes that detect the presence or absence of light, remote optical sensors that use fibers only to transmit and collect light, and purely fiber sensors in which the fiber responds to outside influences.
8. Outside influences such as pressure and temperature can change light intensity transmitted by a fiber, polarization of light in the fiber, or the effective distance light travels through a fiber.
9. Quantities that fibers can sense include rotation, acceleration, pressure, electric current, acoustic waves, and temperature.

Quiz for Chapter 20

1. Which of the following statements is false?
   a. Coherent fiber bundles can transmit images.
   b. Coherent fiber bundles can focus light.
   c. Graded-index fiber segments can focus light.
   d. Imaging fiber bundles contain step-index multimode fibers.

2. A graded-index fiber lens has a pitch of 0.45. How much of a sinusoidal oscillation cycle do light rays experience in passing through it?
   a. 27°.
   b. 45°.
   c. 81°.
   d. 162°.
   e. 180°.

3. What does it mean to say that a fiber bundle has packing fraction of 70%?
   a. 70% of the fibers are intact.
   b. 70% of the input surface is made up of optical fibers.
   c. 70% of the input surface is made up of fiber core.
   d. 70% of the input surface is made up of fiber cladding.
   e. It transmits 70% of the incident light through its entire length.

4. You want to resolve an image with 8 line pairs per millimeter. In theory, what is the largest fiber core size that you could use in a stationary coherent bundle?
   a. 8 μm.
   b. 50 μm.
   c. 62.5 μm.
   d. 100 μm.
   e. 125 μm.

5. Endoscopes used in medicine to view inside the body are:
   a. flexible fiber-optic bundles.
   b. also able to transmit laser beams to treat disease.
   c. rigid fiber-optic bundles.
   d. all of the above.
   e. none of the above.

6. Fiber-optic faceplates are:
   a. specialized sensors that detect temperature variations across a surface.
   b. thin rigid fiber bundles that enhance transmission high-resolution images from a vacuum into air.
   c. assemblies of graded-index fiber lenses that focus light in photocopiers.
   d. used on most television sets.

7. Fiber-optic probes could detect the presence of a part on an assembly line by:
   a. checking if it is blocking light.
   b. sensing its pressure on the assembly-line belt.
   c. measuring its temperature.
   d. pushing against it.

8. Which of the following would not be used in a fiber-optic sensor?
   a. Variation in refractive index of core and cladding with temperature.
   b. Changes in microbending loss when pressure is applied to a fiber confined between plates.
   c. Temperature-induced changes in fluorescence emission.
   d. Diffusion of chemicals into the fiber.

9. What is the most important advantage of interferometric fiber sensors?
   a. Extremely high sensitivity to small effects.
   b. Immunity to temperature variations and electromagnetic interference.
   c. Ease of detecting changes.
   d. They do not require coherent light.

10. How do fiber-optic rotation sensors work?
    a. They use microbending effects to detect changes in pressure caused by rotation.
    b. They interferometrically detect changes in polarization.
    c. They interferometrically detect differences in effective path lengths travelled by light going in opposite directions around a ring.
    d. They are accelerometers that measure rotational acceleration on fibers.
    e. By optical black magic.

Appendix: Laser Safety

With the exercise of reasonable common sense, fiber-optic systems are inherently safe. However, special rules have been developed to cover the safe use of lasers. Their most visible impact is in warning labels printed on many laser data sheets and packages to meet requirements imposed by the Center for Devices and Radiological Health, which is part of the Department of Health and Human Services.

Why are there so many warnings when a couple of milliwatts from a semiconductor laser is not about to burn holes through anything? Because the eye could focus the invisible infrared beam onto a tiny spot on the retina, the light-sensitive layer at the back of the eye. Focusing a 1-mW laser beam onto the retina produces light intensity, in that small spot, similar to that produced on the retina when the eye looks directly at the sun.

Just as with the sun, a momentary glance into a laser beam will not blind you, but staring into it for a long period could leave a blind spot. The beam from a semiconductor laser spreads out so rapidly (by laser standards) that looking at it from a few feet away should pose no threat. (That is not true for gas lasers, which have very narrow beams.) However, you should *never* aim the output of a semiconductor laser directly into your eye. The same applies to the output of an optical fiber connected to a semiconductor laser. Although federal regulations do not apply to LEDs, their output could have the same effect if it was as powerful and as tightly focused as a laser beam (e.g., if it emerged from a fiber right into the eye).

Glossary

Acceptance Angle The angle over which the core of an optical fiber accepts incoming light; usually measured from the fiber axis. Related to numerical aperture (NA).

Active Coupler A coupler that includes a receiver and one or more transmitters to regenerate input signals and send them through output fibers, instead of passively dividing input light.

All-Dielectric Cable Cable made entirely of dielectric (insulating) materials without any metal conductors.

Analog A signal that varies continuously (e.g., sound waves). Analog signals have a frequency and bandwidth measured in hertz.

Ångstrom Symbol (Å) A unit of length, 0.1 nm or 10^{-10} m, often used to measure wavelength but not part of the SI system of units. Often written Angstrom because the special symbol is not available.

Armor A protective layer, usually metal, wrapped around a cable.

Attenuation Reduction of signal magnitude, or loss, normally measured in decibels. Fiber attenuation normally is measured per unit length in decibels per kilometer.

Attenuator An optical element that reduces intensity of a signal passing through it (i.e., attenuates it).

Avalanche Photodiode (APD) A semiconductor photodetector that includes detection and amplification stages. Electrons generated at a p-n junction are accelerated in a region where they free an avalanche of other electrons. APDs can detect faint signals but require higher voltages than other semiconductor electronics.

Average Power The average level of power in signal that varies with time.

Axis The center of an optical fiber.

Backbone System A transmission network that carries high-speed telecommunications between regions (e.g., a nationwide long-distance telephone system).

Backscattering Scattering of light in the direction opposite to that in which it was originally travelling.

Bandwidth The highest frequency that can be transmitted in analog operation.

Baud Strictly speaking, the number of signal-level transitions per second in digital data. For some common coding schemes, this equals bits per second, but this is not true for more complex coding, where it is often misused. Telecommunication specialists prefer bits per second, which is less ambiguous.

Beamsplitter A device that divides incident light into two separate beams.

Bel A relative measurement, denoting a factor of ten change. Rarely used in practice; most measurements are in decibels (0.1 bel).

BER Bit-error rate.

Bidirectional Couplers Couplers that operate in the same way regardless of the direction light passes through them.

Bidirectional Transmission Transmission of signals in both directions, in some cases through the same fiber.

Birefringent Having refractive index that differs for light of different polarizations.

Bistable optics Optical devices with two stable transmission states.

Bit-Error Rate The fraction of bits transmitted incorrectly.

Broadband In general, transmission that covers a wide range of frequencies. The broadband label sometimes is used for a network that carries many different services or for video transmission.

Bundle (of fibers) A group of fibers packaged or manufactured together, with end points in a common plane. The bundle may be rigid or flexible. In coherent fiber bundles, the fibers retain a fixed position relative to each other and can transmit images. Otherwise, they merely carry light between points.

Burrus Diode A surface-emitting LED with a hole etched to accommodate a light-collecting fiber. Named after its inventor, Charles Burrus.

Bypass A circuit that carries telephone signals from a subscriber to another point without the use of local telephone-company circuits.

Byte Eight bits of digital data.

CATV An acronym for cable television derived from Community Antenna TeleVision.

CCITT International Consultative Commission on Telephone and Telegraph, an arm of the International Telecommunications Union, which sets standards.

Central Office A telephone-company facility for switching signals among local telephone circuits; connects to subscriber telephones.

Chromatic Dispersion Pulse spreading caused by variation in light propagation with wavelength measured in picoseconds (of dispersion) per kilometer (of fiber length) per nanometer (of source bandwidth). It is the sum of waveguide and material dispersion.

Cladding The layer of glass or other transparent material surrounding the light-carrying core of an optical fiber. It has a lower refractive index than the core. Coatings may be applied over the cladding.

Coherent Bundle (of fibers) Fibers packaged together in a bundle so they retain a fixed arrangement at the two ends and can transmit an image.

Coherent Communications In fiber optics, a communication system where the output of a local laser oscillator is mixed with the received signal, and the difference frequency is detected and amplified.

Compression Reducing the number of bits needed to encode a signal, typically by eliminating long strings of identical bits or bits that do not change in successive sampling intervals (e.g., video frames).

Connector A device mounted on the end of a fiber-optic cable, light source, receiver, or housing that mates to a similar device to couple light optically into and out of optical fibers. A connector joins two fiber ends or one fiber end and a light source or detector.

Core The central part of an optical fiber that carries light.

Coupler A device that connects three or more fiber ends, dividing one input between two or more outputs or combining two or more inputs into one output.

Coupling Transfer of light into or out of an optical fiber. (Note that coupling does not require a coupler.)

Critical Angle The angle at which light undergoes total internal reflection.

Cut-back Measurements Measurement of optical loss made by cutting a fiber to compare loss of a short segment with loss of a longer one.

Cut-off Wavelength The longest wavelength at which a single-mode fiber can transmit two modes, or (equivalently) the shortest wavelength at which a single-mode fiber carries only one mode.

Cycles per Second Number of oscillations a wave makes or frequency. One cycle per second equals one hertz.

Dark Current The noise current generated by a photodiode in the dark.

Data Link A fiber-optic transmitter, cable, and receiver that transmits digital data between two points.

dBm Decibels below 1 mW.

dBμ Decibels below 1 μW.

Decibel (dB) A logarithmic comparison of power levels, defined as ten times the base-ten logarithm of the ratio of the two power levels. One-tenth of a bel.

Demultiplexer A device that separates a multiplexed signal into its original components; the inverse of a multiplexer.

Detector A device that generates an electric signal when illuminated by light. The most common in fiber optics are photodiodes, photodarlingtons, and phototransistors.

Dielectric Non-conductive.

Digital Encoded as a signal in discrete levels, typically binary ones and zeros.

Diode An electronic device that lets current flow in only one direction. Semiconductor diodes used in fiber optics contain a junction between regions of different doping and include light emitters (LEDs and laser diodes) and detectors (photodiodes).

Diode Laser A semiconductor diode in which the injection of current carriers produces laser light by amplifying photons produced when holes and electrons recombine at the junction between p and n doped regions.

Directional Coupler A coupler in which the way light is transmitted depends on the direction it travels.

Dispersion The spreading out of light pulses as they travel in an optical fiber, proportional to length.

Duplex Dual. A duplex cable contains two fibers; a duplex connector links two pairs of fibers.

Edge-emitting Diode An LED that emits light from its edge, producing more directional output than LEDs that emit from their top surface.

Electromagnetic Interference (EMI) Noise generated when stray electromagnetic fields induce currents in electric conductors.

Electromagnetic Radiation Waves made up of oscillating electrical and magnetic fields perpendicular to one another and travelling at the speed of light. Also can be viewed as photons or quanta of energy. Electromagnetic radiation includes radio waves, microwaves, infrared, visible light, ultraviolet radiation, X rays, and gamma rays.

EMI Electromagnetic interference.

Endoscope A fiber-optic bundle that delivers light and views inside the human body.

Evanescent Wave Light guided in the inner part of an optical fiber's cladding rather than in the core.

Excess Loss Loss of a passive coupler above that inherent in dividing light among the output ports.

Expanded-Beam Connector A connector in which light exiting one fiber is expanded to a diameter much larger than the fiber core, then focused down onto the core of the other fiber.

External Modulation Modulation of a light source by an external device.

Extrinsic Loss Splice losses arising from the splicing process itself.

Eye Pattern A pattern displayed when an oscilloscope is driven by a receiver output and triggered by the signal source that drove the transmitter.

Faceplate A rigid array of short fibers fused together to serve as the face of a cathode-ray tube.

FDDI Fiber-Optic Distributed Data Interface.

Ferrule A tube within a connector with a central hole that contains and aligns a fiber.

Fiber-Optic Distributed Data Interface (FDDI) A standard for a 100-Mbit/s fiber-optic local-area network.

Fiber-Optic Gyroscope A coil of optical fiber that can detect rotation about its axis.

Fluoride Glasses Materials that have the amorphous structure of glass but are made of fluoride compounds (e.g., zirconium fluoride) rather than oxide compounds (e.g., silica).

FOG-M Fiber-Optic Guided Missile.

Frequency Division Multiplexing Multiplexing of analog signals by assigning each a different carrier frequency and combining them in a single signal with a broad range of frequencies.

Fused Fibers A bundle of fibers fused together so they maintain a fixed alignment with respect to each other in a rigid rod.

GaAlAs Gallium aluminum arsenide.

GaAs Gallium arsenide.

Gallium Aluminum Arsenide GaAlAs, a semiconductor compound used in LEDs, diode lasers, and certain detectors.

Gallium Arsenide GaAs, a semiconductor compound used in LEDs, laser diodes, detectors, and electronic components.

Graded-Index Fiber A fiber in which the refractive index changes gradually with distance from the fiber axis, rather than abruptly at the core-cladding interface.

Graded-Index Fiber Lens A short segment of graded-index fiber that focuses light and serves as a lens.

Hard-Clad Silica Fiber A fiber with a hard plastic cladding surrounding a step-index silica core. (Ordinary plastic-clad silica fibers have a soft plastic cladding.)

HDTV High-definition (or high-resolution) television; television with over 1000 lines per screen, about double the resolution of present systems.

Head End The central distribution point in a cable television system.

Hierarchy A set of transmission speeds arranged to multiplex successively higher numbers of circuits.

Hertz Frequency in cycles per second.

High-Resolution Television Television with over 1000 lines per screen, about double the resolution of present systems. Sometimes called HDTV, for high-definition television.

Hydrogen Losses Increases in fiber attenuation that occur when hydrogen diffuses into the glass matrix and absorbs some light.

Index of Refraction The ratio of the speed of light in vacuum to the speed of light in a material, usually abbreviated n.

Index-Matching Fluid A fluid with refractive index close to glass that reduces refractive-index discontinuities.

Indium Gallium Arsenide InGaAs, a semiconductor material used in lasers, LEDs, and detectors.

Indium Gallium Arsenide Phosphide InGaAsP, a semiconductor material used in lasers, LEDs, and detectors.

Infrared Wavelengths longer than 700 nm and shorter than about 1 mm. We cannot see infrared radiation but can feel it as heat. Transmission of glass optical fibers is best in the infrared at wavelengths of 1100–1600 nm.

Infrared Fiber Colloquially, optical fibers with best transmission at wavelengths of 2 μm or longer, made of materials other than silica glass.

InGaAs Indium gallium arsenide.

InGaAsP Indium gallium arsenide phosphide. Properties depend on composition, which is sometimes written $In_{1-x}Ga_xAs_{1-y}P_y$.

Injection Laser Another name for semiconductor or diode laser.

Integrated Optics Optical devices that perform two or more functions integrated on a single substrate; analogous to integrated electronic circuits.

Integrated Optoelectronics Optical and electronic devices (e.g., diode lasers and drive circuitry) integrated on the same semiconductor chip.

Integrated Services Digital Network (ISDN) A network in which many different services are carried in a single digital bit stream.

Intensity Power per unit solid angle.

Interferometric Sensors Fiber-optic sensors that rely on interferometric detection.

Intrinsic Losses Splice losses arising from differences in the fibers being spliced.

Irradiance Power per unit area.

ISDN Integrated Services Digital Network.

Jacket A layer of material surrounding a fiber but not bonded to it—part of the cable, not part of the fiber.

Junction Laser A semiconductor diode laser.

Kevlar A strong synthetic material used in cable strength members; the name is a trademark of the Dupont Company.

LAN Local-area network.

Large-Core Fiber Usually, a fiber with core 200 μm or more, but sometimes applied to 100/140 fiber.

Laser From Light Amplification by Stimulated Emission of Radiation, one of the wide range of devices that generates light by that principle. Laser light is directional, covers a narrow range of wavelengths, and is more coherent than ordinary light. In fiber optics, the most important lasers are semiconductor diode lasers.

LED Light-emitting diode.

Light Strictly speaking, electromagnetic radiation visible to the human eye. Commonly, the term is applied to electromagnetic radiation with properties similar to visible light, including the invisible near-infrared radiation that carries signals in most fiber-optic communication systems.

Light Piping Use of optical fibers to illuminate.

Light-emitting Diode LED, a semiconductor diode that emits incoherent light at the junction between p- and n-doped materials.

Lightguide An optical fiber or fiber bundle.

Lightwave As an adjective, a synonym for fiber-optic, but with somewhat broader connotations (e.g., it can cover optical communication not through fibers).

Local-Area Network LAN, a network that transmits data among many nodes in a small area (e.g., a building).

Local Loop The part of the telephone network extending from the central (switching) office to the subscriber.

Longitudinal Modes Oscillation modes of a laser along the length of its cavity, so twice the length of the cavity equals an integral number of wavelengths. Each longitudinal mode contains only a very narrow range of wavelengths. Distinct from transverse modes.

Loose Tube A protective tube loosely surrounding a cabled fiber, often filled with a gel.

Loss Attenuation of optical signal, normally measured in decibels.

Loss Budget An accounting of overall attenuation in a system.

Margin Allowance for attenuation in addition to that explicitly accounted for in system design.

Mass Splicing Simultaneous splicing of multiple fibers in a cable.

Material Dispersion Pulse dispersion caused by variation of a material's refractive index with wavelength.

Mbit/s Megabit (million bits) per second.

Mechanical Splice A splice in which fibers are joined mechanically (e.g., glued or crimped in place) but not fused together.

Microbending Tiny bends in a fiber that allow light to leak out and increase loss.

Micrometer One-millionth of a meter, abbreviated μm.

Modal Dispersion Dispersion arising from differences in the times that different modes take to travel through multimode fiber.

Mode An electromagnetic field distribution that satisfies theoretical requirements for propagation in a waveguide or oscillation in a cavity (e.g., a laser). Light has modes in a fiber or laser.

Mode Field Diameter The diameter of the one mode of light propagating in a single-mode fiber.

Mode Stripper A device that removes high-order modes in a multimode fiber to give standard measurement conditions.

Multimode Transmits or emits multiple modes of light.

Multiplexer A device that combines two or more signals into a single output.

N Region A semiconductor doped to have an excess of electrons as current carriers.

NA Numerical aperture.

Nanometer A unit of length, 10^{-9} m. It is part of the SI system and has largely supplanted the non-SI Ångstrom (0.1 nm) in technical literature.

Nanosecond One-billionth of a second, 10^{-9} second.

Near Infrared The part of the infrared near the visible spectrum, typically 700–1500 or 2000 nm; it is not rigidly defined.

No Return to Zero NRZ, a digital code in which the signal level is low for a 0 bit and high for a 1 bit, and does not return to zero between successive 1 bits.

Noise Equivalent Power NEP, the optical input power to a detector needed to generate an electrical signal equal to the inherent electrical noise.

Normal (angle) Perpendicular to a surface.

NRZ No return to zero.

Numerical Aperture NA, the sine of half the angle over which a fiber can accept light. Strictly speaking, this is multiplied by the refractive index of the medium containing the light, but that equals 1 for air, the normal medium from which NA is measured.

Optical Time-Domain Reflectometer (OTDR) An instrument that measures transmission characteristics by sending a short pulse of light down a fiber and observing backscattered light.

Optical Waveguide Technically, any structure that can guide light. Sometimes used as a synonym for optical fiber, it also can apply to planar light waveguides.

P Region Part of a semiconductor doped with electron acceptors in which holes (vacancies in the valence electron level) are the dominant current carriers.

Packing Fraction The fraction of the surface area of a fiber-optic bundle that is fiber core.

Passive Coupler A coupler that divides entering light among output ports without generating new light.

PBX Private branch exchange.

PCS Fiber Plastic-clad silica fiber.

Peak Power Highest instantaneous power level in a pulse.

Phase The position of a wave in its oscillation cycle.

Photodarlington A light detector in which a phototransistor is combined in a circuit with a second transistor to amplify its output. Slow but sensitive.

Photodetector A light detector.

Photodiode A diode that can produce an electrical signal proportional to light falling upon it.

Photons Quanta of electromagnetic radiation. Light can be viewed as either a wave or a series of photons.

Phototransistor A transistor that detects light and amplifies the resulting electrical signal. Light falling on the base-collector junction generates a current, which is amplified internally. Simple but slow.

Pin Photodiode A semiconductor detector with an intrinsic (i) region separating the p- and n-doped regions. This design gives fast, linear response and is widely used in fiber-optic receivers.

Planar Waveguide A waveguide fabricated in a flat material such as a film.

Plastic-Clad Silica (PCS) Fiber A step-index multimode fiber in which a silica core is surrounded by a lower-index plastic cladding.

Plenum Cable Cable made of fire-retardant material that generates little smoke, for installation in air ducts.

Point-to-Point Transmission Carrying a signal between two end points, without branching to other points.

Polarization Alignment of the electric and magnetic fields that make up an electromagnetic wave; normally refers to the electric field. If all light waves have the same alignment, the light is polarized.

Polarization-maintaining Fiber Fiber that maintains the polarization of light that enters it.

Preform A cylindrical rod of specially prepared and purified glass from which an optical fiber is drawn.

Pressurization Filling the inside of a cable structure with gas slightly above atmospheric pressure, to prevent entry of moisture.

Pulse Dispersion The spreading out of pulses as they travel along an optical fiber.

Quantum Efficiency The fraction of photons that strike a detector which produce electron-hole pairs in the output current.

Quaternary A semiconductor compound made of four elements (e.g., InGaAsP).

Radiation-hardened Insensitive to the effects of nuclear radiation, usually for military applications.

Radiometer An instrument, distinct from a photometer, to measure power (watts) of electromagnetic radiation.

Rays Straight lines that represent the path taken by light.

Receiver A device that detects an optical signal and converts it into an electrical form usable by other devices.

Recombination Combination of an electron and a hole in a semiconductor that releases energy.

Refraction The bending of light as it passes between materials of different refractive index.

Refractive Index Ratio of the speed of light in vacuum to the speed of light in a material; abbreviated n.

Refractive-Index Gradient The change in refractive index with distance from the axis of an optical fiber.

Regenerator A receiver–transmitter pair that detects a weak signal, cleans it up, then sends the regenerated signal through another length of fiber.

Repeater Often a receiver–transmitter pair that detects, cleans up, and amplifies a weak signal for retransmission through another length of optical fiber. Sometimes a repeater contains multiple regenerators, one for each fiber in a cable.

Responsivity The ratio of detector output to input, usually measured in units of amperes per watt (or microamperes per microwatt).

Return to Zero RZ, a digital coding scheme where signal level is low for a 0 bit and high for a 1 bit during the first half of a bit interval, then in either case returns to zero for the second half of the bit interval.

Ribbon Cables Cables in which many fibers are embedded in a plastic material in parallel, forming a flat ribbon-like structure.

Ring Architecture A network scheme in which a transmission line forms a complete ring.

Rise Time The time it takes output to rise from low levels to peak value. Typically measured as the time to rise from 10 to 90% of maximum output.

RZ Return-to-zero.

Selfoc Lens A tradename used by the Nippon Sheet Glass Company for graded-index fiber lens, a segment of graded-index fiber made to serve as a lens.

Semiconductor Diode Laser A laser in which injection of current into a semiconductor diode produces light by recombination of holes and electrons at the junction between p- and n-doped regions.

Sheath An outer protective layer of a fiber-optic cable.

Signal-to-Noise Ratio The ratio of signal to noise, measured in decibels, an indication of signal quality in analog systems.

Silica Glass Glass made mostly of silicon dioxide, SiO_2, used in conventional optical fibers.

Simplex Single-element (e.g., a simplex connector is a single-fiber connector).

Single-Frequency Laser A laser that emits a range of wavelengths small enough to be considered a single frequency.

Single Mode Containing only one mode. Beware of ambiguities because of the difference between transverse and longitudinal modes. A laser operating in a single transverse mode typically does not operate in a single longitudinal mode.

Single-Polarization Fibers Optical fibers able to carry light in only one polarization.

Splice A permanent junction between two fiber ends.

Splitting Ratio The ratio of power emerging from two output ports of a coupler.

Star Coupler A coupler with more than three ports.

Step-Index Multimode Fiber A step-index fiber with a core large enough to carry light in multiple modes.

Step-Index Single-Mode Fiber A step-index fiber with a small core able to carry light in only one mode.

Submarine Cable A cable laid underwater.

Subscriber Loop The part of the telephone network from a central office to individual subscribers.

Supertrunk A cable that carries several video channels between facilities of a cable television company.

Surface-emitting Diode An LED that emits light from its flat surface rather than its side. Simple and inexpensive, with emission spread over a wide angle.

Switched Network A network that routes signals to their destinations by switching circuits.

T Coupler A coupler with three ports.

T3 Telecommunication transmission at 45 Mbit/s.

T3C Telecommunication transmission at 90 Mbit/s.

TDM Time-division multiplexing, digital multiplexing by taking one pulse at a time from separate signals and combining them in a single bit stream.

Ternary A semiconductor compound made of three elements (e.g., GaAlAs).

III–V (Three–five) Semiconductor A semiconductor compound made of one (or more) elements from the IIIA column of the periodic table (Al, Ga, and In) and one (or more) elements from the VA column (N, P, As, or Sb). Used in LEDs, diode lasers, and detectors.

Threshold Current The minimum current needed to sustain laser action in a laser diode.

Tight Buffer A material tightly surrounding a fiber in a cable, holding it rigidly in place.

Time-Division Multiplexing (TDM) Digital multiplexing by taking one pulse at a time from separate signals and combining them in a single bit stream.

Total Internal Reflection Total reflection of light back into a material when it strikes the interface with a material having a lower refractive index at an angle below a critical value.

Transceiver A combination of transmitter and receiver providing both output and input interfaces with a device.

Transverse Modes Modes across the width of a waveguide (e.g., a fiber or laser). Distinct from longitudinal modes, which are along the length.

Tree A network architecture in which transmission routes branch out from a central point.

Trunk Line A transmission line running between telephone switching offices.

Ultraviolet Electromagnetic waves invisible to the human eye with wavelengths about 10–400 nm.

Videoconferencing Teleconferencing via videophone.

Videophone A telephone-like service with a picture as well as sound.

Visible Light Electromagnetic radiation visible to the human eye at wavelengths of 400–700 nm.

Voice Circuit A circuit able to carry one telephone conversation or its equivalent; the standard subunit in which telecommunication capacity is counted. The U.S. analog equivalent is 4 kHz. The digital equivalent is 56 kbit/s in North America and 64 kbit/s in Europe.

Waveguide A structure that guides electromagnetic waves along its length. An optical fiber is an optical waveguide.

Waveguide Couplers A coupler in which light is transferred between planar waveguides.

Waveguide Dispersion The part of chromatic dispersion arising from the different speeds light travels in the core and cladding of a single-mode fiber (i.e., from the fiber's waveguide structure).

Wavelength The distance an electromagnetic wave travels in the time it takes to oscillate through a complete cycle. Wavelengths of light are measured in nanometers (10^{-9} m) or micrometers (10^{-6} m).

Wavelength-Division Multiplexing (WDM) Multiplexing of signals by transmitting them at different wavelengths through the same fiber.

Y Coupler A variation on the T coupler in which input light is split between two channels (typically planar waveguide) that branch out like a Y from the input.

Zero-Dispersion Wavelength Wavelength at which net chromatic dispersion of an optical fiber is zero. Arises when waveguide dispersion cancels out material dispersion.

Index

Answers to Quizzes

Chapter 1

1. d
2. c
3. a
4. d
5. d
6. a
7. b
8. c
9. a
10. e

Chapter 2

1. b
2. c
3. d
4. c
5. c
6. e
7. b
8. e
9. c
10. d

Chapter 3

1. e
2. e
3. d
4. a
5. b
   c
   d
   e
6. b
7. c
8. a
9. a
10. b

Chapter 4

1. e
2. b
3. e
4. c

5. a
6. d
7. c
8. a
9. d
10. b

Chapter 5

1. b
2. c
3. a
4. b
5. c
6. a
7. c
8. b
9. a
10. d

Chapter 6

1. d
2. a
3. b
4. e
5. a
6. a
7. b
8. c
9. b
10. d

Chapter 7

1. b
2. a
3. d
4. a
5. e
6. b
7. a-A
   b-D
   c-C
   d-E
   e-B
8. d
9. b

10. b

Chapter 8

1. b
2. e
3. c
4. d
5. b
6. a
7. c
8. b
9. a
10. d

Chapter 9

1. b
2. a
3. b
4. d
5. c
6. b
7. c
8. e
9. c
10. b

Chapter 10

1. c
2. b
3. d
4. c
5. d
6. b
7. a
8. a
9. b
10. b

Chapter 11

1. b
2. d
3. c
4. a
5. e
6. c

7. b
8. a
9. c
10. d

Chapter 12

1. a
2. d
3. b
4. c
5. d
6. d
7. a
8. b
9. d
10. c

Chapter 13

1. b
2. e
3. a
4. b
5. b
6. c
7. d
8. c
9. a
10. d

Chapter 14

1. e
2. c
3. d
4. d
5. c
6. c
7. a
8. b
9. d
10. d

Chapter 15

1. e
2. a
3. e

4. b
5. a
6. d
7. a
8. c
9. b
10. b

Chapter 16

1. b
2. d
3. b
4. a
   b
   c
5. a
6. e
7. c
8. c

9. a
   b
   c
   e
10. d

Chapter 17

1. c
2. d
3. c
4. a
5. b
6. c
7. a
8. b
9. a
10. b

Chapter 18

1. b
2. d
3. a
4. c
5. c
6. d
7. b
8. b
9. d
10. d

Chapter 19

1. b
2. c
3. b
4. d
5. c

6. a
7. b
8. b
9. e
10. a

Chapter 20

1. b
2. d
3. c
4. c
5. d
6. b
7. a
8. d
9. a
10. c